Quantum Chemistry of Nanotubes

Electronic Cylindrical Waves

Pavel N. D'yachkov
Quantum Chemistry Laboratory
Kurnakov Institute of General and Inorganic Chemistry
Russian Academy of Sciences
Moscow, Russia

CRC Press is an imprint of the
Taylor & Francis Group, an **informa** business
A SCIENCE PUBLISHERS BOOK

Cover illustration reproduced from 'Electronic Structure of Carbon Nanotubes with Copper Core' by I.A. Bochkov and P.N. D'yachkov in the "Russian Journal of Inorganic Chemistry", Jan. 2013, Springer Nature. Reproduced by permission.

CRC Press
Taylor & Francis Group
6000 Broken Sound Parkway NW, Suite 300
Boca Raton, FL 33487-2742

First issued in paperback 2020

ISBN-13: 978-1-138-59887-4 (hbk)
ISBN-13: 978-0-367-77946-7 (pbk)

Library of Congress Cataloging-in-Publication Data

Names: D'yachkov , P. N. (Pavel Nikolaevich), author.
Title: Quantum chemistry of nanotubes : electronic cylindrical waves / Pavel N. D'yachkov (Quantum Chemistry Laboratory, Kurnakov Institute of General and Inorganic Chemistry, Russian Academy of Sciences, Moscow, Russia).
Description: Boca Raton, FL : CRC Press, 2019. | "A science publishers book." | Includes bibliographical references and index.
Identifiers: LCCN 2019015120 | ISBN 9781138598874 (hardback)
Subjects: LCSH: Nanotubes. | Quantum chemistry.
Classification: LCC TA418.9.N35 D5275 2019 | DDC 620.1/93--dc23
LC record available at https://lccn.loc.gov/2019015120

Visit the Taylor & Francis Web site at
http://www.taylorandfrancis.com

and the CRC Press Web site at
http://www.crcpress.com

I was attracted to the idea that even in the smallest particles of matter can be found mathematical formulas.
Werner Heisenberg

Preface

Quantum chemistry is the quantum mechanics of atoms, molecules, and solids. Until recently, the quantum chemistry of crystals was regarded as part of a solid-state physics and was a theoretical foundation of materials sciences. Now in the time of nanotechnology, when individual molecules have become the building block of electronic devices, molecular quantum chemistry has become an important part of materials science. Nanotubes lie halfway from the molecules to crystals and they are at a center progress in nanoelectronics.

Quantum chemistry of nanotubes is being developed by the efforts of the molecular and solid-state theorists. At first, the electronic structure of carbon nanotubes was described using the simplest quantum chemical Hückel molecular orbital theory elaborated for π-conjugated organic compounds. The qualitatively correct predictions obtained of the relationships between the geometry and band structure of carbon nanotubes engendered a large number of experimental studies of their electronic properties and design of nanoelectronics elements. There was a need for more accurate calculations of nanotubes, and an alternative and somewhat theoretically more substantiated and detailed description of the nanotubes electron states came from the solid-state quantum chemistry when Slater's augmented plane wave method was taken as a starting point for tubules studies. However, the cylindrical (tubular) geometry of nanotubes determines the formation of cylindrical electron waves in tubules, rather than plane waves as in crystals. This book provides a detailed exposition of nonrelativistic and relativistic linearized augmented cylindrical wave technique and demonstrates its applications to the various carbon and non-carbon, achiral and chiral, pure, intercalated and doped single-walled, double-walled, and embedded nanotubes and nanowires. We hope to convince the reader that the use of cylindrical waves for nanotubes offers a great advantages in the studies of their properties.

Pavel N. D′yachkov

Contents

Introduction

Nanotubes are the nanometer-scale chemical compounds with cylindrical geometry (Iijima 1991). All the single-walled carbon nanotubes can be constructed by rolling up a single graphite sheet, and the structures of tubules can be labeled by the pair of integers n_1 and n_2 (where $n_1 \geq n_2 \geq 0$), which, together with C-C bond length $d_{C\text{-}C} = 1.42$ Å determine a geometry of the tubule. They are currently the focus of intense multidisciplinary studies because of their unique physical and chemical properties and their prospects for practical applications in molecular electronics (Saito et al. 1998, De Volder et al. 2013, Laird et al. 2015, Dekker 2018, Peng 2018, Bai et al. 2018, He et al. 2018, Gupta et al. 2018, Cao et al. 2017) that is perhaps the most intriguing area of nanotechnology. Chemically, the carbon nanotubes are the aromatic conjugated compounds. The theoretical simulation of the nanotubes electronic structure have received much attention since 1992, when the first calculations for the band structures of all-carbon nanotubes were done using the Linear Combination of Atomic Orbitals (LCAO) and π-electron tight-binding techniques (Mintmire et al. 1992, Saito et al. 1992a, Saito et al. 1992b, Hamada et al. 1992, White et al. 1993). Particularly, it was shown that the electronic structure of any tubule in a region of the occupied and unoccupied π states can be obtained within the tight-binding model by folding the graphene π bands along a certain direction in the two-dimensional Brillouin zone.

The zone-folding technique has proven immensely successful in providing physically relevant information on the nature of electronic interactions in nanotubes. In this approach, it was predicted that a nanotube is a metal if $n_1 - n_2$ is a multiple of 3, or a semiconductor otherwise. Later on, the modified tight-binding π electron and all-valence orbital band structure models were developed that account for the effects of the $\sigma\pi$-hybridization, misalignment of carbon p_π-orbitals on the curved graphene surface, chirality, and diameter dependence of nearest-neighbor hopping integral (Saito et al. 2001, Mintmire et al. 1993, Mintmire and White 1998, Rubio et al. 1999, Hagen and Hertel 2003, Vadapalli and Mintmire 2006, Bulusheva et al. 1998, Bussi et al. 2005). This leads particularly to modifications of the gap energies, which reveals that the single-walled carbon nanotubes with $n_1 - n_2 = 3n$ ($n = 1,2,\ldots$) can be very-small-gap semiconductors, and that these effects are of great importance in the case of the small- and moderate radius nanotubes.

These include terms causing the trigonal warping effects, similar to the zone folding technique results for the highest occupied and lowest unoccupied π states of the single-walled carbon nanotubes that were obtained using an effective-mass $k{\cdot}p$ approximation (Ajiki and Ando 1993, 1996).

Going beyond the semi empirical tight-binding methods or $k{\cdot}p$ approximation, one can perform the first-principles calculations for actual curved-surfaced tubule. A version of LCAO pseudopotential method, in which the core electrons are replaced by the nonlocal norm-conserving pseudopotentials and the valence electrons are treated using a linear combination of multiple-ζ and polarized atomic orbitals, was applied in the calculation of the electronic dispersion energies and density of states in the Fermi energy region for several small tubules (Christ and Sadeghpour 2007, Reich et al. 2002). In the Fermi energy region, the electronic structure of nanotubes was also studied using the plane-wave basic and *ab initio* pseudopotential Local-Density-Functional theory (DFT) (Blasé et al. 1994, Charlier et al. 1996 a,b, Machón et al. 2002, Marinopoulos et al. 2003, Zólyomi and Kürti 2004, Cabria et al. 2003, Dubay et al. 2002, Li et al. 2001, Liu and Chan 2002, Kürti et al. 2004, Miyake and Saito 2005). One assumes that predictive power and accuracy of the tight-binding calculations are less robust compared with the *ab initio* DFT methods (Christ and Sadeghpour 2007). Using this *ab initio* method, it was possible to study in more detail the curvature-induced π- and σ-band mixing and the deviations of chemical bonding from the sp^2 hybridization. In particular, these effects were found to significantly alter the electronic structure of narrow nanotubes compared to the predictions of the tight-binding model. In the case of small-diameter insulating nanotubes, where the diameter is less than 1 nm, the strong π-σ rehybridization modifies the low-lying nondegenerate conduction band states (Marinopoulos et al. 2003, Zólyomi and Kürti 2004). The pseudopotential calculations on the narrow tubes such as (5, 0) revealed total closure of the gap.

Since the 90s, we are developing a Linearized Augmented Cylindrical Wave (LACW) method for the nanotubes band structure (D'yachkov et al. 1998, D'yachkov et al. 1999, D'yachkov and Kirin 1999). The LACW method is an extension of the one-dimensional multiatomic systems with tubular structure of the Augmented Plane Wave (APW) theory suggested for the bulk materials (Slater 1937, Slater 1974) and developed the latter in a form of more efficient linearized APW (LAPW) technique (Andersen 1975, Koelling and Arbman 1975). During the past decades, the techniques for APW and LAPW calculations have reached the point at which, with the aid of large computers, a solution of the band structure problem may be obtained for particular crystals, including those with heavy metals (Singh 1994, Singh and Nordstrom 2006). There is good reason to believe that the LAPW method is one of the most accurate computational schemes in the bulk solid-state electronic structure theory. These developments have challenged an extension of the LAPW approach to low-dimensional systems such as nanofilms (Jepsen et al. 1978, Krakauer et al. 1979), nanotubes and nanowires (D'yachkov 1997, 2004, 2016, Mokrousov et al. 2005, 2006) and fullerene-type spherical clusters (D'yachkov and Kuznetsov 2004).

The LACW method, as applied to the nanotubes, has an advantage over the conventional LCAO and plane-wave pseudopotential methods. While low level LCAO calculations are relatively easy to do, improving the accuracy quickly becomes both technically demanding and computationally very expensive. The main concern with approaches of the LCAO method is the transferability of the basic set. Moreover, it is well known from the band structure of bulk materials that the LCAO basis is adequate to achieve good results for the valence band, but sometimes not for the conduction band, because this basis does not include the delocalized conducting plane-wave-type functions. The plane-wave pseudopotential calculations suffer from a slow convergence and an unfavorable scaling: The number of basic functions and the time taken to perform such a calculation on a computer increase asymptotically with the cube of the number of atoms (Skylaris et al. 2005). This method is computationally quite cumbersome for calculating the band structure of the chiral tubes without rotational symmetry and very large translational unit cells; hence, the availability of such results in the literature is very limited. The purely delocalized nature of the plane-wave basic set puts obstacles in the way of calculating the inner low-energy states of the valence band, e.g., the carbon nanotubes p_σ and *s* bands. The basis of the LACW method has both localized and delocalized components. Finally, the main argument for using cylindrical waves is to account for the cylindrical geometry of the nanotubes in an explicit form that offers the obvious advantages.

The purpose of this book is two-fold. First of all, it provides the physical basis and gives the detailed exposition of the LACW method, so that newcomers to nanomaterial's studies can quickly learn the technique, and if desired construct a working code. Connections are made between the standard LAPW method for solids and the LACW for nanotubes that must make the LACW approach more understandable for the readers experienced in the solid-state LAPW methods. The experimental detection of spin-orbit gaps in the carbon nanotubes stimulated an interest to the spin-dependent band structure calculations of tubules. In the case of crystals, the LAPW band structure calculations with the relativistic effects had already become routine work some decades ago, and a relativistic version of LACW theory developed is just an obvious extension of the relativistic LAPW technique to the cylindrical multiatomic systems. The nanotubes may have various atomic-scale point impurities, which can appear during the nanotube growth or can be created by external action and change the nanotubes electron properties. Based on the LACW and Green's function techniques, a method for calculating the electronic structure of the point substitutional impurities in nanotubes was developed. Again, this approach is closely related to the Green's function method designed for the point defects in the bulk materials. The LACW method can be used to study the electronic structure of double-walled carbon nanotubes consisting of two concentric graphene cylinders with extremely strong covalent bonding of atoms within the individual graphitic sheets, very weak van der Waals type interaction between them and electron tunneling between the walls. Moreover, the electronic structure of single-walled carbon nanotubes embedded in a crystal matrix can be simulated by means of the LACW method. It should be noted that similar to the LAPW method, the LACW technique is also applicable to non-carbon inorganic compounds up to the gold and platinum tubules; the calculations of non-carbon systems do not require any new methodical receptions.

Secondly, we summarize the results of application of the LACW method to the various carbon and non-carbon, achiral and chiral, pure, intercalated and doped single-walled, double-walled, and embedded nanotubes and nanowires. These results can be useful for a deeper understanding of the properties of specific materials and their practical use in nanoelectronics. We hope to convince the reader that the use of cylindrical waves for the nanotubes offers the great advantages in the studies of their properties.

The structure of this book is as follows. A presentation of the LACW method begins in Chapter One with a description of the method for constructing the muffin-tin electron potentials for the tubular and cylindrical polyatomic systems. The next sections are devoted to development of the method to the achiral single-walled nanotubes. Here, we also present the results of its application to numerical calculations of nanotubes with armchair and zigzag geometries and different chemical composition. The two sections of Chapter Two are devoted, respectively, to the LACW investigations of the effects of the interlayer tunneling of electrons on the electronic characteristics of the double-walled nanotubes and to the analysis of the electronic properties of samples in a form of achiral single-walled nanotubes embedded in a semiconductor matrix. In Chapter Three, we develop a symmetry-adapted version of a LACW method. In this case, the cells contain only two carbon atoms, and the theory becomes applicable to any single-walled nanotube independent of the number of atoms in a translational unit cell. The calculated total band structures and densities of states of the chiral and achiral, semiconducting, semimetallic, and metallic carbon and non-carbon nanotubes containing up to the 118804 atoms per translational unit cell are presented. In Chapter Four, the point defects in carbon nanotubes are described. The method described avoids using the supercell and superlattice geometries and combines the advantages of density-functional ab initio theory with the Green's function approach. The point defects in the nanotubes are illustrated by examples of the density of electronic states in the vicinity of nitrogen and boron substitutional impurities in chiral and achiral tubules. The last chapter of this book, presents the relativistic version of the LACW method and its applications.

The LACW method was emerged and developed in our laboratory; perhaps, this justifies an unusually large number of references to our publications in the bibliography. These results were obtained in collaboration with the former PhD students of our laboratory Dr. O.M. Kepp, Dr. D.V. Kirin, Dr. D.V. Makaev, Dr. B.S. Kuznetsov, Dr. D.Sh. Kutlubaev, Dr. E.P. Dyachkov, V.A. Zaluev, L.O. Khorshavin, D.O. Krasnov, who bore the bulk of the work associated with the computer implementation of the method. Some of the data presented here are the results of collaboration with Prof. H. Hermann from the Institute of Solids and Materials (Dresden), with Prof. Yu.F. Zhukovsky and Dr. S.N. Piskunov from the University of Latvia, as well as with Prof. A.V. Nikolayev from Moscow State University. My sincere appreciation to all of them.

Finally, my wife Ludmila D'yachkova created comfortable conditions for living and working.

Those who are interested in obtaining computer programs can contact.

Pavel N. D'yachkov

e-mail: p_dyachkov@rambler.ru

CHAPTER

1

Augmented Cylindrical Waves for Nonchiral Nanotubes and Wires

Let us start from the simplest case of LACW calculations of nonchiral single-walled systems considering their translational symmetry only (D'yachkov et al 1998, 1999).

1.1. Geometry of Carbon Nanotubes

All the perfect carbon single-walled nanotubes can be constructed by rolling up a single graphene sheet, and the structures of tubule can be visualized as a conformal mapping of a two-dimensional graphene lattice onto the surface of a cylinder. One can make such a seamless tubule without any special distortion of their bonding angles other than the introduction of curvature to the carbon hexagons through the rolling process. Each tubule can be labeled by the pair of integers (n_1, n_2) (where $n_1 \geq n_2 \geq 0$) which, together with C–C bond length $d_{C\text{-}C}$, determine nanotubes, geometry (Fig. 1.1).

The nanotubes generated by mapping are translationally periodic along the tubule's axis. The tubules (n, n) and (n, 0), called the armchair and zigzag tubules based on the appearance of their cross-sections, are the achiral compounds. In armchair and zigzag nanotubes, the two bonds of each hexagon are oriented perpendicular and parallel to the nanotube symmetry Z axis, respectively; in the chiral tubes, hexagons orientation with respect to Z axis is intermediate. The armchair and zigzag nanotubes have relatively small numbers of atoms in the minimum translational unit cell, $N_{Tr} = 4n$. In the case of the chiral tubules even with small diameters, the minimum number of atoms per unit cell can be very large. For example, the translational unit cells of the achiral (10, 10) and chiral (10, 9) single-walled nanotubes with virtually equal diameters contain 40 and 1084 carbon atoms, respectively.

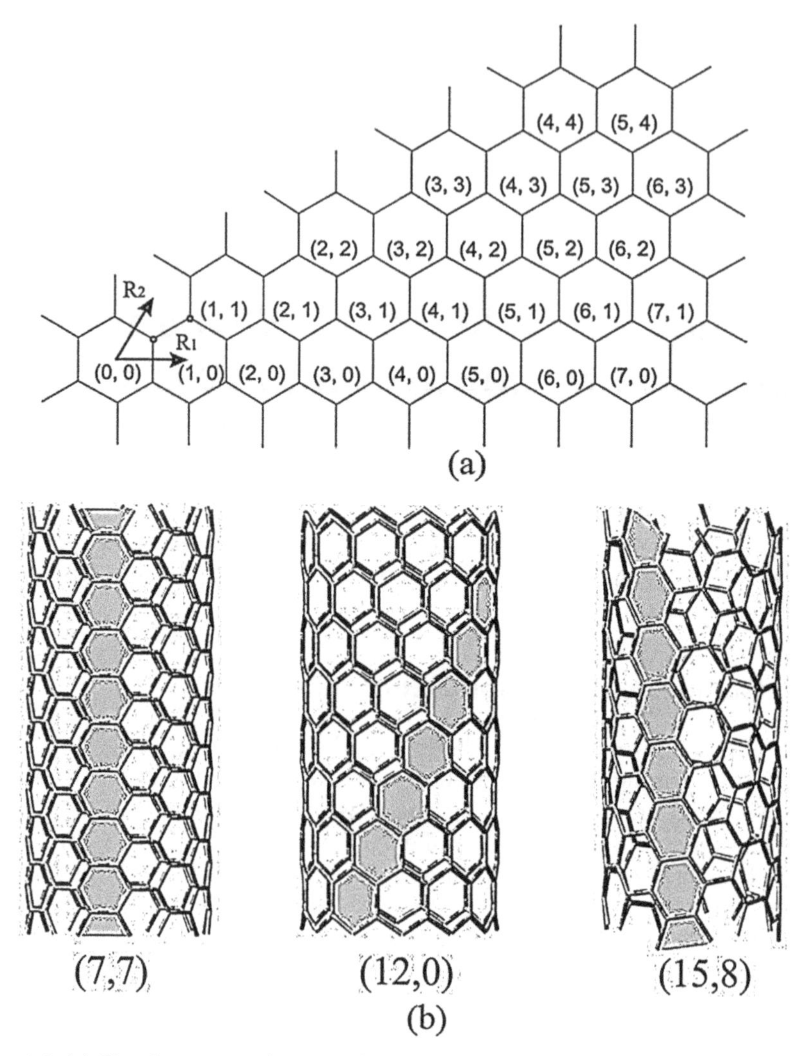

Fig. 1.1. (a) Graphene monolayer with indexed lattice sites. Rolling the layer from (0, 0) to (n_1, n_2) point gives a single-walled (n_1, n_2) tubule. (b) Examples of armchair (7, 7), zigzag (12, 0), and chiral (15, 8) nanotubes.

1.2. Theory

1.2.1. One-Electron Orbitals and Secular Equations

In the LAPW and LACW methods, the concept of one-electron orbitals is used. It is assumed that separate electrons in a polyatomic system are characterized by wave functions of their own, or spin orbitals. Each spin orbital ψ_i represents a function of spatial coordinates of the electron and its spin. It is assumed in the simplest approach that spin orbitals can be written as products of spatial and spin functions $\psi_i(\mathbf{r})\alpha$ and $\psi_i(\mathbf{r})\beta$, where α and β are the spin functions of electrons with spins up and down, respectively. The spatial function $\psi_i(\mathbf{r})$ is called an orbital. A study of the electronic structure of a multiatomic system therefore reduces to a study of its orbitals if spin-orbit coupling is neglected. These orbitals are the functions defined in three-dimensional space and are therefore easy to visualize. The $\psi_i(\mathbf{r})$ function, together with the corresponding one-electron energies E_i are found by solving the one-electron Schrödinger equation

$$H\psi_i(\mathbf{r}) = E_i\psi_i(\mathbf{r}) \tag{1.1}$$

with effective one-electron Hamiltonian written in atomic Rydberg units (Plank constant $\hbar = 1$, electron mass $m = 1/2$, electron charge $e = \sqrt{2}$)

$$H = -\left(\frac{\partial^2}{\partial x^2} + \frac{\partial^2}{\partial y^2} + \frac{\partial^2}{\partial z^2}\right) + U \equiv -\Delta + U. \tag{1.2}$$

This Hamiltonian contains the electron kinetic energy operator, $-\Delta$, and the operator U describing the summed action on the electron in consideration of all the other electrons in the system and all its nuclei α.

In all the LAPW and LACW methods, the basic functions $\phi_n(\mathbf{r})$ are continuous and differentiable everywhere, and the Rayleigh-Ritz variational principle is then easily applied. Expanding the electronic wave functions

$$\psi_i(\mathbf{r}) = \sum_n a_{i,n}\varphi_n(\mathbf{r}) \tag{1.3}$$

and applying the variational principle then yield the secular equations:

$$\det\left\| \langle\varphi_n|H|\varphi_m\rangle - E\langle\varphi_n|\varphi_m\rangle \right\| = 0, \tag{1.4}$$

$$\sum_n \left[\langle\varphi_n|H|\varphi_m\rangle - E_i\langle\varphi_n|\varphi_m\rangle\right] a_{i,n} = 0, \tag{1.5}$$

where $\langle\varphi_n|H|\varphi_m\rangle$ and $\langle\varphi_n|\varphi_m\rangle$ are the Hamiltonian H and overlap matrix elements, respectively. As different from the original Slater's APW approach, the most important feature of the LAPW and LACW secular equations (1.4) and (1.5) is that the Hamiltonian and overlap matrices are energy independent, which permits the simultaneous determination of the eigenvalues E_i and eigenvectors $a_{i,n}$.

1.2.2. Cylindrical Muffin-Tin Potential

Finding the U is a nontrivial task, which is solved differently in different quantum chemical methods. For instance, in the Hartree-Fock approximation, U is a nonlocal operator. In the LAPW method, the local density approximation $U = U(\mathbf{r})$ is commonly used:

$$U(\mathbf{r}) = -2\sum_{\alpha}\frac{z_{\alpha}}{|\mathbf{r}-\mathbf{R}_{\alpha}|}+2\int\frac{\rho(\mathbf{r}')}{|\mathbf{r}-\mathbf{r}'|}d\mathbf{r}'+V_{X\alpha}(\mathbf{r}), \quad (1.6)$$

where z_{α} and R_{α} are the nuclear charge and coordinate of atom α and $\rho(\mathbf{r})$ is the electron density. The first term in (1.6) describes attraction between the electrons and nuclei, the second one corresponds to the mutual repulsion of electrons, and $V_{X\alpha}(\mathbf{r})$ is the exchange correlation potential (Hohenberg and Kohn 1964, Kohn and Sham 1965). A frequently used equation for $V_{X\alpha}(\mathbf{r})$ called the Slater's potential is:

$$V_{X\alpha}(\mathbf{r}) = -6\alpha\left(\frac{3}{8\pi}\rho(r)\right)^{1/3}, \quad \alpha = \frac{2}{3}. \quad (1.7)$$

However, the application of Eq. (1.6) for practical calculations of the band structure of complex materials is sometimes too time-consuming, and developing the APW method Slater proposed a simpler approach for calculating the local potential $U(\mathbf{r})$. He clearly and concisely states the main idea of the method: Near an atomic nucleus the potential and wave functions are similar to those in an atom; they vary strongly, but near the spherical ones. On the contrary, between the atoms both the potential and wave functions are smoother. In the Muffin-Tin (MT) approximation, as it is called, the electron potential is suggested to be constant in the interstitial region and spherically symmetric in the MT-spheres. In the MT approximation in Eq. (1.6), z_{α} is substituted for the MT-sphere effective charge z_{α}^{MT}. The constant potential can be taken as the origin for measurements of energy. The radii of the MT spheres are selected so that the spheres of neighboring atoms are in contact, which corresponds to the maximal volume of nonoverlapping MT regions. Such a choice is physically rather evident: information (electron density and potential) on the chemical nature of atoms constituting a polyatomic system is contained only inside the MT spheres. The largest possible volume of the MT spheres corresponds to the maximal amount of such information. In this model, the electronic band structure of a crystal is determined by the free electron movement in the interspherical region and by the electron scattering on the MT spheres (Fig. 1.2).

The LAPW method with MT approximation has been extensively tested and applied in the numerous calculations and shown to be the good technique for quantitative electronic band structure calculations of closed packed materials, but it becomes less reliable as the site symmetry and coordination decreases and a volume of interspherical region increases.

The high accuracy and computational efficiency of the MT-LAPW method in the theory of electron properties of crystals suggest that this

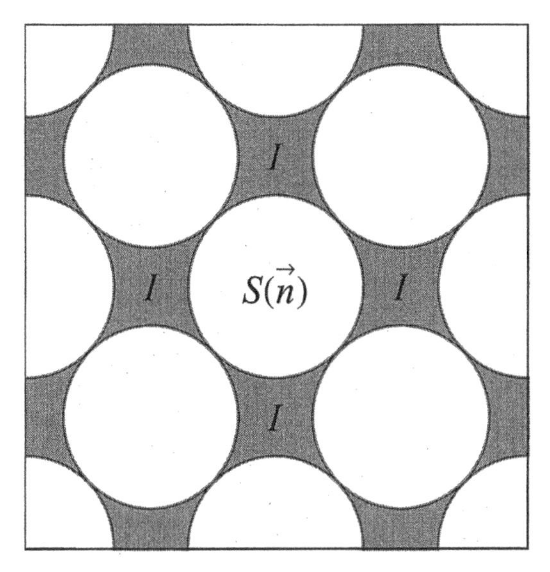

Fig. 1.2. Dual partitioning of space in crystal: region inside the nonoverlapping MT spheres *S* and the interstitial region *I*. The MT spheres touch but do not overlap.

method can be the good starting point for development of the quantitative technique for calculating the nanotubes electron properties. In common with the standard and most simple bulk LAPW technique, the one-electron potential is used and is constructed in terms of the MT and local density approximations in the LACW theory too. In our LACW model, we also apply the $\rho^{1/3}$ local density exchange potential (1.7). The electron density $\rho(\mathbf{r})$ of tubule is calculated as a superposition of atomic densities, and its spherically symmetric part $\rho(r)$ is taken in the MT spheres. However, the electronic potential of the nanomaterials differs drastically from that of the bulk system. An infinite motion of electrons is possible in any direction in crystals, but it is obviously limited in the case of nanomaterials by their size and shape. Nanotubes are the giant cage molecules looking like closed hollow cylindrical shells. The electron motion in tubules is confined by an approximately cylindrical layer with a thickness of an order of the doubled van der Waals radius of the atom. In the case of the nanotube, there are two vacuum regions inside and outside the tubule (Fig. 1.3).

The nanotube and vacuum regions are separated by the essentially impenetrable approximately cylindrical potential barriers Ω_a and Ω_b that are obviously absent in the case of bulk crystal. Therefore, the MT method must be slightly adapted to the cylindrical structures. The term cylindrical MT potential is taken to mean that the one-electron potential is spherically symmetric in the atomic regions $\Omega_{I\alpha}$ and constant in the interstitial space Ω_{II} up to the two impenetrable cylindrical potential barriers. Note that

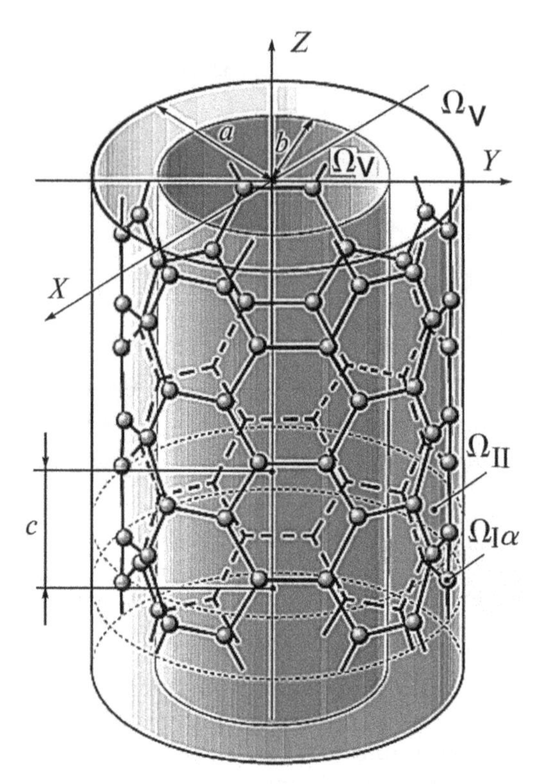

Fig. 1.3. Nanotube in a tubular potential. The nanotube is positioned between infinite barriers Ω_a and Ω_b, which separate the multiatomic system from the two vacuum regions Ω_v outside and inside of tubule.

the nanotubes structure virtually approach the closed packed one due to these barriers, and the MT approximation becomes applicable. The radii a and b depend on the radius of tubule R_{NT} and distance δ between the Ω_a and Ω_b barriers that is a parameter of the cylindrical MT theory.

In the LACW method, as different from the LAPW approach to bulk materials, the MT approximation requires an introduction of free parameter, namely, the width of cylindrical layer. This complicates the self-consistent calculations of nanotubes in the terms of the LACW method. Both for the self-consistent and non-self-consistent calculations, the results depend on the choice of this parameter, the best fitted parameters being different in these cases. Fortunately, our test self-consistent calculations of the carbon tubules using a regular k-point mesh, where seven points were equally spaced between Brillouin zone center and boundary, show that the effect of self-consistency is virtually equivalent to that of a change of the width δ of a cylindrical layer. Therefore, we perform the non-self-consistent calculations of the band structure of nanotubes in order to avoid this ambiguity in the choice of the width of the cylindrical layer, the electron density of the nanotubes being constructed as the superposition of the atomic ones.

Thus, in the cylindrical MT approach, the electronic bands of the nanotube are determined by the free electron movement in the cylindrical interspherical region, by the electron scattering on the MT spheres, and by the electron reflection from the inner and outer barriers. In order to realize the LACW method of calculation of the electronic structure of nanotubes, it is now necessary to translate these words into the language of quantum mechanics:

(i) We must solve the Schrödinger equation for the interstitial region. The cylindrical waves are the solutions of this problem, they are known from text books on quantum mechanics;
(ii) We must solve this equation for the MT sphere regions. Here, the atomic orbitals are the solutions due to the spherical symmetry of potential;
(iii) Next, we have to sew together the obtained wave functions on the boundaries of MT spheres so that the resulting basic functions are continuous and differentiable. This is not a completely trivial algebraic problem, since the cylindrical waves have a simple form in the general cylindrical coordinate system, and solutions of the atomic problem, in the local spherical coordinate system. Fortunately, using the addition theorem for cylindrical functions, they can be represented in terms of the local spherical coordinate system too;
(iv) It remains only to calculate the Hamiltonian and overlap matrix elements in terms of these basic functions having a rather simple analytical form; and
(v) Substituting the matrix elements into the secular equations and solving these equations one determines the required electronic dispersion curves and the wave functions of the nanotube.

We begin to implement this program, starting with the simplest case of nonchiral nanotubes (D'yachkov and Kirin 1999, 2001).

1.2.3. Solution of the Schrödinger Equation for Interspherical Region

In the interspherical region, the basic functions are the solutions of the Schrödinger equation for the free electron movement inside an infinite tube with the outer and inner radii a and b, respectively (Galitski et al. 1984, 2013). When expressed in the atomic Rydberg units and cylindrical coordinates Z, Φ and R, this equation takes the form

$$\left\{-\left[\frac{1}{R}\frac{\partial}{\partial R}\left(R\frac{\partial}{\partial R}\right)+\frac{1}{R^2}\frac{\partial^2}{\partial \Phi^2}+\frac{\partial^2}{\partial Z^2}\right]+U(R)\right\}\Psi(Z,\Phi,R)=E\Psi(Z,\Phi,R). \quad (1.8)$$

The potential $U(R)$, determining the region in which electrons of an isolated nanotube are allowed to move, takes the form

$$U(R) = \begin{cases} 0, & b \le R \le a \\ \infty, & R < b,\ R > a. \end{cases} \quad (1.9)$$

Equation (1.8) is derived from the one-electron Hamiltonian $H = -\Delta + U(R)$ by substituting the kinetic energy operator $-\Delta$ by the corresponding expression in cylindrical coordinate system.

Because of the cylindrical symmetry of the $U(R)$ potential (1.9), the solution of Eq. (1.8) has the form $\Psi(Z, \Phi, R) = \Psi_P(Z,k)\ \Psi_M(\Phi)\Psi_{MN}(R)$. Here,

$$\Psi_P(Z,k) = (2/\sqrt{c})\exp[i(k+k_P)Z],\ k_P = (2\pi/c)P,\ P = 0, \pm1, \pm2, \ldots \quad (1.10)$$

is the wave function that describes the free motion of an electron along the translational symmetry axis Z with the period c. The wave vector k belongs to the one-dimensional Brillouin zone: $-\pi/c \le k \le \pi/c$. The function

$$\Psi_M(\Phi) = \frac{1}{\sqrt{2\pi}}e^{iM\Phi},\ M = 0, \pm1, \pm2, \ldots \quad (1.11)$$

describes the rotation of electron about the symmetry axis of system. The $\Psi_{|M|N}(R)$ function, determining the radial motion of the electron, is the solution of the equation

$$\left[-\frac{1}{R}\frac{d}{dR}R\frac{d}{dR}+\frac{M^2}{R^2}\right]\Psi_{|M|N}(R)+U(R)\Psi_{|M|N}(R)=E_{|M|N}\Psi_{|M|N}(R). \quad (1.12)$$

Here, N is the radial quantum number and $E_{|M|,N}$ is the energy spectrum; the energy

$$E = K^2_P + E_{|M|,N} \quad (1.13)$$

corresponds to the wave function $\Psi(Z, \Phi, R)$, where $K_P = k + k_P$.

At $b \le R \le a$, $U(R) = 0$ and Eq. (1.12) is written as

$$\left[\frac{d^2}{dR^2}+\frac{1}{R}\frac{d}{dR}+\kappa^2_{|M|,N}-\frac{M^2}{R^2}\right]\Psi_{|M|N}(R)=0, \quad (1.14)$$

where $\kappa_{|M|,N} = (E_{|M|,N})^{1/2}$. After substituting $\kappa R = x$ and $\Psi(R) = y(x)$ into Eq. (1.14), it reduces to the canonical Bessel equation $x^2y'' + xy' + (x^2 - M^2)y = 0$. Its solutions are referred to as cylindrical functions of the Mth order (Watson 1966, Korn and Korn 1961).

Any solution of the Bessel equation can be represented as a linear combination of its partial solutions (Fig. 1.4), cylindrical Bessel functions of the first J_M and second Y_M kinds (Neumann functions):

$$\Psi_{|M|N}(R) = C_{MN}^{J} J_M\left(\kappa_{|M|,N} R\right) + C_{MN}^{Y} Y_M\left(\kappa_{|M|,N} R\right), \tag{1.15}$$

where C_{MN}^{J} and C_{MN}^{Y} are the constants that should be chosen in such a way as to ensure the normalization of the wave function $\Psi_{M,N}(R)$

$$\int_b^a \left|\Psi_{|M|N}(R)\right|^2 R dR = 1 \tag{1.16}$$

and its vanishing at the interior and exterior potential barriers:

$$C_{MN}^{J} J_M\left(\kappa_{|M|,N} a\right) + C_{MN}^{Y} Y_M\left(\kappa_{|M|,N} a\right) = 0, \tag{1.17}$$

$$C_{MN}^{J} J_M\left(\kappa_{|M|,N} b\right) + C_{MN}^{Y} Y_M\left(\kappa_{|M|,N} b\right) = 0. \tag{1.18}$$

From the set of Eqs. (1.17, 1.18), the relationship between C_{MN}^{J} and C_{MN}^{Y}

$$C_{MN}^{Y} = -C_{MN}^{J} \frac{J_M\left(\kappa_{|M|,N} a\right)}{Y_M\left(\kappa_{|M|,N} a\right)} \tag{1.19}$$

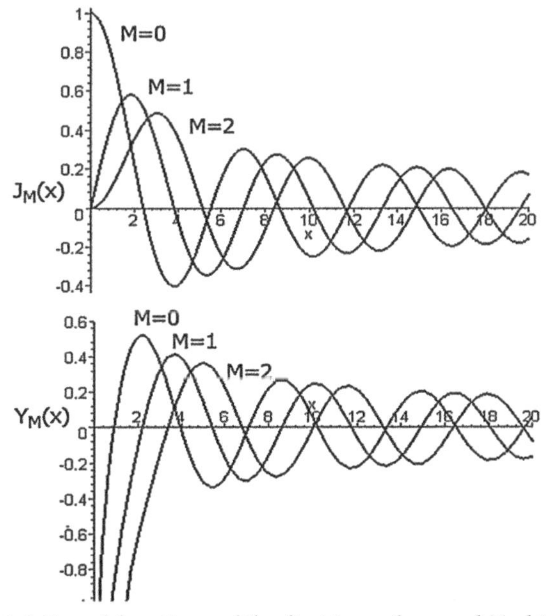

Fig. 1.4. Bessel functions of the first J_M and second Y_M kinds.

and equation for calculation of $\kappa_{|M|,N}$

$$J_M\left(\kappa_{|M|,N}a\right)Y_M\left(\kappa_{|M|,N}b\right)=J_M\left(\kappa_{|M|,N}b\right)Y_M\left(\kappa_{|M|,N}a\right) \tag{1.20}$$

are obtained. To calculate integral (1.16), let us use the equation

$$\int^{z} zF_M(\kappa z)G_M(\kappa z)dz=\frac{z^2}{4}\left[2F_M(\kappa z)G_M(\kappa z)-F_{M-1}(\kappa z)G_{M+1}(\kappa z)\right.$$
$$\left.-F_{M+1}(\kappa z)G_{M-1}(\kappa z)\right] \tag{1.21}$$

for an indefinite integral and the recurrence formulas

$$zF_{M-1}(z)=zF'_M(z)+MF_M(z), \tag{1.22}$$

$$-zF_{M+1}(z)=zF'_M(z)-MF_M(z), \tag{1.23}$$

where F_M and G_M are any two cylindrical functions, in particular, J_M and Y_M (Watson 1966, Korn and Korn 1961). Then,

$$\int^{z} zF_M(\kappa z)G_M(\kappa z)dz=\frac{z^2}{2}\left[F'_M(\kappa z)G'_M(\kappa z)+\left(1-\left(\frac{M}{\kappa z}\right)^2\right)F_M(\kappa z)G_M(\kappa z)\right], \tag{1.24}$$

where F'_M and G'_M are the derivatives of the cylindrical functions. Finally, from normalization integral (1.16), we obtain the equation

$$\int_b^a \Psi^*_{|M|N}(R)\Psi_{|M|N}(R)RdR=\frac{a^2}{2}\Psi'^*_{|M|N}(a)\Psi'_{|M|N}(a)-\frac{b^2}{2}\Psi'^*_{|M|N}(b)\Psi'_{|M|N}(b)$$
$$=\frac{a^2}{2}\left[C^J_{MN}J'_M\left(\kappa_{|M|,N}a\right)+C^Y_{MN}Y'_M\left(\kappa_{|M|,N}a\right)\right]^2-$$
$$\frac{b^2}{2}\left[C^J_{MN}J'_M\left(\kappa_{|M|,N}b\right)+C^Y_{MN}Y'_M\left(\kappa_{|M|,N}b\right)\right]^2=1 \tag{1.25}$$

for the coefficients C^J_{MN} and C^Y_{MN}. Thus, the basic function in the Ω_{II} region in the general cylindrical coordinate system takes the form

$$\Psi^{PMN,k}_{II}=\frac{1}{\sqrt{2\pi c}}\exp\left[i(K_PZ+M\Phi)\right]\left[C^J_{MN}J_M\left(\kappa_{|M|,N}R\right)+C^Y_{MN}Y_M\left(\kappa_{|M|,N}R\right)\right]. \tag{1.26}$$

1.2.4. Muffin-Tin Regions

In LAPW and LACW models, the MT regions are treated similarly (Andersen 1975, Koelling and Arbman 1975). Namely, inside the MT sphere, the Hamiltonian H_{MT} is spherically symmetric. The radial wave functions $u_{l\alpha}(r)$ are taken to be the solutions of the radial Schrödinger equation with energy $E_{l\alpha}$

$$H_{MT\alpha}u_{l\alpha}(r) = E_{l\alpha}u_{l\alpha}(r). \tag{1.27}$$

The $E_{l\alpha}$values are kept fixed within some energy region depending on the nature of atom. In our calculations of nanotubes, for the non-transition elements α we take the $E_{0\alpha}$ and $E_{1\alpha}$ values to be equal to the atomic energies of the valence *ns* and *np* electrons (e.g., –14.1 and –5.9 eV for the C atom). For $8 \geq l \geq 2$, we arbitrarily take the large values of $E_{l\alpha}$ between 10 and 30 eV, which corresponds to a weak participation of these states in the chemical bonding in the compounds. In the case of transition metals atoms α, the $E_{0\alpha}$ and $E_{2\alpha}$ values were equal to atomic energies of the valence $(n + 1)s$ and *nd* electrons; other $E_{l\alpha}$energies for $l \leq 8$ were between 10 and 30 eV. [The results of the linear augmented wave calculations are known to be stable with respect to reasonable variations of the constant-energy parameters $E_{l\alpha}$ (Andersen 1975, Koelling and Arbman 1975)]. The contribution of states with $l > 8$ was neglected.

In Rydberg units, Eq. (1.27) takes the form

$$\frac{1}{r}\frac{d^2 r u_{l\alpha}(r)}{dr^2} + \left(E_{l\alpha} - V_{MT\alpha}(r) - \frac{l(l+1)}{r^2}\right)u_{l\alpha}(r) = 0. \tag{1.28}$$

Here, $V_{MT\alpha}(r)$ is the local density spherically symmetric potential in the region of the MT sphere α. The functions $u_{l\alpha}$ are normalized inside the MT spheres as

$$\int_0^{r_\alpha} [u_{l\alpha}(r)]^2 r^2 dr = 1. \tag{1.29}$$

Here, r_α is a radius of MT sphere of atom α. Differentiating the Eqs. (1.27) and (1.28) with respect to energy yields a differential equation for the energy derivative of radial functions $\dot{u}_{l\alpha} = [\partial u_{l\alpha} / \partial E]_{E_{l\alpha}}$

$$H\dot{u}_{l\alpha}(r) = u_{l\alpha}(r) + E_l \dot{u}_{l\alpha}(r) \tag{1.30}$$

or

$$\frac{1}{r}\frac{d^2 r\dot{u}_{l\alpha}(r)}{dr^2} + u_{l\alpha}(r) + \left(E_{l\alpha} - V_{MT\alpha}(r) - \frac{l(l+1)}{r^2}\right)\dot{u}_{l\alpha}(r) = 0 \tag{1.31}$$

The functions $\dot{u}_{l\alpha}(r)$ and $u_{l\alpha}$, are orthogonal as can be seen by differentiating Eq. (1.29) with respect to energy,

$$\int_0^{r_\alpha} \dot{u}_{l\alpha}(r)u_{l\alpha}(r)r^2 dr = 0, \int_0^{r_\alpha} \dot{u}_{l\alpha}(r)Hu_{l\alpha}(r)r^2 dr = 0 \tag{1.32}$$

and

$$\int_0^{r_\alpha} u_{l\alpha}(r) H \dot{u}_{l\alpha}(r) r^2 dr = 1, \int_0^{r_\alpha} \dot{u}_{l\alpha}(r) H \dot{u}_{l\alpha}(r) r^2 dr = E_{l\alpha} N_{l\alpha}, \quad (1.33)$$

where

$$N_{l\alpha} = \int_0^{r_\alpha} |\dot{u}_{l\alpha}(r)|^2 r^2 dr.$$

Multiplying Eq. (1.28) by $r^2 \dot{u}_{l\alpha}(r)$ and Eq. (1.31) by $r^2 u_{l\alpha}(r)$, subtracting Eq. (1.31) from Eq. (1.28), and integrating then yield the identity

$$\int_0^{r_\alpha} r \left(\dot{u}_{l\alpha} \frac{d^2 r u_{l\alpha}}{dr^2} - u_{l\alpha} \frac{d^2 r \dot{u}_{l\alpha}}{dr^2} \right) dr = 1. \quad (1.34)$$

Moreover, we have

$$\int_0^{r_\alpha} r \left(\dot{u}_{l\alpha} \frac{d^2 r u_{l\alpha}}{dr^2} - u_{l\alpha} \frac{d^2 r \dot{u}_{l\alpha}}{dr^2} \right) dr = r^2 \left(\dot{u}_{l\alpha} \frac{du_{l\alpha}}{dr} - u_{l\alpha} \frac{d\dot{u}_{l\alpha}}{dr} \right). \quad (1.35)$$

Finally, from Eqs. (1.34) and (1.35), we obtain

$$\left[r^2 \left(\dot{u}_{l\alpha} u'_{l\alpha} - u_{l\alpha} \dot{u}'_{l\alpha} \right) \right]_{r=r_\alpha} = 1, \quad (1.36)$$

where $u'_{l\alpha} = du_{l\alpha}/dr$ and $\dot{u}'_{l\alpha} = d\dot{u}_{l\alpha}/dr$ are radial derivatives of the $u_{l\alpha}(r)$ and $\dot{u}_{l\alpha}(r)$ functions.

As in the LAPW method, in terms of the local spherical coordinate system (r, θ, φ) inside the MT spheres ($\Omega_{I\alpha}$), the basic function $\Psi_{k,PMN}$ is expanded in spherical harmonics $Y_{lm}(\theta, \varphi)$:

$$\Psi_{I\alpha} = \sum_{l=0}^{\infty} \sum_{m=-l}^{l} \left[A_{lm\alpha}^{MNP} u_{l\alpha}(r, E_{l\alpha}) + B_{lm\alpha}^{MNP} \dot{u}_{l\alpha}(r, E_{l\alpha}) \right] Y_{lm}(\theta, \varphi). \quad (1.37)$$

Here

$$Y_{lm}(\theta, \phi) = N_{lm} P_l^{|m|}(\cos\theta) e^{im\phi}, N_{lm} = (-1)^{\frac{m+|m|}{2}} i^l \left[\frac{2l+1}{4\pi} \frac{(l-|m|)!}{(l+|m|)!} \right]^{1/2}, \quad (1.38)$$

and $P_l^{|m|}$ are the augmented Legendre polynomials.

Finally, the coefficients $A_{lm\alpha}^{MNP}$ and $B_{lm\alpha}^{MNP}$ are to be selected so that both the basic functions $\Psi_{k,PMN}$ and their derivatives have no discontinuities at the boundaries of the MT spheres.

1.2.5. Sewing Conditions

To do this, let us express the Ψ_{II} function (Eq. 1.26) through the cylindrical coordinates Z_α, Φ_α, and R_α of the center of the αth sphere and through the local spherical coordinate system r, θ, and φ with the origin on the nucleus of the atoms (Fig. 1.5).

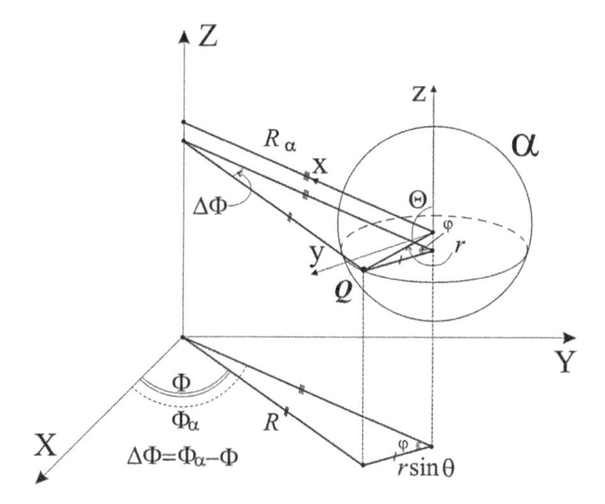

Fig. 1.5. General cylindrical and local spherical coordinate systems.

Let $Q = Q(Z, \Phi, R)$ be the point with coordinates Z, Φ, and R in the general cylindrical coordinate system and with coordinates r, θ, and φ in the local spherical coordinate system. Then, considering $Z = Z_\alpha + \rho \cos\theta$ and $\Delta\Phi = \Phi_\alpha - \Phi$, we have

$$\Psi_{\mathrm{II}\alpha}(k,P,M,N) = \frac{1}{\sqrt{2\pi c}} \exp\left[i\left(K_P Z_\alpha + M\Phi_\alpha\right)\right] \exp\left[iK_P r\cos\theta\right] \times \left[C^J_{MN} J_M\left(\kappa_{|M|,N} R\right) + C^Y_{MN} Y_M\left(\kappa_{|M|,N} R\right)\right] e^{-iM\Delta\Phi} \quad (1.39)$$

In order to finally write this equation in terms of the r, θ, and φ coordinates, one has to eliminate the cylindrical coordinates R and $\Delta\Phi$. To achieve this objective, one has to use the theorem of addition (expansion) for cylindrical functions F_M (Fig. 1.6), according to which

$$e^{iM\cdot\cdot} F_M(\kappa r_1) = \sum_{m=-\infty}^{\infty} J_m(\kappa r_3) F_{m+M}(\kappa r_2)\, e^{im\phi}, \quad (1.40)$$

where F_M is the cylindrical Bessel functions of the first (J_M) or second (Y_M) kind and κ is an arbitrary complex number (Watson 1966).

Substituting $-M$ for M in the general Eq. (1.40) for the theorem of addition and using the relationship $F_{-M} = (-1)^M F_M$, we can write:

$$e^{-iM\Delta\Phi} F_M\left(\kappa_{|M|,N} R\right) = (-1)^M \sum_{m=-\infty}^{+\infty} J_m\left(\kappa_{|M|,N} r \sin\theta\right) F_{m-M}\left(\kappa_{|M|,N} R_\alpha\right)\, e^{im\phi}. \quad (1.41)$$

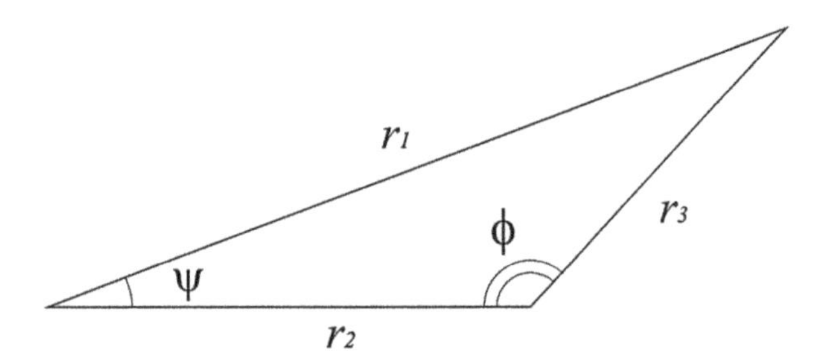

Fig. 1.6. Illustration to the theorem of addition.

Finally, the cylindrical waves in terms of local spherical coordinates take the form

$$\Psi_{II\alpha,MNP} = \frac{1}{\sqrt{2\pi c}} \exp\left[i\left(K_P Z_\alpha + M\Phi_\alpha\right)\right] \exp\left(iK_P r\cos\theta\right)(-1)^M \times$$
$$\sum_{m=-\infty}^{+\infty} \left[C^J_{MN} J_{m-M}\left(\kappa_{|M|,N} R_\alpha\right) + C^Y_{MN} Y_{m-M}\left(\kappa_{|M|,N} R_\alpha\right)\right] J_m\left(\kappa_{|M|,N} r\sin\theta\right) e^{im\varphi}. \quad (1.42)$$

Now, both the $\Psi_{I\alpha,PMN}$ (1.37) and $\Psi_{II\alpha,PMN}$ (1.42) functions are written in terms of the same local spherical coordinate systems, and from the equality of these functions and of their radial derivatives at the MT sphere boundary, we have

$$\sum_{l=|m|}^{\infty} \left[A^{MNP}_{lm\alpha} u_{l\alpha}\left(r_\alpha, E_{l\alpha}\right) + B^{MNP}_{lm\alpha} \dot{u}_{l\alpha}\left(r_\alpha, E_{l\alpha}\right)\right] N_{lm} P_l^{|m|}(\cos\theta)$$
$$= \frac{1}{\sqrt{2\pi c}} \exp\left[i\left(K_P Z_\alpha + M\Phi_\alpha\right)\right] \times (-1)^M \exp\left(iK_P r_\alpha \cos\theta\right)$$
$$\times \left[C^J_{MN} J_{m-M}\left(\kappa_{|M|,N} R_\alpha\right) + C^Y_{MN} Y_{m-M}\left(\kappa_{|M|,N} R_\alpha\right)\right] J_m\left(\kappa_{|M|,N} r_\alpha \sin\theta\right), \quad (1.43)$$

$$\sum_{l=|m|}^{\infty} \left[A^{MNP}_{lm\alpha} u'_{l\alpha}\left(r_\alpha, E_{l\alpha}\right) + B^{MNP}_{lm\alpha} \dot{u}'_{l\alpha}\left(r_\alpha, E_{l\alpha}\right)\right] N_{lm} P_l^{|m|}(\cos\theta) =$$
$$\frac{1}{\sqrt{2\pi c}} (-1)^M \exp\left[i\left(K_P Z_\alpha + M\Phi_\alpha\right)\right] \exp\left[iK_P r_\alpha \cos\theta\right] \times$$
$$\left[C^J_{MN} J_{m-M}\left(\kappa_{|M|,N} R_\alpha\right) + C^Y_{MN} Y_{m-M}\left(\kappa_{|M|,N} R_\alpha\right)\right] \times$$
$$\left[\kappa_{|M|,N} \sin\theta J'_m\left(\kappa_{|M|,N} r_\alpha \sin\theta\right) + iK_P \cos\theta J_m\left(\kappa_{|M|,N} r_\alpha \sin\theta\right)\right]. \quad (1.44)$$

We now multiply both sides of Eq. (1.44) by $P_l^{|m|}(\cos\theta)\sin\theta$, integrate over θ from 0 to π, and consider the orthogonality of the Legendre polynomials

$$\int_0^{\pi} P_l^m(\cos\theta)P_{l'}^m(\cos\theta)\sin\theta\, d\theta = \delta_{ll'}\frac{2}{2l+1}\frac{(l+m)!}{(l-m)!}. \tag{1.45}$$

Then, with account of Eqs. (1.43) and (1.44), we finally obtain the sewing conditions

$$A_{lm\alpha}^{MNP} = D_{lm\alpha}^{MNP} r_\alpha^2 a_{lm\alpha}^{MNP},\quad B_{lm\alpha}^{MNP} = D_{lm\alpha}^{MNP} r_\alpha^2 b_{lm\alpha}^{MNP}. \tag{1.46}$$

Here, the following designations are used

$$D_{lm\alpha}^{MNP} = \frac{1}{\sqrt{2c}}(-1)^{\frac{1}{2}(m+|m|)+l}i^{l}\left[(2l+1)\frac{(l-|m|)!}{(l+|m|)!}\right]^{1/2}\exp\left[i\left(K_P Z_\alpha + M\Phi_\alpha\right)\right]\times$$
$$(-1)^M\left[C_{MN}^J J_{m-M}\left(\kappa_{|M|,N}R_\alpha\right)+C_{MN}^Y Y_{m-M}\left(\kappa_{|M|,N}R_\alpha\right)\right], \tag{1.47}$$

$$a_{lm\alpha}^{MNP} = I_2^{m\alpha}\, \dot{u}_{l\alpha}(r_\alpha, E_{l\alpha}) - I_1^{m\alpha}\, \dot{u}'_{l\alpha}(r_\alpha, E_{l\alpha}), \tag{1.48}$$

$$b_{lm\alpha}^{MNP} = I_1^{m\alpha}\, u'_{l\alpha}(r_\alpha, E_{l\alpha}) - I_2^{m\alpha}\, u_{l\alpha}(r_\alpha, E_{l\alpha}). \tag{1.49}$$

$$I_1^{m\alpha} = 2\int_0^{\pi/2}\exp\left(iK_P r_\alpha\cos\theta\right)J_m\left(\kappa_{|M|,N}r_\alpha\sin\theta\right)P_l^{|m|}(\cos\theta)\sin\theta d\theta, \tag{1.50}$$

$$I_2^{m\alpha} = 2\int_0^{\pi/2}\exp\left(iK_P r_\alpha\cos\theta\right)\left[iK_P\cos\theta J_m(\kappa_{|M|,N}r_\alpha\sin\theta)+(1/2)\kappa_{|M|,N}\sin\theta\right]\times$$
$$[J_{m-1}(\kappa_{|M|,N}r_\alpha\sin\theta)-J_{m+1}(\kappa_{|M|,N}r_\alpha\sin\theta)]P_l^{|m|}(\cos\theta)\sin\theta d\theta. \tag{1.51}$$

1.2.6. Overlap Integrals

The integral of the product of the basic functions Ψ_{M2N2P2} and Ψ_{M1N1P1} over the unit cell Ω is equal to the integral of the product of cylindrical waves $\Psi^*_{II,M_2N_2P_2}$ and $\Psi_{II,M_1N_1P_1}$ over the interspherical regions Ω_{II} plus the sum of the integrals of the product of spherical parts of the basic functions $\Psi^*_{I\alpha,M_2N_2P_2}$ and $\Psi_{I\alpha,M_1N_1P_1}$ over the MT regions. The integral over interspherical region Ω_{II} is equal to the integral over unit cell Ω minus the sum of the integrals over the MT regions:

$$\left\langle\Psi^*_{M_2N_2P_2}\middle|\Psi_{M_1N_1P_1}\right\rangle = \int_\Omega \Psi^*_{II,M_2N_2P_2}\Psi_{II,M_1N_1P_1}dV$$
$$-\sum_\alpha\int_{\Omega_\alpha}\Psi^*_{II\alpha,M_2N_2P_2}\Psi_{II\alpha,M_1N_1P_1}dV$$
$$+\sum_\alpha\int_{\Omega_\alpha}\Psi^*_{I\alpha,M_2N_2P_2}\Psi_{I\alpha,M_1N_1P_1}dV. \tag{1.52}$$

Due to the fact that the cylindrical waves as the solutions of Schrödinger equation are orthonormalized, the integral over Ω is equal to the product of the δ functions. As a result, the last equation takes the form

$$\begin{aligned}\langle P_2M_2N_2 \mid P_1M_1N_1\rangle = \delta_{P_2P_1}\delta_{M_2M_1}\delta_{N_2N_1} \\ -\sum_{\alpha}\int_{\Omega_\alpha}\Psi^*_{II\alpha,M_2N_2P_2}\Psi_{II\alpha,M_1N_1P_1}dV \\ +\sum_{\alpha}\int_{\Omega_\alpha}\Psi^*_{I\alpha,M_2N_2P_2}\Psi_{I\alpha,M_1N_1P_1}dV. \qquad (1.53)\end{aligned}$$

With the use of Eqs (1.37, 1.42) for the $\Psi_{I\alpha,MNP}$ and $\Psi_{II\alpha,MNP}$ functions, we finally obtain after simple algebra the following equation for the overlap matrix elements:

$$\begin{aligned}\langle P_2M_2N_2 | P_1M_1N_1\rangle = \delta_{P_2P_1}\delta_{M_2M_1}\delta_{N_2N_1} - \frac{1}{c}(-1)^{M_1+M_2} \\ \times\sum_{\alpha}\exp\left\{ i\left[\left(K_{P_1}-K_{P_2}\right)Z_\alpha+\left(M_1-M_2\right)\Phi_\alpha\right]\right\} \\ \times\sum_{m=-\infty}^{+\infty}\left[C^J_{M_2N_2}J_{m-M_2}\left(\kappa_{|M_2|,N_2}R_\alpha\right)+\right. \\ \left. C^Y_{M_2N_2}Y_{m-M_2}\left(\kappa_{|M_2|,N_2}R_\alpha\right)\right]\times \\ \left[C^J_{M_1N_1}J_{m-M_1}\left(\kappa_{|M_1|,N_1}R_\alpha\right)+C^Y_{M_1N_1}Y_{m-M_1}\left(\kappa_{|M_1|,N_1}R_\alpha\right)\right] \\ \times\left[I_3^{m\alpha}-r_\alpha^4\sum_{l=|m|}^{\infty}\frac{(2l+1)(l-|m|)!}{2(l+|m|)!}S_{lm\alpha}\right], \qquad (1.54)\end{aligned}$$

where

$$\begin{aligned}I_3^{m\alpha} = 2\int_0^{\pi/2}\int_0^{r_\alpha}\cos\left[r\left(K_{P_1}-K_{P_2}\right)\cos\theta\right]J_m\left(\kappa_{|M_2|,N_2}r\sin\theta\right) \\ \times J_m\left(\kappa_{|M_1|,N_1}r\sin\theta\right)r^2\sin\theta\, d\theta dr, \qquad (1.55)\end{aligned}$$

$$S_{lm\alpha} = \left(a^{M_2N_2P_2}_{lm\alpha}\right)^* a^{M_1N_1P_1}_{lm\alpha} + N_{lm\alpha}\left(b^{M_2N_2P_2}_{lm\alpha}\right)^* b^{M_1N_1P_1}_{lm\alpha}. \qquad (1.56)$$

1.2.7. Hamiltonian Matrix Elements

Similar to the overlap elements (1.53), the Hamiltonian matrix elements of single-walled nanotubes are written as

$$\begin{aligned}\left\langle\Psi^*_{M_2N_2P_2}\middle|H\middle|\Psi_{M_1N_1P_1}\right\rangle = \int_\Omega\Psi^*_{II,M_2N_2P_2}(-\Delta)\Psi_{II,M_1N_1P_1}dV - \\ \sum_\alpha\int_{\Omega_\alpha}\Psi^*_{II\alpha,M_2N_2P_2}(-\Delta)\Psi_{II\alpha,M_1N_1P_1}dV + \sum_\alpha\int_{\Omega_\alpha}\Psi^*_{I\alpha,M_2N_2P_2}H_{MT_\alpha}\Psi_{I\alpha,M_1N_1P_1}dV. \qquad (1.57)\end{aligned}$$

Again, the integral over Ω_{II} is equal to the integral over Ω minus the sum of the integrals over the MT regions Ω_α. The cylindrical waves are eigenfunctions of the kinetic energy operator of electrons in the cylindrical potential well; therefore,

$$\int_\Omega \Psi^*_{II,M_2N_2P_2}(-\Delta)\Psi_{II,M_1N_1P_1}dV = \left(K_{P_2}K_{P_1} + \kappa_{|M_2|N_2}\kappa_{|M_1|N_1}\right)\delta_{M_2M_1}\delta_{N_2N_1}\delta_{P_2P_1}. \tag{1.58}$$

When calculating the integral of the cylindrical waves over MT Ω_α regions, we use their representation in terms of spherical coordinate system (1.42), apply the equation

$$\int_{\Omega_\alpha} \Psi^*_{II\alpha,M_2N_2P_2}(-\Delta)\Psi_{II\alpha,M_1N_1P_1}dV = -\int_{\Omega_\alpha}\left(i\nabla\Psi_{II\alpha,M_2N_2P_2}\right)^*\left(i\nabla\Psi_{II\alpha,M_1N_1P_1}\right)dV, \tag{1.59}$$

and formula for operator ∇ in spherical coordinates:

$$\nabla = \frac{\partial}{\partial r}\boldsymbol{i}_r + \frac{1}{r}\frac{\partial}{\partial\theta}\boldsymbol{i}_\theta + \frac{1}{r\sin\theta}\frac{\partial}{\partial\phi}\boldsymbol{i}_\phi, \tag{1.60}$$

where $\boldsymbol{i}_r$, $\boldsymbol{i}_\theta$ and $\boldsymbol{i}_\varphi$ are orthonormal unit vectors.

The last integral on the right-hand side of Eq. (1.57) corresponding to the V_{MT_α} regions is calculated taking into account the form of the basic functions $\Psi_{I\alpha,MNP}$ (1.57) and the values of the integrals of the radial wave functions and their derivatives (1.32, 1.33)

$$\int_{\Omega_\alpha} \Psi^*_{I\alpha,M_2N_2P_2}H_{MT_\alpha}\Psi_{I\alpha,M_1N_1P_1}dV =$$
$$\sum_{l=0}^{\infty}\sum_{m=-l}^{l}\left\{E_{l\alpha}\left[\left(A_{lm\alpha}^{M_2N_2P_2}\right)^* A_{lm\alpha}^{M_1N_1P_1} + N_{lm\alpha}\left(B_{lm\alpha}^{M_2N_2P_2}\right)^* B_{lm\alpha}^{M_1N_1P_1}\right]\right.$$
$$\left. + \left(A_{lm\alpha}^{M_2N_2P_2}\right)^* B_{lm\alpha}^{M_1N_1P_1}\right\}. \tag{1.61}$$

It remains to substitute the equations (1.46) and (1.51) for $A_{lm\alpha}^{MNP}$ and $B_{lm\alpha}^{MNP}$.

The ultimate expression for the Hamiltonian matrix elements takes the form

$$\langle P_2M_2N_2|H|P_1M_1N_1\rangle = \left(K_{P_1}K_{P_2} + \kappa_{|M_1|,N_1}\kappa_{|M_2|,N_2}\right)\delta_{P_2P_1}\delta_{N_2N_1}\delta_{M_2M_1} -$$
$$\frac{1}{c}(-1)^{M_1+M_2}\sum_\alpha \exp\left\{i\left[\left(K_{P_1} - K_{P_2}\right)Z_\alpha + (M_1 - M_2)\Phi_\alpha\right]\right\}\times$$
$$\sum_{m=-\infty}^{+\infty}\left[C^J_{M_2N_2}J_{m-M_2}\left(\kappa_{|M_2|,N_2}R_\alpha\right) + C^Y_{M_2N_2}Y_{m-M_2}\left(\kappa_{|M_2|,N_2}R_\alpha\right)\right]\times$$

$$\left[C^{J}_{M_1N_1}J_{m-M_1}\left(\kappa_{|M_1|,N_1}R_\alpha\right)+C^{Y}_{M_1N_1}Y_{m-M_1}\left(\kappa_{|M_1|,N_1}R_\alpha\right)\right]\times$$
$$\left\{K_{P_1}K_{P_2}I_3^{m\alpha}+\kappa_{|M_2|,N_2}\kappa_{|M_1|,N_1}I_3^{\prime m\alpha}+m^2 I_4^{m\alpha}-\right.$$
$$\left. r_\alpha^4\sum_{l=|m|}^{\infty}\frac{(2l+1)(l-|m|)!}{2(l+|m|)!}\left[E_{l\alpha}S_{lm\alpha}+\gamma_{lm\alpha}\right]\right\}. \tag{1.62}$$

Here

$$I_3^{\prime m\alpha}=2\int_0^{\pi/2}\int_0^{r_\alpha}\cos\left[r\left(K_{P_1}-K_{P_2}\right)\cos\theta\right]J_m^{'}\left(\kappa_{|M_2|,N_2}r\sin\theta\right)$$
$$\times J_m^{'}\left(\kappa_{|M_1|,N_1}r\sin\theta\right)r^2\sin\theta d\theta dr, \tag{1.63}$$

$$I_4^{m\alpha}=2\int_0^{r_\alpha}\int_0^{\pi/2}\cos\left[r(K_{P_1}-K_{P_2})\cos\theta\right]J_m(\kappa_{|M_2|,N_2}r\sin\theta)$$
$$\times J_m(\kappa_{|M_1|,N_1}r\sin\theta)(\sin\theta)^{-1}drd\theta, \tag{1.64}$$

$$\gamma_{lm\alpha}=\left\{I_2^*(P_2,M_2,N_2)I_1(P_1,M_1,N_1)+I_1^*(P_2,M_2,N_2)I_2(P_1,M_1,N_1)\right\}$$
$$\times\dot{u}_{l\alpha}(r_\alpha)u'_{l\alpha}(r_\alpha)-I_2^*(P_2,M_2,N_2)I_2(P_1,M_1,N_1)\dot{u}_{l\alpha}(r_\alpha)u_{l\alpha}(r_\alpha)$$
$$-I_1^*(P_2,M_2,N_2)I_1(P_1,M_1,N_1)\dot{u}'_{l\alpha}(r_\alpha)u'_{l\alpha}(r_\alpha). \tag{1.65}$$

As in the LAPW method, when calculating $\gamma_{lm,\alpha}$ the product $\dot{u}'_{l\alpha}(r_\alpha,E_{l\alpha})u_{l\alpha}(r_\alpha,E_{l\alpha})$ was replaced by $\dot{u}_{l\alpha}(r_\alpha,E_{l\alpha})u'_{l\alpha}(r_\alpha,E_{l\alpha})$ to provide the Hermitian conjugacy of the Hamiltonian matrix. The integrals I_1–I_4 of the Bessel functions of the first kind and of the Legendre polynomials are calculated numerically.

1.2.8. Dispersion Curves and Densities of States

Finally, using the expressions obtained for the overlap integrals (1.54) and the matrix elements of the Hamiltonian (1.61), one can determine the electronic dispersion curves $E_n(k)$ of the nanotube from the secular equation

$$\det\left|\,|\langle P_2M_2N_2|H|P_1M_1N_1\rangle|_k-E_n(k)\langle P_2M_2N_2|P_1M_1N_1\rangle|_k\,\right|=0, \tag{1.66}$$

while the Bloch functions of the nanotube are written as

$$\Psi_{kn}(\mathbf{r})=\sum_{MNP}a_{MNP}^{kn}\Psi_{MNP}(\mathbf{r}). \tag{1.67}$$

In addition to the dispersion curves $E_n(k)$, the electronic structure of nanotubes can be characterized by energy dependences of the total and partial densities of states (DOS). As to the total DOS, it is determined by the dispersion curves $E_n(k)$ only

$$N(E) = \sum_n \left[dE_n(k)/dk\right]^{-1}. \tag{1.68}$$

In order to obtain the partial DOS in the MT sphere

$$N_l^\alpha(E) = \sum_n Q_{nl}^\alpha(k)\left[dE_n(k)/dk\right]^{-1}, \tag{1.69}$$

one has to calculate additionally the components Q_{nl}^α of expansion of electron density of doubly occupied orbital $\Psi_{k,n}(\mathbf{r})$ with respect to l

$$2\int_{\Omega_\alpha} \left|\Psi_{k,n}(\mathbf{r})\right|^2 d\mathbf{r} = \sum_l Q_{nl}^\alpha(k). \tag{1.70}$$

In Eqs. (1.69) and (1.70), $Q_{nl}^\alpha(k)$ the partial charges are

$$\begin{aligned}Q_l^\alpha(k,n) = {} & \frac{2r_\alpha^4}{c}\sum_{P_2M_2N_2}\sum_{P_1M_1N_1}\bar{a}_{P_2M_2N_2}^{kn}a_{P_1M_1N_1}^{kn} \\ & \times\exp\left\{i\left[\left(K_{P_1}-K_{P_2}\right)Z_\alpha+\left(M_1-M_2\right)\Phi_\alpha\right]\right\}\times(-1)^{M_1+M_2} \\ & \times\sum_{m=-l}^{l}\left[C_{M_2N_2}^J J_{m-M_2}\left(\kappa_{|M_2|,N_2}R_\alpha\right)+C_{M_2N_2}^Y Y_{m-M_2}\left(\kappa_{|M_2|,N_2}R_\alpha\right)\right]\times \\ & \left[C_{M_1N_1}^J J_{m-M_1}\left(\kappa_{|M_1|,N_1}R_\alpha\right)+C_{M_1N_1}^Y Y_{m-M_1}\left(\kappa_{|M_1|,N_1}R_\alpha\right)\right] \\ & \times\frac{(2l+1)(l-|m|)!}{2(l+|m|)!}S_{lm,\alpha}.\end{aligned} \tag{1.71}$$

One can also determine the partial DOS of the interspherical region from the equation

$$N_{IS}(E) = \sum_n Q_n^{IS}(k)\left[dE_n(k)/dk\right]^{-1}, \tag{1.72}$$

where

$$\begin{aligned}Q^{IS}(k,n) = {} & \frac{2}{c}\sum_{P_2M_2N_2}\sum_{P_1M_1N_1}\bar{a}_{P_2M_2N_2}^{kn}a_{P_1M_1N_1}^{kn} \\ & \times\exp\left\{i\left[\left(K_{P_1}-K_{P_2}\right)Z_\alpha+\left(M_1-M_2\right)\Phi_\alpha\right]\right\}\times(-1)^{M_1+M_2} \\ & \times\sum_{m=-\infty}^{+\infty}\left[C_{M_2N_2}^J J_{m-M_2}\left(\kappa_{|M_2|,N_2}R_\alpha\right)+C_{M_2N_2}^Y Y_{m-M_2}\left(\kappa_{|M_2|,N_2}R_\alpha\right)\right]\times \\ & \left[C_{M_1N_1}^J J_{m-M_1}\left(\kappa_{|M_1|,N_1}R_\alpha\right)+C_{M_1N_1}^Y Y_{m-M_1}\left(\kappa_{|M_1|,N_1}R_\alpha\right)\right]I_3^{m\alpha}.\end{aligned} \tag{1.73}$$

1.3. Applications

1.3.1. Carbon Nanotubes

Carbon nanotubes have attracted great attention due to their good geometry and physical properties. They have been discussed, for example, as promising material for applications as building blocks for molecular electronics, field-emission displays diodes, electromechanical devices, and computer (Dekker 1999, Dekker 2018, Peng 2018, He et al. 2018, Gupta et al. 2018, De Volder et al. 2013, Cao et al. 2017, Laird et al. 2015, de Heer et al. 1995, Bonard et al. 1998, Chung et al. 2002, Collins et al. 1997, Shulaker et al. 2013). Since the geometry and, therefore, the physical properties of the single-walled nanotubes are determined by their diameters and chiralities, careful determination of the nanotubes structure is required. For individual, isolated tubules, scanning tunnel microscopy and resonance Raman scattering (Wildöer et al. 1998, Odom et al. 1998, Wirth et al. 2000, Jorio et al. 2001, Bai et al. 2018) can be used successfully for structure elucidation. In macroscopic samples, the nanotubes can be synthesized both as randomly oriented and aligned arrangements of the tubes with different diameters and chirality (Ebbesen and Ajayan 1992, Li et al. 2001, Li et al. 1996, Tu et al. 2002, Zhang et al. 2017). Therefore, the experimental data typically represent an average over the distribution of type and diameter of the tubules. If electronic applications are considered, methods such as optical absorption, electron energy loss, photoemission, X-ray absorption, and scanning tunneling spectroscopy are most suitable for characterization of such samples (Bachilo et al. 2002, Knupfer 2001). In these methods, the experimental electronic DOS of the tubules are used to determine their structure. The energy separation E_{11} between the lowest conduction band and highest valence band singularities seems to be the most important parameter of the DOS. The value of E_{11} is much larger in metallic nanotubes compared with semiconducting ones, and can be used to distinguish metallic from semiconducting tubes.

Let us discuss the band structures of the achiral metallic, semiconducting, and quasi-metallic carbon nanotubes as calculated in terms of the LACW method. The results are used particularly to correlate the minimum direct gaps between the conduction and valence band singularities with the nanotube diameter d and optical absorption spectra.

It should be remembered that the Hückel-type zone-folding approach shows that armchair (n, n) nanotubes should have the metallic-type band structures, in which π bands overlap at the Fermi level. The $(n, 0)$ tubules with n indivisible by three are to be the semiconductors having a few tenths of eV energy gaps. Finally, if $n = 3q$ (with integer q), the tubules should have the quasi-metallic band structure with zero energy gap between valence and conduction bands, in which the dispersion curves touch, but do not overlap at the Fermi level. The π-electronic zone-folding approach was fairly successful in determining the families of single-walled nanotubes, but it fails in predicting the exact gaps, and the first-principle calculations of band structures are necessary to study their electronic behavior.

1.3.1.1. Metallic Nanotubes

In our calculations of all the carbon single-walled nanotubes, we take δ = 4.3 a.u. that is $a = R_{NT} + 2.3$ a.u. and $b = R_{NT} - 2.3$ a.u., where R_{NT} is radius of the tubule. This value of δ is equal to the half-sum of the covalent (0.7 Å) and van der Waals (1.7 Å) radii of the atom C; the LACW calculations with this value of δ reasonably predict the valence band width equal to about 22 eV for all carbon tubules.

As the first example of application of LACW method, Fig. 1.7 shows the calculated LACW band structure and DOS for the armchair (12,12) carbon nanotube that can be compared with purely π-electronic data (D'yachkov et al. 2002).

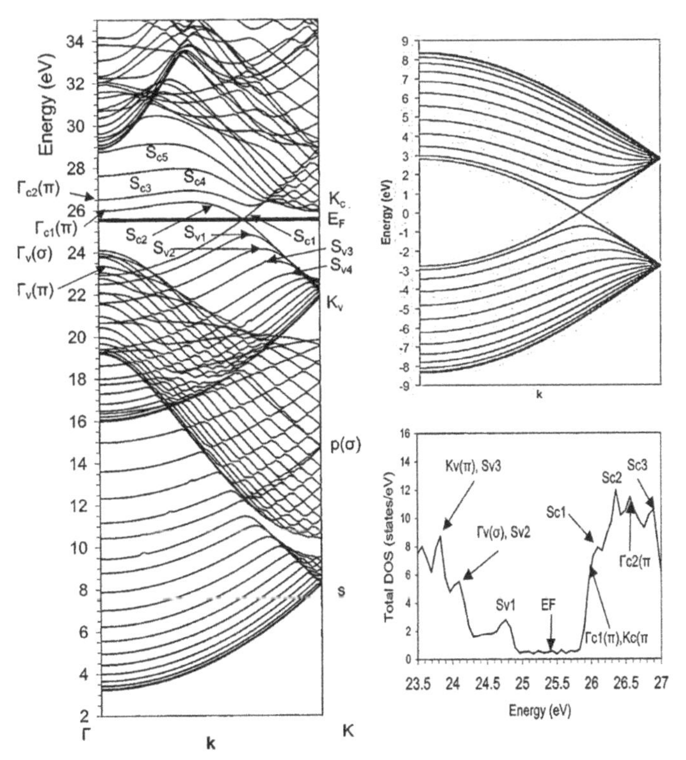

Fig. 1.7. Band structure of (12, 12) carbon nanotube calculated using the LACW (left) and Hückel (right top) methods and LACW total DOS in the Fermi energy region (right bottom). Here and below, the energy is measured from the potential of the interspherical region.

Both the LACW and Hückel computation results show that the nanotube has a metal-type band structure with the Fermi level located at the intersection of two π-bands at the point $k = (2/3)(\pi/c)$. The density of states near the Fermi level between the first singularities of the valence band and conduction band is constant. However, the LACW calculation gives more informative band structure. In the center of the Brillouin zone Γ, the upper occupied σ level $\Gamma_v(\sigma)$ is located above the upper occupied π level $\Gamma_v(\pi)$. Note, that in the case of in graphite, the upper occupied π-state at the Γ-point is below the upper occupied σ-state too (Tatar and Rabii 1982, Fink and Lambin 2001). In π-electron models, S_{v1} and S_{c1} states are the boundary singularities of the valence and conduction bands, respectively, and the minimal gap is $E_{11}(\pi\pi^*) = E(S_{v1}) - E(S_{c1})$. However, as can be seen in Fig. 1.7, in the center Γ and at the boundary K of the Brillouin zone, the lower $\Gamma_{c1}(\pi)$ and $K_{c1}(\pi)$ states are located below the S_{c1}. The $S_{v1} - S_{c1}$ gap still corresponds to the direct transition with the minimal energy. The gap $E_{11}(\sigma\pi^*) = E[\Gamma_{c1}(\pi)] - E[\Gamma_v(\sigma)]$ corresponds to the second direct transition.

For (n, n) carbon nanotubes with n = 8–10, the $E_{11}(\pi\pi^*)$ and $E_{11}(\sigma\pi^*)$ energies are almost the same, and for tubes of smaller diameter $E_{11}(\sigma\pi^*)$ is 0.2–0.5 eV smaller than $E_{11}(\pi\pi^*)$ (Fig. 1.8).

The plots of the direct transition energies $E_{11}(\sigma\pi^*)$ and $E_{11}(\pi\pi^*)$ versus the diameter d of the armchair nanotubes show significant deviations from the relationship $E_{11} \sim d^{-1}$ predicted by the simple π electron model. The situation is complicated by the close $E_{11}(\pi\pi^*)$ and $E_{11}(\sigma\pi^*)$ values and by intersection of these characteristics in the range n = 8–10.

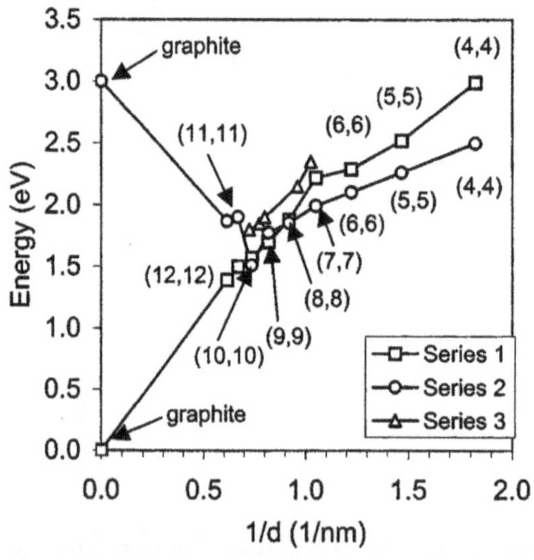

Fig. 1.8. Theoretical direct optical gaps for ππ* (series 1) and σπ* (series 2) transitions in zone center and experimental data (Liu et al. 2001) (series 3) versus diameters, d, of the (n, n) tubes.

The electronic structure of the (3,3) nanotube is sharply different (Fig. 1.9).

Here, the S_{c1} singularity coincides with the Fermi level, which leads to a sharp increase in the density of states at this level and correlates with the experimentally observed superconductivity of the (3, 3) nanotube (Tang et al. 2001).

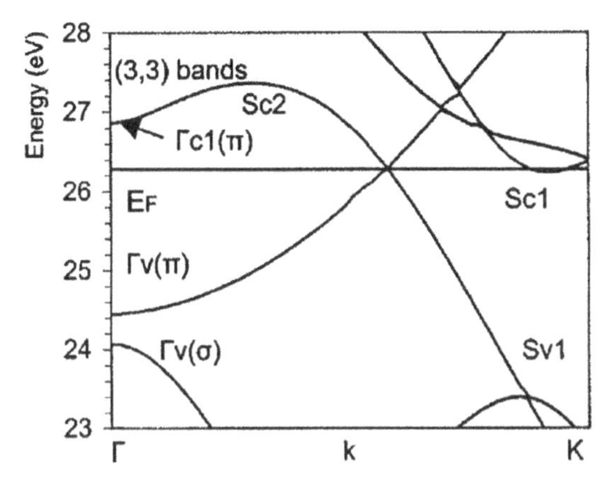

Fig. 1.9. LACW band structure of (3, 3) nanotube.

1.3.1.2. Semiconducting Nanotubes

We also studied the electron properties of the semiconducting $(n, 0)$ zigzag nanotubes in terms of the LACW method (D'yachlov and Hermann 2004). We particularly considered the relationships between the E_{ii} energies for i =1 and 2 and diameter d for the values of d between 4 and 20 Å. For this purpose, we have calculated the complete band structures of the $(n, 0)$ tubules with $5 \leq n \leq 26$; we omit the quasi-metallic structures having n evenly divisible by 3, which will be discussed below. As a typical example, the band structure and the DOS of the semiconducting (13, 0) tubules with d = 10 Å are presented in Fig. 1.10.

As expected, the zigzag tubes with $10 \leq n \leq 26$ have a semiconducting electronic structure with minimum energy gap E_{11} between the valence and conduction bands located at the Brillouin zone center Γ. In agreement with the simple π-electron models, both the highest occupied (S_{v1}) and the lowest unoccupied (S_{c1}) singularities correspond to the π-bands located at the Γ point, and $E_{11} = E(S_{c1}) - E(S_{v1})$. The relationship between the gap energies E_{11} and diameters of tubules is presented in Fig. 1.11.

The gap energy E_{11} is an oscillating function of d^{-1}. The $E_{11}(d^{-1})$ values alternate between the two curves corresponding to the zigzag nanotubes with n mod 3 = 1 and n mod 3 = 2. The curve mod 3 = 1 is located totally above the curve mod 3 = 2. The maximum values of E_{11} = 0.90 eV for the series mod 3 = 1 and E_{11} = 0.56 eV for the series mod 3 = 2 correspond to

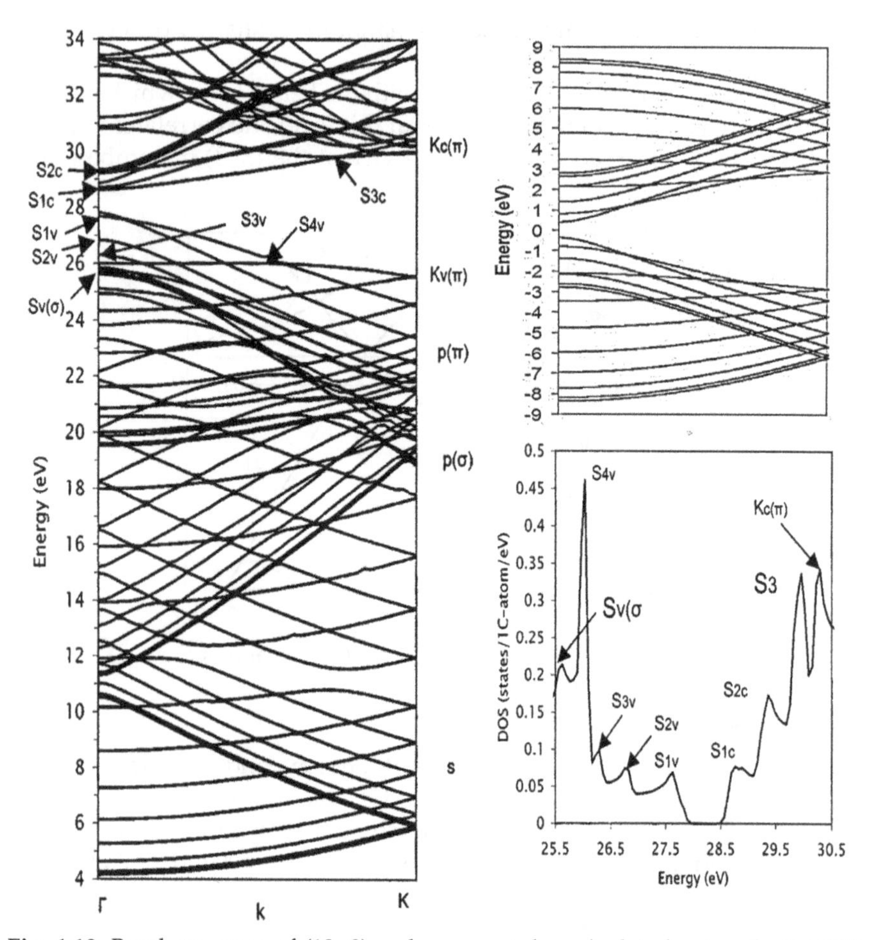

Fig. 1.10. Band structure of (13, 0) carbon nanotube calculated using the LACW (left) and Hückel (right top) methods and LACW total DOS in the Fermi energy region (right bottom).

the values of d equal to 12.5 ($n = 16$) and 11 Å ($n = 14$), respectively. Further decrease of the tubules diameter results in an abrupt descent of E_{11} due to the increasing curvature of the tubules surface and the progressive mixing of the π- and σ-states. For the small-diameter nanotubes with $n \leq 8$ ($d \leq$ 6.3 Å), the gap at the Γ point is closed up and the simple π-electron tight-binding predictions about the semiconducting behavior break down.

Figure 1.11 shows clearly that there is no one-to-one correspondence between the E_{11} and d values even for the nanotubes with the same chirality. For example, the energy gap E_{11} equal to about 0.3 eV is matched by four zigzag nanotubes with d = 7.8, 8.6, 20.4, and 40.8 Å (n = 10, 11, 26, and 52, respectively). This ambiguity must add complexity to the nanotubes structure determination on the basis of optical gap E_{11}. The same is true

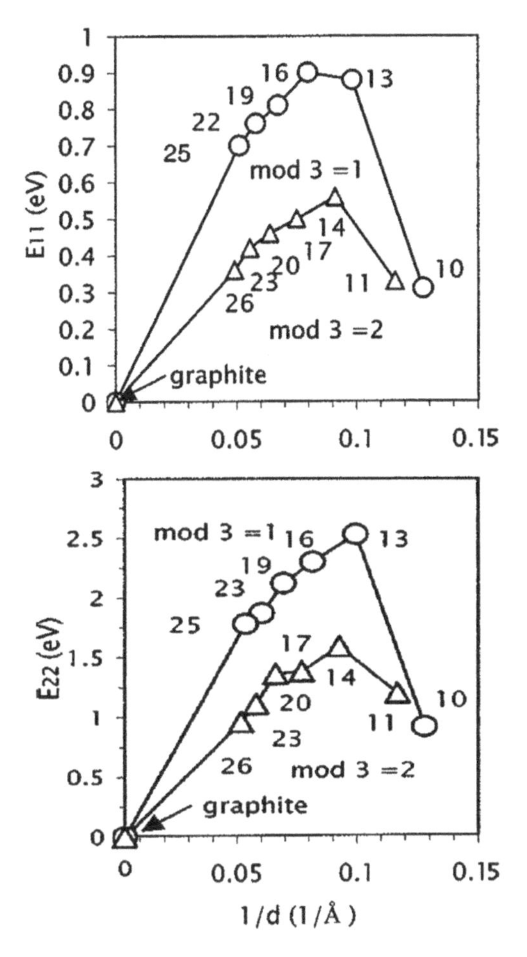

Fig. 1.11. Minimum optical gaps E_{11} and E_{22} versus diameter d of the semiconducting $(n, 0)$ tubules.

for the second direct gap $E_{22} = E(S_{c2}) - E(S_{v2})$. It was suggested that the gap energy E_{22} could be used as an additional important parameter of the DOS for the determination of the semiconducting nanotubes diameters in terms of the relationships between the E_{22} and d^{-1}. The amplitudes of the $E_{22}(d^{-1})$ function oscillations are about three times greater than those of the $E_{11}(d^{-1})$ function (Fig. 1.11). This behavior gives rise to complications in the analysis of experimental data, but it is also possible to consider it as an additional criterion for classifying the carbon tubules.

1.3.1.3. Quasi-metallic Nanotubes

If n is a multiple of three, the simple π-electronic model predicts that in the carbon $(n, 0)$ nanotubes the top of the valence band and the bottom of the conductivity band touch each other at $k = 0$. Figure 1.12 shows an example of band structure of such quasi-metallic nanotube.

The LACW data confirm this qualitative model, but in reality the band structure is much more complex. A special section of Chapter 3 is devoted to a more detailed description of the electronic properties of such quasi-metallic nanotubes, where a symmetrized version of the LACW technique is developed and applied to the chiral tubules.

1.3.1.4. N, B, and O Doping

The use of chemically modified nanotubes instead of pure carbon tubules increases their utility in molecular electronics (Bobenko et al. 2014). Boron and nitrogen are the most suitable dopants for an experimental study of doping of carbon nanotubes (Komorowski and Cottam 2017, Stephan et al. 1994, Zhang et al. 1997, Terrones et al. 1996, Sen et al. 1998, Miyamoto et al. 1996, Satishkumar et al. 1999). For example, the boron-doped nanotubes of an average composition $C_{35}B$ were produced by the pyrolysis of a mixture of acetylene and diborane. Pyridine pyrolysis with the use of cobalt as a catalyst yielded nitrogen-doped nanotubes of similar composition $C_{33}N$. Oxygen can also substitute for some carbon atoms in a nanotube, e.g., during oxidative purification of nanotubes or during nanotube manufacture in an oxygen-containing medium (He et al. 1998,

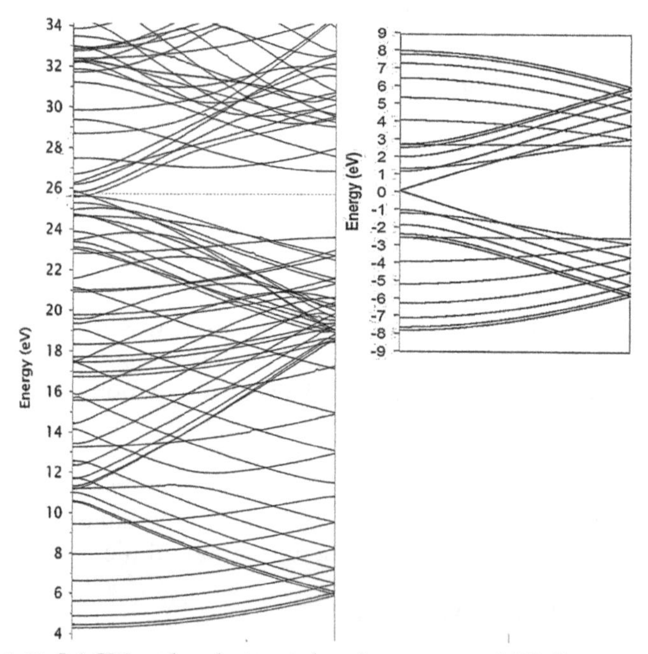

Fig. 1.12. LACW and π-electronic band structures of (12, 0) nanotube.

Kuznetsova et al. 2000). Evidently, such impurity atoms should change the electronic structure of a nanotube. Boron should render the nanotube p-conductive; nitrogen results in n-conductivity. The LACW method together with super-cell model was used to investigate the electronic properties of the (5, 5), (15, 0), and (16, 0) carbon nanotubes containing the boron, nitrogen, or oxygen atoms as substitutional impurities (Nikulkina and D'yachkov 2004).

Let us start with doped A-(5, 5) nanotubes containing one non-carbon atom per unit cell. Figure 1.13 shows a structure of such a nanotube.

The substitution of a nitrogen atom for one carbon atom in the unit cell of carbon nanotube perturbs the band structure and densities of states (Fig. 1.14).

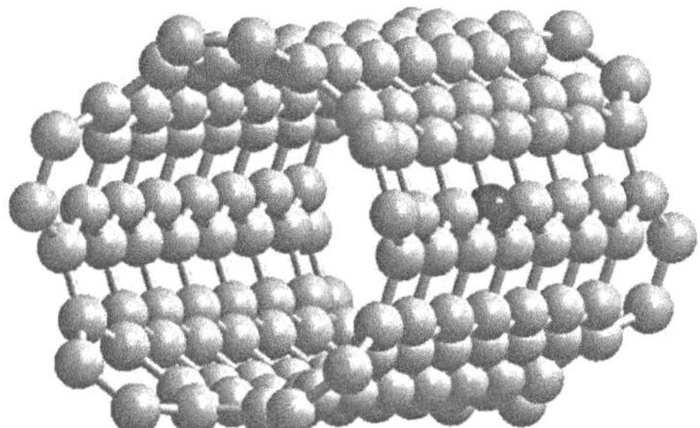

Fig. 1.13. Geometry of doped nanotube.

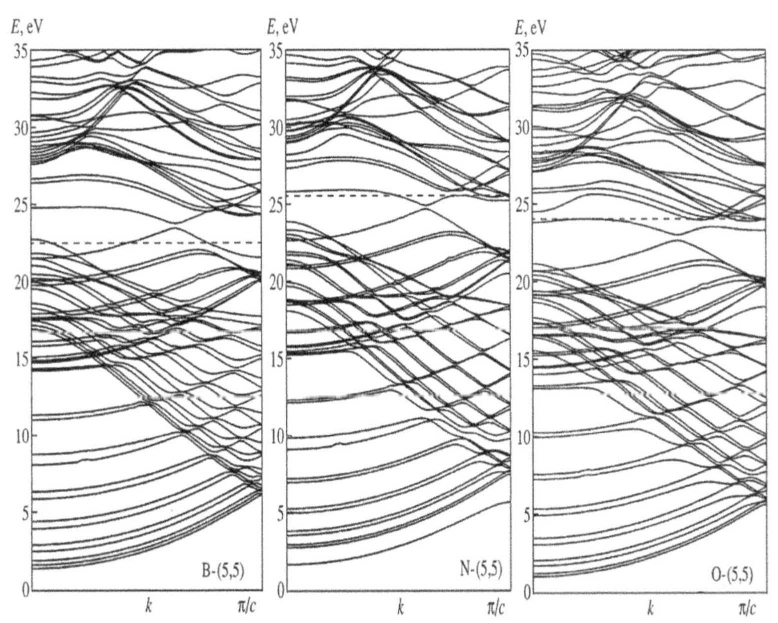

Fig. 1.14. Band diagrams for carbon doped (5, 5) nanotubes containing one N, O, or B atom per unit cell.

The incorporation of the nitrogen atom, which has one more electron than carbon, elevates the Fermi level and shifts it to the conduction band of the pure carbon (5, 5) nanotube. This correlates, in particular, with the rigid band model. The density of states near the Fermi level changes significantly: in a pure carbon (5, 5) nanotube, the Fermi level occurs in the region of a DOS valley; in a nitrogen-doped nanotube, it lies near a peak (Fig. 1.15).

The abrupt increase in the DOS at the Fermi level can significantly change the physical properties of the nanotube, such as electrical conductivity and electronic heat capacity. The valence band width in the N-doped (5, 5) nanotube is 1 eV greater than in the carbon (5, 5) nanotube as a result of the appearance of a 2*s*(N) band at the valence band bottom; an extra peak, associated with the nitrogen 2*s* states, appears in the low-energy region of the DOS.

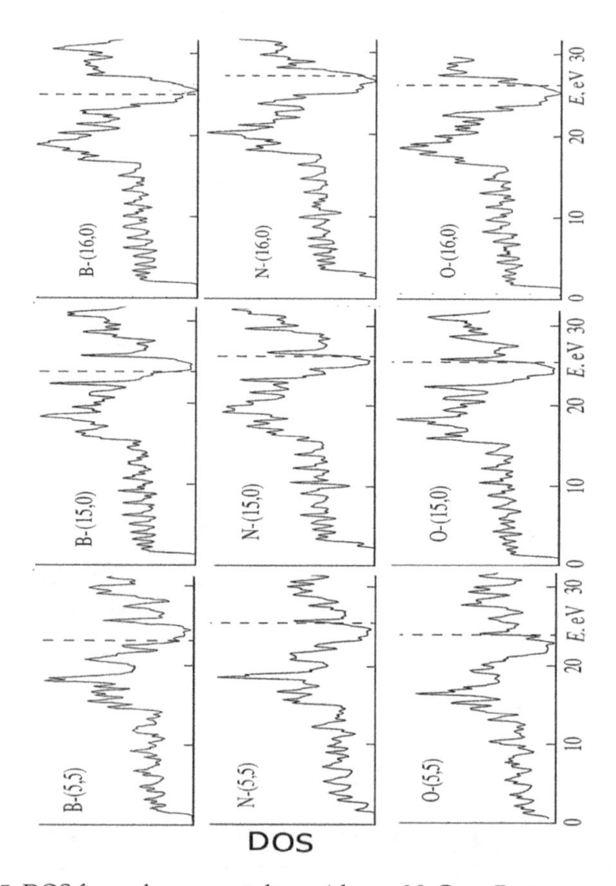

Fig. 1.15. DOS for carbon nanotubes with one N, O, or B atom per unit cell (Gaussian broadening with a half width of 0.1 eV).

Pristine carbon (5, 5) nanotube has a fivefold axis; some of its bands are doubly degenerate. As a result of symmetry lowering caused by the substitution, all the doubly degenerate bands are split. This splitting averages 0.2 eV. Peaks in the DOS curves are broadened for the same reason. Even more significant changes in the electronic structure of the (5, 5) nanotube are caused by the substitution of an oxygen atom for one carbon atom. The transfer to the oxygen doped (5, 5) nanotube shifts the Fermi level further to the conduction band of the starting carbon nanotube. Here, the Fermi level appears precisely at the DOS peak.

The boron atom has one electron less than the carbon. Therefore, boron doping of a pure carbon (5, 5) nanotube decreases the Fermi level and shifts it to the valence band of pristine carbon nanotube. The DOS at the Fermi level again increases, since this level leaves the valley in the DOS curve of the pure carbon system. The splitting of degenerate energy levels of carbon nanotube on the boron doping is more significant than on the nitrogen doping.

According to the LACW data, the pristine zigzag (16,0) carbon nanotube is a semiconductor with the band gap of 0.57 eV with Fermi level located in the band gap and zero DOS at this level. In doped nanotubes, the band diagram changes from a semiconductor to a metallic character: Fermi level shifts from the bandgap to the conduction band (upon nitrogen or oxygen doping) or to the valence band (upon boron doping).

The effects of the analogous doping of the quasi-metallic (15, 0) tubule are similar to those observed in the metallic armchair (5, 5) tubule. They appear as the shifts of the Fermi level to the conduction or valence band of the pristine nanotube in the cases of acceptor and donor impurities accompanied by a rise in the DOS at the Fermi level.

1.3.1.5. 3d-Metal Dopants in Nanotubes

Carbon nanotubes are synthesized by the Chemical Vapor Deposition (CVD) and laser ablation methods. In the CVD method, a hydrocarbon (most frequently, acetylene or methane) is passed over a cobalt- or nickel-containing catalyst. In the laser ablation method, transition metals are also used as catalysts. The method consists of bombarding a graphite target placed into a vacuum chamber by a pulse laser. Carbon knocked out of the target by a laser pulse is deposited on a cold substrate as nanotubes. Introducing different catalysts into the target enables the synthesis of different types of nanotubes; as a rule, 1–2 at % Ni or Co is introduced. Since transition metals play a significant role in synthesis of nanotubes, the question of their effect on the properties of end product is of importance (Avramescu et al. 2016, Ajayan 2004). The role of the catalyst has been comprehensively studied. A catalytic reaction of methane with supported nickel nanoparticles of 5–20 nm in diameter has been carried out at 500°C directly in a transmission electron microscope. It has been demonstrated that hydrocarbon molecules decompose on the catalyst surface. Nanotubes grow on the catalyst as on the template and literally dangle from tiny catalyst particles. When transition metals are used in the synthesis, their atoms can replace some of the carbon atoms. This uncontrollable substitution can significantly modify the electronic properties of a nanomaterial.

We calculated the electronic properties of a semiconducting (13, 0) and metallic (5, 5) nanotubes with Co or Ni dopant atoms. We calculated (5, 5) nanotubes with one metal atom per three translation cells, which corresponded to the composition $C_{59}Co$ and a metal concentration of ~1.7 at %, and (13, 0) nanotubes with one metal atom per two cells, which corresponded to the composition $C_{51}Co$ and a metal composition of ~2 at %. Figure 1.16 shows the effect of the cobalt dopant on the total and partial densities of states of the (5, 5) and (13, 0) nanotubes.

Comparison with data for the pristine (5, 5) tubule shows that introduction of Co does not break the metal-like character of the (5, 5) nanotube; moreover, the electron concentration at the Fermi level increases by one order of magnitude: from 2 to 20 states/eV per C_{60} and $C_{59}Co$ fragments. The introduction of Co into a semiconducting (13, 0) carbon nanotube leads to its metallization; the DOS at the Fermi level increases from 0 up to 6.5 states/eV. The partial DOS show that both the

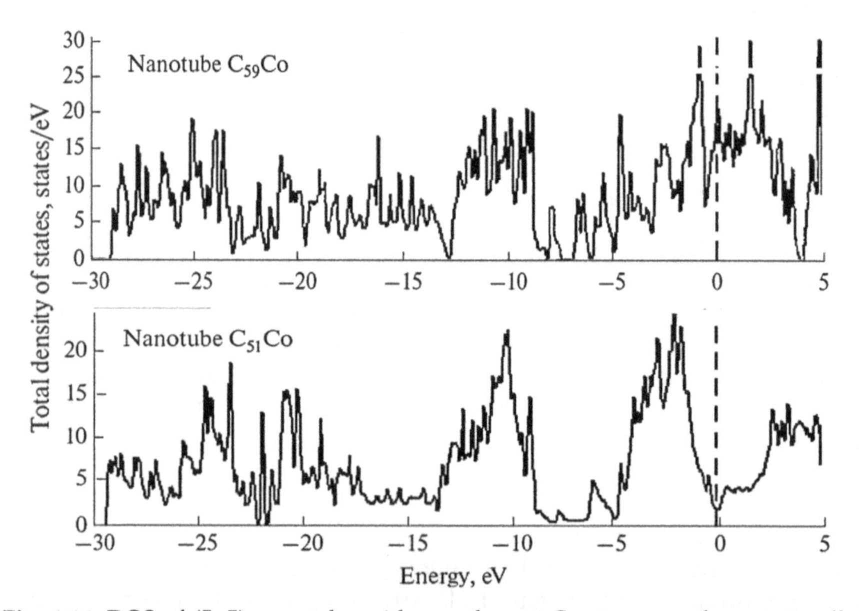

Fig. 1.16. DOS of (5, 5) nanotube with one dopant Co atom per three unit cells ($C_{59}Co$) and DOS of (13, 0) nanotube with one dopant Co atom per two unit cells ($C_{51}Co$).

metal *d* electrons and carbon *p* electrons fill this gap. The *s* electrons of metal form a strong and broad band at a distance of 23–29 eV from the Fermi level, i.e., noticeably below the valence band bottom of the pristine carbon nanotube. This inner band can be identified by spectral methods and can be used to reveal doped nanotubes. An introduction of nickel into nanotubes leads to the similar modifications of the electronic structure of these nanotubes.

1.3.2. BN Nanotubes

Advances in carbon nanomaterials science stimulated interest in noncarbon analogues of nanotubes. As early as in 1994, quantum mechanical tight-binding calculations predicted the existence, stability, and basic electronic properties of tubules based on hexagonal Boron Nitride (BN) (Rubio et al. 1994). A year later, BN nanotubes were synthesized in a laboratory (Chopra et al. 1995). The BN nanotubes have the same atomic structure as their carbon counterparts, but alternating B and N atoms substitute for C atoms in graphene-like planes with almost no change in atomic spacing (Kim et al. 2018, Zhi et al. 2010, Soares and Guerini 2011). The band structures of ideal BN nanotubes have been studied in many works using the tight-binding and pseudo-potential theories (Osadchii et al. 2003, Xiang et al. 2003, Guo and Lin 2005). In complete agreement with the quantum-mechanical prediction, BN nanotubes appear to be wide-gap materials, in which the width of the forbidden gap is equal to 4.5–5.5 eV and is almost independent of their diameter and chirality. This is a considerable difference of BN nanotubes from carbon nanotubes, because, the latter nanotubes may be both metallic and semiconducting depending on their structure. A large optical gap and a weak dependence of the gap on the structure of BN nanotubes make use of BN nanotubes in nanoelectronic devices favorable over carbon nanotubes in some cases. In particular, due to the large band gap, the molecular transistors based on BN nanotubes could operate at higher temperatures. Moreover, BN nanotubes can be used as an outer insulating shell in nanocables with a conducting metallic core (Enyashin et al. 2004). The BN nanotubes are stable under the action of oxidants (Chen et al. 2004, Piquini et al. 2005) even at high temperatures and this property also expands the potential field of their application.

Let us discuss the pristine and doped BN nanotubes electronic properties as calculated in terms of LACW method (Golovacheva and D'yachkov 2005, D'yachkov and Makaev 2009). In the MT approximation, the results of calculations depend on one parameter, the cylindrical layer thickness δ. In this case, the value δ = 4.6 a.u. was obtained from the condition that the width of the bandgap E_g of ideal BN nanotubes of a sufficiently large radius approaches the value $E_g = 4.5$ eV for the hexagonal boron nitride. Note that this is the same δ value as for the carbon tubules.

1.3.2.1. Ideal Nanotubes

We first discuss the band structures of ideal nanotubes of the armchair type (5, 5) and zigzag type (9, 0), which have almost identical diameters d = 6.6 and 6.8 Å (Fig. 1.17).

In the region of the bottom of the valence band in both nanotubes, there is a band of hybridized s(B) and s(N) states with a width of about 5 eV. This band is separated by a gap of about 5 eV from the sp_σ band, where the contribution of s states decreases as energy increases. A band

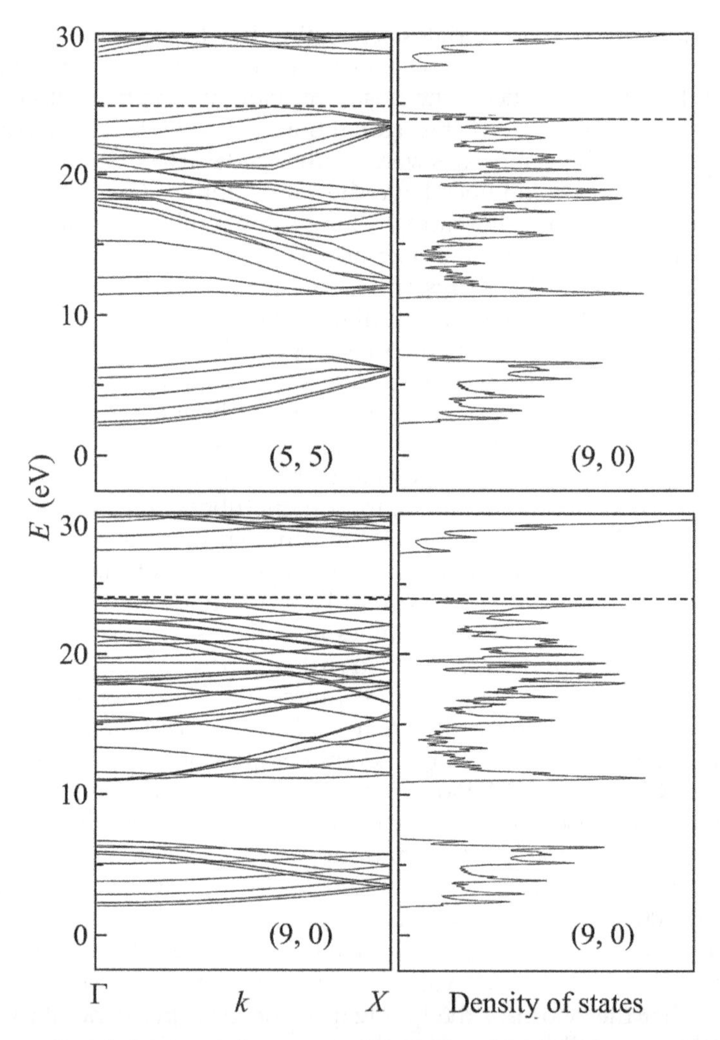

Fig. 1.17. Band structures of BN nanotubes (5, 5) and (9, 0). The energy is measured from the potential of the region between spheres; the dashed straight line is the position of the upper filled level.

of predominantly p_σ states is located above, and its high-energy part overlaps with low-energy states of the π band. The top of the valence band in armchair nanotubes is located near the point $k = (2/3)(\pi/c)$. The minima $E_c(\Gamma)$ and $E_c(X)$ of the conduction band are located at the points Γ and X of the Brillouin zone, and their energies almost coincide with each other. In all zigzag nanotubes calculated, the bottom of the conduction band and the top of the valence band are at the point Γ.

1.3.2.2. Packing Defects

All the BN nanotubes are the wide-gap materials, in which the width of the forbidden gap is equal to about 4 eV and is almost independent of their diameter and chirality; however, this statement is valid for ideal BN nanotubes, whereas the presence of packing defects may significantly change their physical properties, limiting and in some cases extending the possibilities of their applications. Intrinsic substitution defects corresponding to the presence of a boron atom B_N in a nitrogen position or, vice versa, nitrogen atom N_B in the boron position are the simplest and most important defects that are easily formed in the processes of synthesizing nanotubes. We calculated nanotubes that contain one defect per one, two, and three translational unit cells corresponding to an impurity concentration between 5 and 1.5 at. %. Figures 1.18 and 1.19 show the band structures and densities of states of the nanotubes (5, 5) and (9, 0) with such defects.

It is seen that presence of B_N or N_B impurities significantly affects the band structure and densities of states for BN nanotubes of both types. The main feature of the band structure of nanotubes with such defects is the appearance of a new band—the defect band D^π(B) or D^π(N)—in the bandgap of the ideal nanotube. In the (5, 5) nanotube with one defect per unit cell, the defect band D^π(B, N) is located in the ranges 25.2–27.05 and 24.1–26.9 eV for the substitution of nitrogen by boron and vice versa, respectively. When boron is in excess, the defect band D^π(B) is completely unfilled and forms the conduction band bottom. When nitrogen is in excess, the band D^π(N) is completely filled and forms the valence-band top. As a result of the formation of such defects, the BN nanotube is transformed from a wide-band gap material to a semiconductor. In the presence of B_N defect, the minimum of the conduction band is located at the point Γ for 25.2 eV and the valence band has two maxima with almost identical energies of 24.59 and 24.53 eV at the point Γ and near $k = (2/3)$ (π/c), so that the energies of the direct and indirect transitions virtually coincide with each other and are equal to 0.61 and 0.67 eV, respectively. In the presence of an N_B defect, the top of the valence band is located at $k \leq (1/2)(\pi/c)$, the bottom of the conduction band has two maxima at the point Γ and near $k = (1/3)(\pi/c)$, and the minimum gap is equal to 0.52 eV.

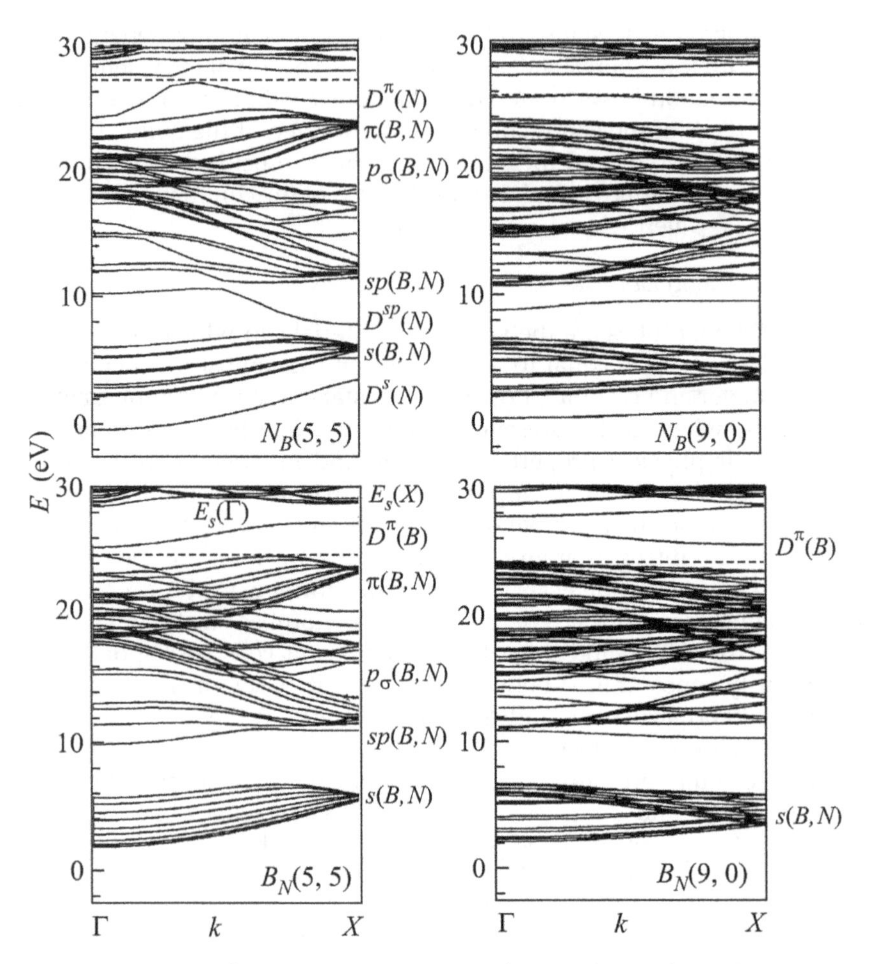

Fig. 1.18. Band structure of BN (5, 5) and (9, 0) tubes with one B_N or N_B defect per cell.

In the (9, 0) nanotube with one defect per unit cell, the width of the defect bands D^{π}(B) or D^{π}(N) is half the value for the (5, 5) nanotube. This relation is attributed to the approximately double difference in the defect concentration in the (9, 0) and (5, 5) nanotubes with 36 and 20 atoms per unit cells, respectively. The minimum of the defect band D^{π}(B) in the BN (9, 0) system is located at the edge of the Brillouin zone at 25.25 eV and forms the bottom of the conduction band, and its maximum is located at the center of the Brillouin zone and corresponds to the energy of 26.46 eV. The minimum optical band corresponds to indirect $E_v(\Gamma)$– D^{π}(B) transition from the center to the edge of the Brillouin zone with an energy of 1.36 eV, which is low as compared to a gap of 3.6 eV in an ideal nanotube. For the N_B (9, 0) system, the filled defect band D^{π}(N) is characterized by a smooth

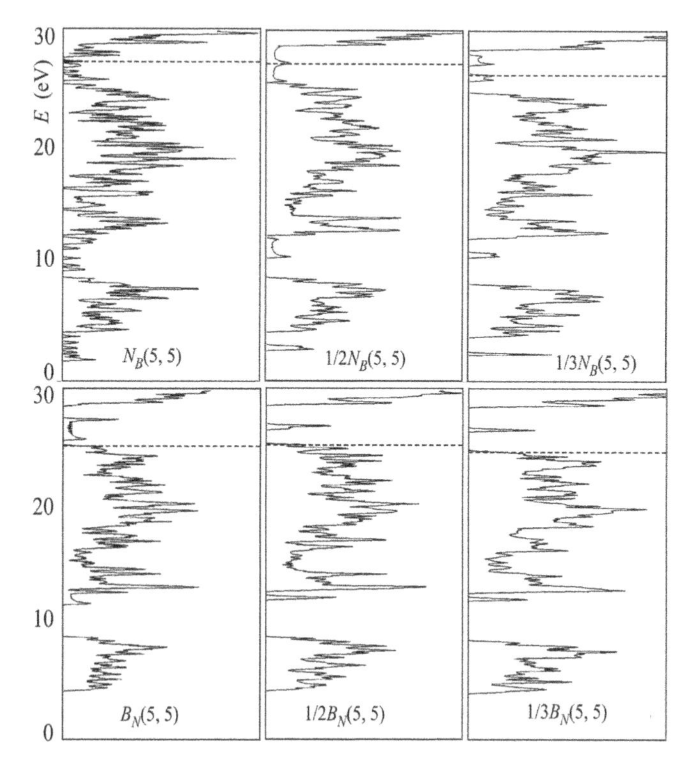

Fig. 1.19. DOS of (5, 5) BN nanotube with defects N_B or B_N.

maximum near $k = (1/3)(\pi/c)$ at 25.8 eV and a minimum at 25.1 eV and the minimum optical gap is almost the same, 1.56 eV. It is seen that a decrease in the symmetry of nanotubes due to the appearance of defects gives rise to the splitting of all doubly degenerate bands and broadening of the DOS. If nitrogen is in excess, the structure of the valence band of nanotubes changes more significantly. First, the width of the valences band increases by more than 2 eV due to a splitting of one D^s(N) band. Second, the gap between the *s* and *sp* bands is filled due to the separation of the D^{sp}(N) band. When boron is in excess, the gap between the *s* and *sp* bands decreases by only 2 eV and the width of the *s* band almost does not change.

In the dependence of the DOS on the concentration of defects (Fig. 1.19), it is clearly seen that a decrease in the concentration results in a fast decrease in the widths of the D^π(B, N) bands located in the gap of ideal nanotubes. The center of these bands is almost not shifted. For the lowest calculated concentration, the D^π(B, N) bands are degenerate virtually into discrete levels and BN nanotubes with intrinsic defects appear to be semiconductors with a bandgap width of 1 to 2 eV independent on the nanotubes structure and type of the impurity atoms. Similar narrowing and smoothing of the bands that accompany a decrease in the defect concentration is also observed for the D^{s}(N) and D^{sp}(N) band of the valence band of nanotubes with the excess of nitrogen atoms.

Thus, on the basis of the above LACW calculations, one may conclude that the presence of intrinsic packing defects significantly affects the band structure of BN nanotubes, which must be taken into account when constructing devices based on these nanotubes.

1.3.2.3. Isoelectronic Impurities

The possible modifications of the structure and physical properties of BN nanotubes are not completely reduced to intrinsic packing defects only. Impurities can be introduced to the nanotubes, changing their chemical composition and, as a consequence, electronic properties. Let us investigate the influence of isoelectronic impurity atoms of III and V elements on the electronic structure of the BN nanotubes. The study was carried out for the nanotubes with the substitutions of a nitrogen atom by phosphorus (P_N), arsenic (As_N) or antimony (Sb_N) atoms and the substitution of a boron atom by indium (In_B), gallium (Ga_B) or aluminum (Al_B) atoms. It is shown below that the presence of impurities significantly affects the band structures and densities of electronic states of BN tubules. We calculated the nanotubes with one impurity per one, two, and three unit cells.

Figure 1.20 shows that the substitution of one nitrogen atom per unit cell of the (5, 5) BN nanotube by a phosphorus atom results in substantial changes in the DOS: the pure nanotube has an energy gap of 3.5 eV, whereas a phosphorous-doped nanotube, 2.8 eV.

In the valence band region, the D^{s}(P) band splits from the *s*(B, N) band. This band produces a single peak in the DOS separated by 1.3 eV from the *s*(B, N) band. The shape of the upper dispersion curve in the valence band changes, leading to the shift of the states to a higher-energy region and to the appearance of a new π(P) peak in the DOS, which forms the top of the valence band in a doped system. An impurity band π*(P) appears in the low-energy region of the conduction band. A decrease of concentration of the P_N defects results in low-energy shifts of the occupied π(P) and *s*(P) bands and in a high-energy shift of the unoccupied π*(P) band. Figure 1.20 shows that the influence of the heavier impurity atom As on the electronic structure of the (5, 5) BN nanotube is qualitatively similar to the case of phosphorus. The major differences are the following: a peak in the density of *s*(As) states of impurity is located closer to the *s*(B, N) band and the width of the upper filled π band is equal to 2.0 eV, which is twice as large as in the phosphorous-doped nanotube. The energy gap, in this case, is smaller than that in the phosphorous-doped nanotube and is equal to 2.3 eV. Due to the introduction of one Sb atom per unit cell of the (5, 5) BN nanotube and formation of broad π(Sb) and π*(Sb) bands, the energy gap

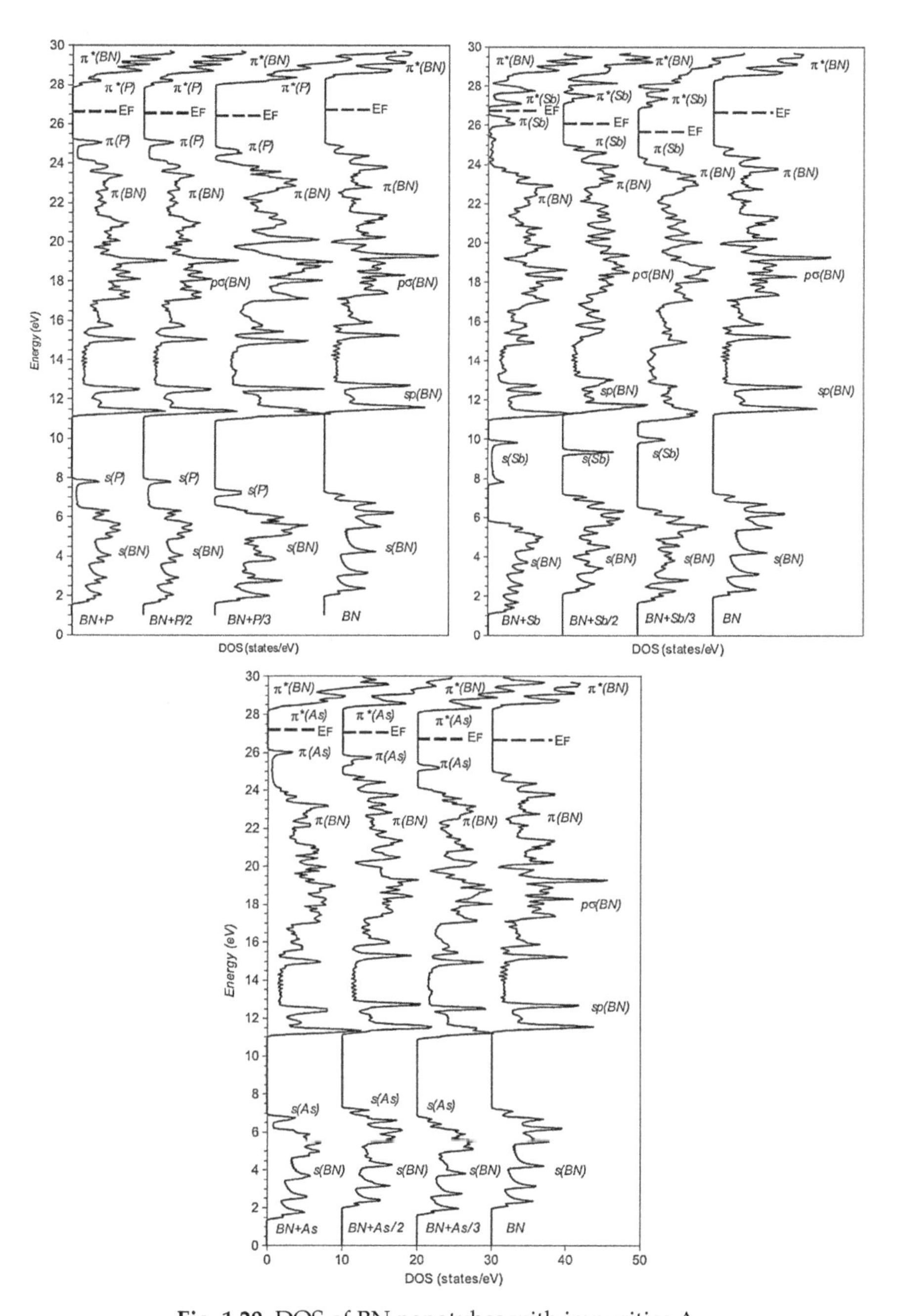

Fig. 1.20. DOS of BN nanotubes with impurities A_N.

is virtually closed, the optical gap E_g being equal to 0.5 eV only (Fig. 1.20). However, the electronic structure changes of the (5, 5) BN nanotube are not so drastic for the lower impurity concentrations corresponding to one Sb atom per two or three translational unit cells.

The substitution of one boron atom by an indium atom results in a rather weak perturbation of the band structure of the BN nanotube (Fig. 1.21).

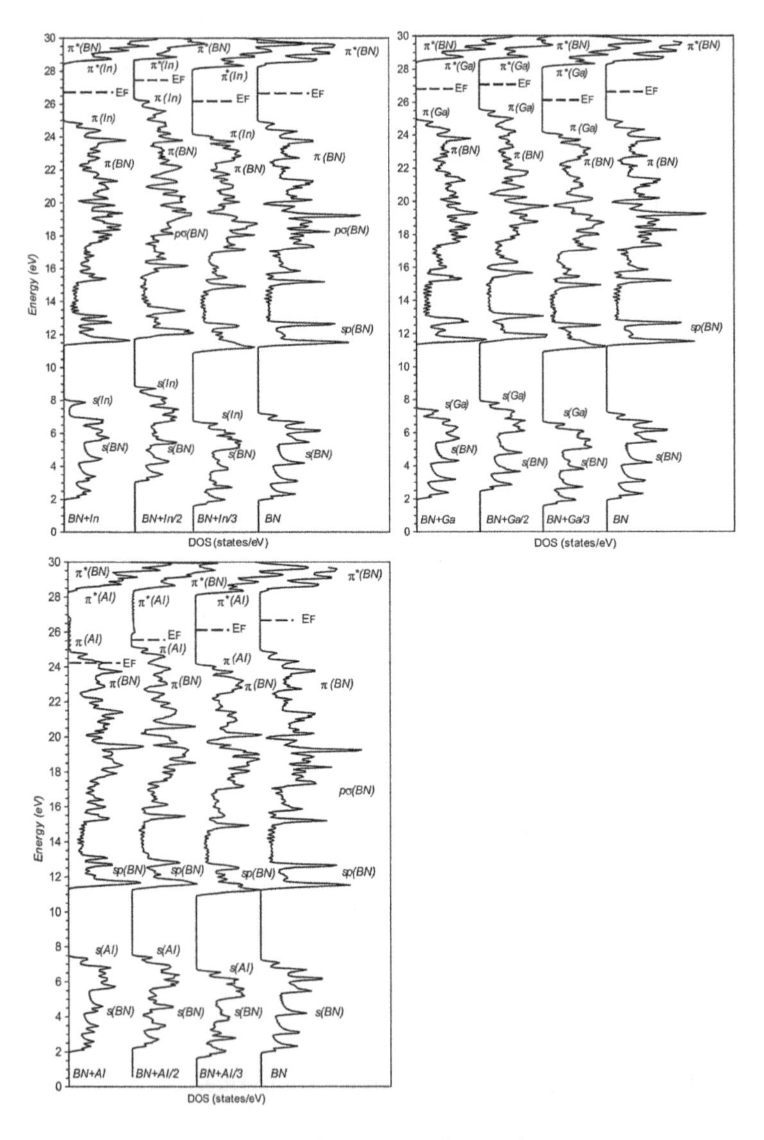

Fig. 1.21. DOS of BN nanotubes with impurities A_B.

In this case, the maximum of the upper dispersion curve of the valence band produced by the π(In) band is located near the $k = (2/3)(\pi/c)$ point of the Brillouin zone, the width of the energy gap being equal to 2.9 eV. The effect of Ga impurity atom is qualitatively the same as in the case of In. The nanotube with one Al atom per unit cell is metallic.

All the mentioned defects can be observed by optical and photoelectron spectroscopy methods, as well as by measuring electrical properties of the nanotubes. They can be used to create electronic devices based on such nanotubes.

1.3.3. Silicon Carbide

Silicon carbide nanotubes are currently synthesized by using chemical vapor deposition and carbothermal reduction of the silica by using conventional heating or by microwave heating of SiO_2 and carbon tubular structures (Tony et al. 2017). The atomic structure of such nanotubes results from replacement of a half of the C atoms in the single-walled carbon tube by the Si atoms so that each C atom is only coordinated by Si atoms, and vice versa. The one-dimensional nanostructured SiC such as SiC nanorods, nanotubes and nanowires have been studied extensively during the last decade owing to their versatile application in fabrication of optoelectronic, electronic, and sensor devices on nanometer scale. The SiC nanotubes have been proven as suitable materials for wide applications in high power, elevated temperature and harsh environment.

Using the LACW method, the band structure of (n, n) and $(n, 0)$ silicon carbide nanotubes for $n = 5$–10 was calculated (Larina et al. 2008). The Si–C bond length was equal to 1.78 Å, as in the aromatic system $H_3C_3Si_3H_3$, an analog of benzene. The width of cylindrical layer was equal to arithmetic average of the silicon covalent and Van der Waals radii—the largest of the two radii of atoms in silicon carbide.

As an example, Fig. 1.22 shows the band structure of the (7, 7) SiC nanotube. It can be seen that it is a semiconductor, with its valence band top E_V near $k = (2/3)(\pi/c)$ and its conduction band bottom is located at the center of the Brillouin zone, $k = 0$. The SiC nanotube differs in electronic structure from its carbon analog. Armchair carbon nanotubes are known to be metallic, with an overlap of the valence and conduction bands at $k = (2/3)\pi/c$. Similar to the BN nanotubes, this distinction is related to the antisymmetric component of the electronic potential in partially ionic compounds splitting the π-bands near $k = (2/3)\pi/c$. In the range $n = 7$–10, the SiC (n, n) nanotubes are semiconductors, and their band gap decreases steadily with increasing n: $E_g = 0.28$ eV at $n = 7$, 0.26 eV at $n = 8$, 0.19 eV at $n = 9$, and 0.11 eV at $n = 10$. The Si-C chemical bond is less ionic than the B-N bond; therefore, the gaps in the SiC nanotubes are small in comparison with gaps of the BN tubules.

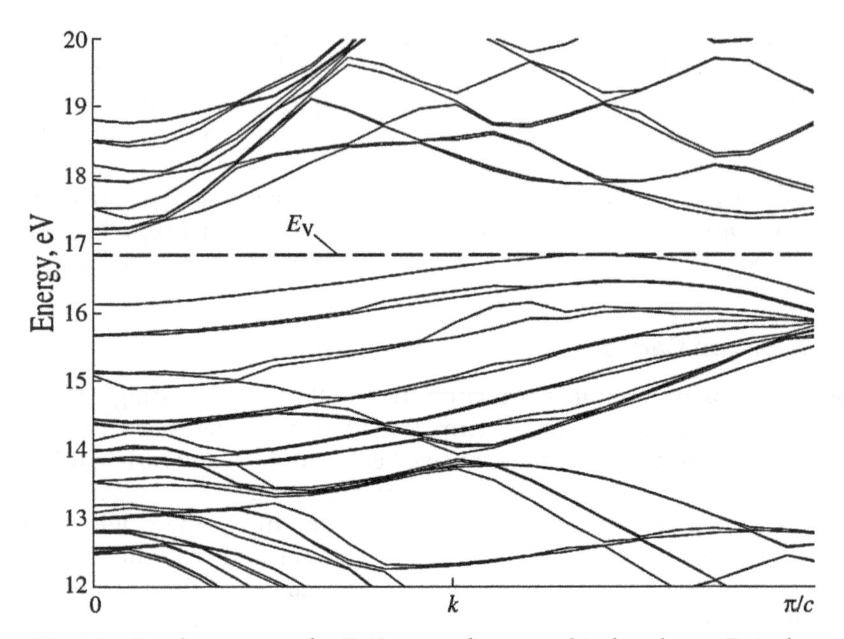

Fig. 1.22. Band structure of a (7, 7) nanotube around its band gap; E_V is the valence band top.

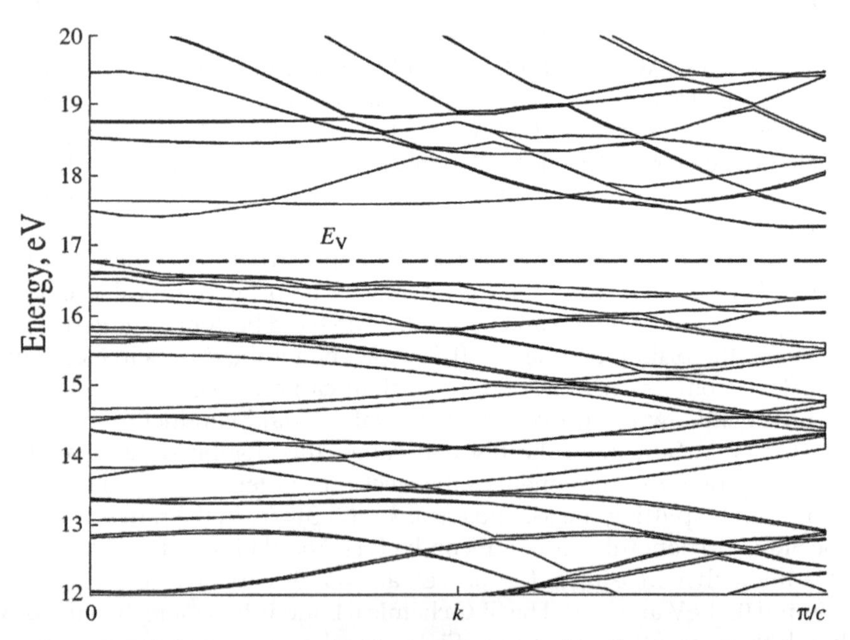

Fig. 1.23. Band structure of a (9, 0) nanotube around its band gap; E_V is the valence band top.

To exemplify calculation results for $(n, 0)$ zigzag nanotubes, Fig. 1.23 shows the band structure of the (9, 0) system. At $n = 7–9$, the $(n, 0)$ SiC nanotubes are the semiconductors, and their band gap increases steadily with n: $E_g = 0.39$ eV at $n = 7$, 0.46 eV at $n = 8$, and 0.62 eV at $n = 9$. In all $(n, 0)$ semiconducting nanotubes, the valence band top is located at $k = 0$, and the conduction band bottom, near the edge of the Brillouin zone. The minimum gap corresponds to a transition at the center of the Brillouin zone. Due to the great curvature, the armchair and zigzag SiC nanotubes with $n = 5$ and 6 have metallic conductivity according to these results.

1.3.4. Gold Au_{16} Nanotube

Gold nanoparticles have physical and chemical properties that noticeably differ from the characteristics of the bulk material and depend on their size and geometry (Yarzhemsky and Battocchio 2011). Gold nanoparticles are better catalysts than the bulk material (Haruta 2005); they can be used in design of complex functional clusters (Daniel and Astruc 2004) and exhibit quantum properties related to their small size (Burda et al. 2005). The Au–Au chemical bonds in nanoparticles essentially differ from the bonds in the bulk phase (Hutchings and Hashmi 2006). In particular, it has been shown by spectroscopy that, on average, the shortest distances between the Au atoms in gold nanoparticles are 0.042 Å longer than in the metal (Szczerba et al. 2011). Until recently, the known gold nanoparticles were the clusters containing from a few atoms to several tens of atoms and to the ordered nanocrystalline structures in the form of decahedra containing up to 250 gold atoms (Yarzhemsky and Battocchio 2011). Later, linear one atom thick nanowires (Yanson et al. 1998), nanotubes with different wall thickness (Kondo and Takayanagi 2000, Bi 2008), nanorods (Roy et al. 2013), and bundles of nanorods up to 250 nm long (Bulusu and Zeng 2006) have been fabricated. Interest in such one-dimensional gold nanomaterials is created, in particular, by the fact that they are very promising candidates for use as contacts between the elements of molecular electronics.

We calculated the band structure of an achiral gold nanotube, a translational unit cell of which is shown in Fig. 1.24 (D'yachkov 2015).

It consists of two parallel planes each of which contains eight gold atoms so that the entire tube is generated by translation of two layers, i.e., 16 gold atoms. Each next layer is rotated by an angle of $\pi/16$ around the nanotube axis with respect to the previous layer. Figure 1.24 shows three layers, the first layer being translationally equivalent to the third layer. Information on the structure of the Au_{16} nanotube was obtained by molecular dynamics methods. The average Au–Au bond length is equal to 2.53 Å, and the nanotube radius $R_{NT} = 3.84$ Å. The radii of the potential barriers were determined as follows: $a = R_{NT} + \delta$ and $b = R_{NT} - \delta$, where δ

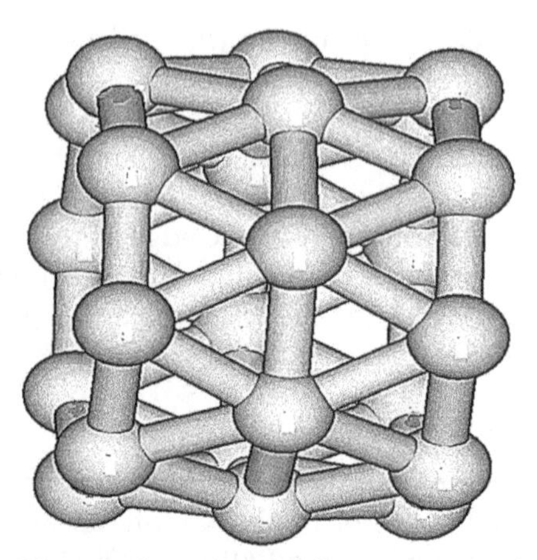

Fig. 1.24. Fragment of the Au_{16} nanotube.

= 3.0 Å is the arithmetic mean of the covalent and van der Waals radii of the gold atom.

Figure 1.25 shows the total and partial *d*(Au) and *s*(Au) DOS of electrons in the nanotube.

The figures demonstrate that the nanotube have a metal type electronic structure; the band gap is absent. The Fermi level is located at the peak of the total DOS, which should contribute to the high electron tunneling conductance of the system. The valence band width is equal to 11 eV. The *s*(Au) states are located completely in the valence band and they are not involved in electron transport. The top of the valence band and the conduction band are made of the *d*(Au) states and, thus, electron transport in the nanotube is determined mainly by the *d*(Au) electrons. Near the Fermi level, the curves of total and partial *d*(Au) DOS are qualitatively coincident. The important for heavy metals data on the spin-dependent band structures of gold nanotubes are presented in Chapter 5.

1.3.5. Cylindrical Nanorods

Inner cavity of a nanotube can be filled, for example, with metals as shown in Fig. 1.26.

In such a nanorod, there is no inner vacuum region; hence there is no inner impermeable potential barrier. Geometrically, the nanorods are systems with a simpler geometry in comparison with nanotubes. The LACW technique is of course applicable to nanorods, only a few clarifications are required (D'yachkov et al. 1998, 1999)

1.3.5.1. Method of Calculation

The potential of the interspherical region now depends only on the radius *a* of the external potential barrier

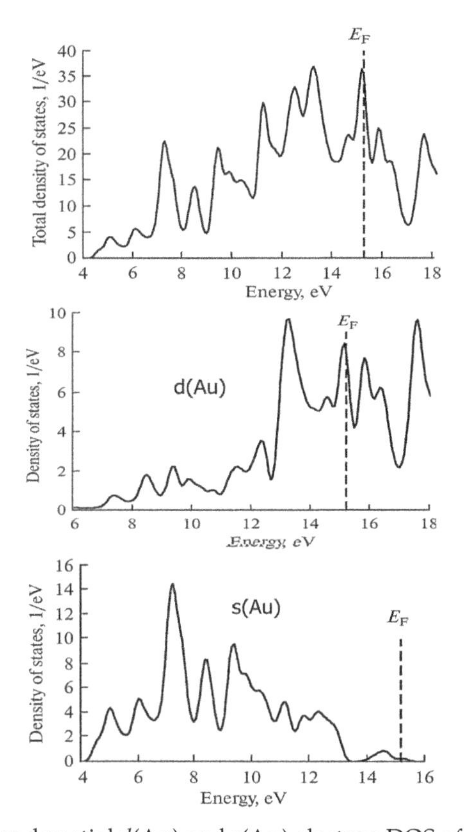

Fig. 1.25. Total and partial d(Au) and s(Au) electron DOS of Au_{16} nanotube.

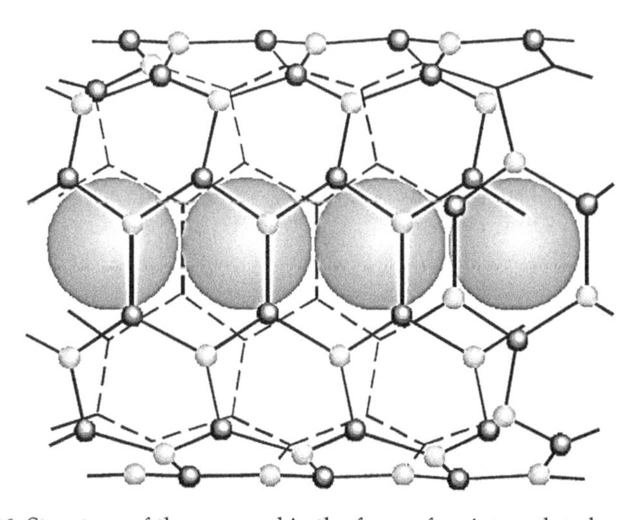

Fig. 1.26. Structure of the nanorod in the form of an intercalated nanotube.

$$U(R)=\begin{cases}0, & R\le a\\ \infty, & R>a.\end{cases} \tag{1.74}$$

Since we are interested in the finite solutions of the Schrödinger equation for R = 0, and the Bessel functions of the second kind tend to infinity as $R \to 0$, the radial functions of the interspherical region coincide with the Bessel functions of the first kind

$$\Psi_{|M|,N}(R)=C^{J}_{MN}\,J_{|M|}(\kappa_{|M|,N}R), \tag{1.75}$$

with the normalizing factor

$$C^{J}_{MN}=\sqrt{2}\Big/\left\{a\left|J'_{|M|}(\kappa_{|M|,N}a)\right|\right\} \tag{1.76}$$

determined from condition

$$\int_0^a\left|\Psi_{|M|,N}(R)\right|^2 R\,dR=1. \tag{1.77}$$

Now, $\kappa_{|M|,N}a$ is the Nth root of the function $J_{|M|}(x)$

$$\Psi_{M,N}(a)=C^{J}_{MN}J_{|M|}(\kappa_{|M|,N}a)=0\,. \tag{1.78}$$

As a result, in a local spherical coordinate system, a cylindrical wave takes the form

$$\Psi_{\mathrm{II}\alpha}(k,P,M,N)=\frac{1}{\sqrt{\Omega}\left|J'_{M}(\kappa_{|M|,N}a)\right|}\exp i(K_P Z_a+M\Phi_\alpha)(-1)^M\times$$

$$\exp\left(iK_P\,r\cos\,\theta\right)\sum_{m=-\infty}^{\infty}J_m(\kappa_{|M|,N}r\sin\theta)\,J_{m-M}(\kappa_{|M|,N}R_\alpha)\exp(im\varphi), \tag{1.79}$$

where $\Omega = \pi a^2 c$ is the volume of the unit cell.

Inside the MT spheres, the basic function has the previous form of the expansion in spherical harmonics (Eq. 1.37). A comparison of Eqs. (1.42) and (1.79) shows that absence of the cavity affects only the structural part of the wave function $\Psi_{\mathrm{II}\alpha}$ depending on the coordinates R_α, Z_α, and Φ_α of the center of the sphere α. The factors depending on the current local coordinates r, θ, and ϕ, on which the integration is performed in calculating the overlap $<P_2M_2N_2\,|\,P_1M_1N_1>$ and Hamiltonian $<P_2M_2N_2\,|\,H\,|\,P_1M_1N_1>$ matrix elements, do not change. Therefore, only the structural part changes when passing from a nanotube to a nanorod in these matrix elements. Namely, the Hamiltonian and overlap integrals for the nanorod are obtained from analogous expressions for the nanotube by replacing c^{-1} by $2\pi\left\{\Omega\left|J'_{M_2}\left(\kappa_{|M_2|,N_2}a\right)J'_{M_1}\left(\kappa_{|M_1|,N_1}a\right)\right|\right\}^{-1}$ and replacing the product of the $\left[C^{J}_{M_2N_2}J_{m-M_2}\left(\kappa_{|M_2|,N_2}R_\alpha\right)+C^{Y}_{M_2N_2}Y_{m-M_2}\left(\kappa_{|M_2|,N_2}R_\alpha\right)\right]$

and $\left[C^{J}_{M_1N_1}J_{m-M_1}\left(\kappa_{|M_1|,N_1}R_\alpha\right)+C^{Y}_{M_1N_1}Y_{m-M_1}\left(\kappa_{|M_1|,N_1}R_\alpha\right)\right]$ by the product $J_{m-M_2}(\kappa_{|M_2|,N_2}R_\alpha)$ and $J_{m-M_1}(\kappa_{|M_1|,N_1}R_\alpha)$. Thus, we have

$$\langle P_2M_2N_2 \mid P_1M_1N_1\rangle = \delta_{P_1P_2}\delta_{M_1M_2}\delta_{N_1N_2} - \frac{2\pi}{\Omega\left|J'_{M_1}(\kappa_{|M_1|,N_1}a)\right|\left|J'_{M_2}(\kappa_{|M_2|,N_2}a)\right|}\times$$
$$(-1)^{M_1+M_2}\sum_\alpha \exp\left\{i\left[(K_{P_1}-K_{P_2})Z_\alpha+(M_1-M_2)\Phi_\alpha\right]\right\}\times$$
$$\sum_{m=-\infty}^{\infty} J_{m-M_2}(\kappa_{|M_2|,N_2}R_\alpha)J_{m-M_1}(\kappa_{|M_1|,N_1}R_\alpha)\left[I_3^{m\alpha}-r_\alpha^4\sum_{l=|m|}^{\infty}\frac{(2l+1)(l-|m|)!}{2(l+|m|)!}S_{lm\alpha}\right] \quad (1.80)$$

and

$$\left\langle P_2M_2N_2 \mid \hat{H} \mid P_1M_1N_1\right\rangle = \left(K_{P_1}K_{P_2}+\kappa_{|M_1|,N_1}\kappa_{|M_2|,N_2}\right)\delta_{P_1P_2}\delta_{M_1M_2}\delta_{N_1N_2} -$$
$$\frac{2\pi}{\Omega\left|J'_{M_1}(\kappa_{|M_1|,N_1}a)\right|\left|J'_{M_2}(\kappa_{|M_2|,N_2}a)\right|}(-1)^{M_1+M_2}$$
$$\times\sum_\alpha \exp\left\{i\left[(K_{P_1}-K_{P_2})Z_\alpha+(M_1-M_2)\Phi_\alpha\right]\right\}\times$$
$$\sum_{m=-\infty}^{\infty} J_{m-M_2}(\kappa_{|M_2|,N_2}R_\alpha)J_{m-M_1}(\kappa_{|M_1|,N_1}R_\alpha)\times$$
$$\left\{K_{P_1}K_{P_2}I_3^{m\alpha}+\kappa_{|M_2|,N_2}\kappa_{|M_1|,N_1}I_3^{\prime m\alpha}+m^2I_4^{m\alpha}-\right.$$
$$\left. r_\alpha^4\sum_{l=|m|}^{\infty}\frac{(2l+1)(l-|m|)!}{2(l+|m|)!}\left[E_{l\alpha}S_{lm\alpha}+\gamma_{lm\alpha}\right]\right\}. \quad (1.81)$$

Analogously, using Eqs. (1.71) and (1.73), one obtains the partial charges for nanorods

$$Q_l^\alpha(k,n) = 2\sum_{P_2M_2N_2}\sum_{P_1M_1N_1}\bar{a}^{kn}_{P_2M_2N_2}a^{kn}_{P_1M_1N_1}\times$$
$$\frac{2\pi}{\Omega\left|J'_{M_1}(\kappa_{|M_1|,N_1}a)\right|\left|J'_{M_2}(\kappa_{|M_2|,N_2}a)\right|}r_\alpha^4\times$$
$$(-1)^{M_1+M_2}\exp\left\{i\left[(K_{P_1}-K_{P_2})Z_\alpha+(M_1-M_2)\Phi_\alpha\right]\right\}$$
$$\times\sum_{m=-l}^{l} J_{m-M_2}(\kappa_{|M_2|,N_2}R_\alpha)J_{m-M_1}(\kappa_{|M_1|,N_1}R_\alpha)\times$$
$$\sum_{l=|m|}^{\infty}\frac{(2l+1)(l-|m|)!}{2(l+|m|)!}S_{lm\alpha}, \quad (1.82)$$

$$Q_{IS}(k,n) = 2 \sum_{P_2 M_2 N_2} \sum_{P_1 M_1 N_1} \bar{a}^{kn}_{P_2 M_2 N_2} a^{kn}_{P_1 M_1 N_1} \frac{2\pi}{\Omega \left| J'_{M_1}(\kappa_{|M_1|,N_1} a) \right| \left| J'_{M_2}(\kappa_{|M_2|,N_2} a) \right|} \times$$

$$(-1)^{M_1+M_2} \sum_{\alpha} \exp\left\{ i\left[(K_{P_1} - K_{P_2}) Z_\alpha + (M_1 - M_2)\Phi_\alpha \right] \right\} \times$$

$$\sum_{m=-\infty}^{\infty} J_{m-M_2}(\kappa_{|M_2|,N_2} R_\alpha) J_{m-M_1}(\kappa_{|M_1|,N_1} R_\alpha) I_3^{m\alpha}. \tag{1.83}$$

1.3.5.2. Monoatomic Chains

Within the current trends of miniaturization of electronic devices, the problems of interconnects between nanodevices attract growing interest in the stability, band structure, and conductivity of the single-atom width chains [Tsukagoshi et al. 1999, Ohnishi et al. 1998, Hu 2011, Qiu et al. 2011, Ataca et al. 2008]. It still remains a challenge to fabricate stable atomic chains; however, there is great progress in this field now. A linear chain with 44 C atoms in compound Tr–C_{44}–Tr with bulky terminal groups Tr = tris(3,5-di-t-butylphenyl)methyl) [Chalifoux and Tykwinski 2010], as well as the similar shorter chains connecting two graphene species have been detected [Jin et al. 2009, Cretu et al. 2013]. The ligand supported linear chains of transition metal atoms with well-defined organic ends are also known [Chen et al. 2006]. Carbon nanotubes can be utilized as sheaths to stabilize monoatomic chains, which are unstable alone, and can be applied in many fields. A carbon chain with a length of 20 nm containing more than 100 atoms enclosed in a shell of multi-walled carbon nanotubes was found [Zhao et al. 2003]. The entrapped La atoms arrange linearly with a typical chain length equal to 10 nm inside a nanotube of a suitable diameter [Guan et al. 2008]. The incorporation of molten iodine into single-walled carbon nanotubes with a diameter of 1 nm generates iodine chains longer than 10 nm [Rodrigues et al. 2003]. Ultrafine metal nanowires were obtained by introducing metallic atoms into the zeolite channels. Information on the electronic structure of purely metallic chains can be important for the interpretation of the band structure of metal-doped carbon nanotubes.

The monoatomic linear chains are well-known as text-book examples studied using the simple tight-binding or free electron models. Such chains are the simplest cylindrical systems too. The arrangement of atoms on the symmetry axis facilitates their calculation using the LACW technique since in this case $R_\alpha = 0$ and $\Phi_\alpha = 0$ and $J_M = \delta_{M,0}$ in Eqs. (1.80) - (1.83), which take the form

$$\langle P_2M_2N_2 \mid P_1M_1N_1\rangle = \delta_{P_1P_2}\delta_{M_1M_2}\delta_{N_1N_2} - \delta_{M_1M_2}\frac{2\pi}{\Omega\left|J'_{M_1}(\kappa_{|M_1|,N_1}a)\right|\left|J'_{M_2}(\kappa_{|M_1|,N_2}a)\right|}\times \times\sum_{\alpha}\exp\left[i(K_{P_1}-K_{P_2})Z_\alpha\right]\left[I_3^{M_1,\alpha} - r_\alpha^4\sum_{l=|m|}^{\infty}\frac{(2l+1)(l-|M_1|)!}{2(l+|M_1|)!}S_{lM_1\alpha}\right], \tag{1.84}$$

$$\left\langle P_2M_2N_2 \mid \hat{H} \mid P_1M_1N_1\right\rangle = \left(K_{P_1}K_{P_2} + \kappa_{|M_1|,N_1}\kappa_{|M_2|,N_2}\right)\delta_{P_1P_2}\delta_{M_1M_2}\delta_{N_1N_2} - \frac{2\pi}{\Omega\left|J'_{M_1}(\kappa_{|M_1|,N_1}a)\right|\left|J'_{M_2}(\kappa_{|M_1|,N_2}a)\right|} \times\sum_{\alpha}\exp\left[i(K_{P_1}-K_{P_2})Z_\alpha\right]\delta_{M_1,M_2}\times \left\{K_{P_1}K_{P_2}I_3^{M_1\alpha} + \kappa_{|M_1|,N_2}\kappa_{|M_1|,N_1}I_3'^{M_1\alpha} + M_1^2 I_4^{M_1\alpha} - r_\alpha^4\sum_{l=|m|}^{\infty}\frac{(2l+1)(l-|M_1|)!}{2(l+|M_1|)!}\left[E_{l\alpha}S_{lM_1\alpha} + \gamma_{lM_1\alpha}\right]\right\}, \tag{1.85}$$

$$Q_l^{\alpha}(k,n) = 2\sum_{P_2M_2N_2}\sum_{P_1M_1N_1}\bar{a}^{kn}_{P_2M_2N_2}a^{kn}_{P_1M_1N_1}\times \frac{2\pi}{\Omega\left|J'_{M_1}(\kappa_{|M_1|,N_1}a)\right|\left|J'_{M_1}(\kappa_{|M_1|,N_2}a)\right|}r_\alpha^4\delta_{M_1,M_2} \times\exp\left[i(K_{P_1}-K_{P_2})Z_\alpha\right]\sum_{l=|M_1|}^{\infty}\frac{(2l+1)\left(l-|M_1|\right)!}{2\left(l+|M_1|\right)!}S_{lM_1\alpha}, \tag{1.86}$$

$$Q_{IS}(k,n) = 2\sum_{P_2M_2N_2}\sum_{P_1M_1N_1}\bar{a}^{kn}_{P_2M_2N_2}a^{kn}_{P_1M_1N_1}\times \frac{2\pi}{\Omega\left|J'_{M_1}(\kappa_{|M_1|,N_1}a)\right|\left|J'_{M_2}(\kappa_{|M_2|,N_2}a)\right|}\times \delta_{M_1,M_2}\sum_{\alpha}\exp i\left[(K_{P_1}-K_{P_2})Z_\alpha\right]I_3^{M_1\alpha} \tag{1.87}$$

and were used in the LACW calculation of chains.

1.3.5.3. Carbynes

Monoatomic carbon chains, called the carbynes, are the simplest carbon molecular systems that may be of interest as the potential materials for nanoelectronics. According to a valence of the carbon atoms, formation of the two structural forms with the translational unit cells having one and two carbon atoms is possible for the linear carbon chains. In a cumulenic form of carbyne, the C atoms form the double bonds (...=C=C=...), but the single and triple bonds alternate in the polynic structure (...–C≡C–C≡C–...). Figure 1.27 shows the energy bands of cumulenic and polyynic carbynes.

Because of the cylindrical symmetry of carbynes, there are σ- and π-type dispersion curves in these systems. The σ-bands are occupied. The twofold orbitally degenerate π bands correspond to the semiclassical clockwise and anticlockwise rotational motion of electrons around the symmetry axis. The polyynic carbyne has the semiconducting-type band structure with a gap between bonding (π) and antibonding (π*) states equal to 1.14 eV. The metallic cumulenic carbyne has one π band crossing the Fermi level at the center of the Brillouin zone.

1.3.5.4. 3d Metal Chains

Figure 1.28 shows the band structures of the linear metallic chains for atoms from K to Zn.

It can be seen that an infinite $[K]_\infty$ chain is a one-dimensional metal with a half-occupied nondegenerate 4*s* band $E_1(1)$ (D'yachkov and Kepp 2000). The $[Ca]_\infty$ chain is a one-dimensional insulator with the occupied 4*s* band and gap of 0.7 eV corresponding to an indirect transition to a doubly degenerate band $E_2(1)$ formed mainly by $4p_x$ and $4p_y$ orbitals. All the transition metal chains from $[Sc]_\infty$ to $[Zn]_\infty$ have the metallic type band structure. In the low-energy region, the dispersion curves of all the

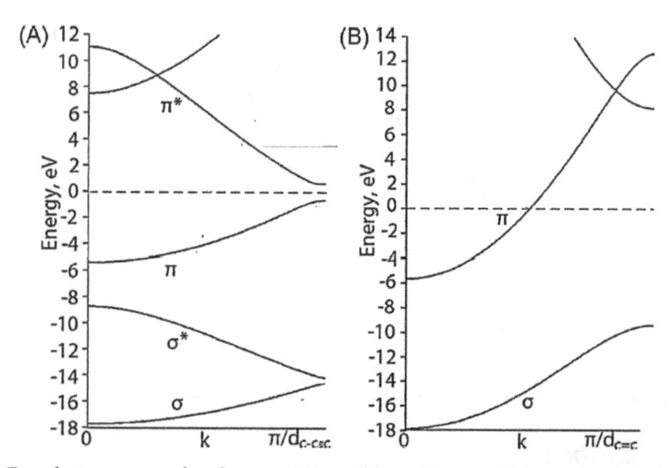

Fig. 1.27. Band structure of polyynic (A) and cumulenic (B) carbynes. Calculations are performed for the experimental bond lengths $d_{C-C} = 1.360$ Å, $d_{C\equiv C} = 1.205$ Å and $d_{C=C} = 1.26$ Å in polyynic and cumulenic forms. Radius $a = 1.2$ Å of the cylindrical barrier is equal to half-sum of the covalent and van der Waals radii of carbon.

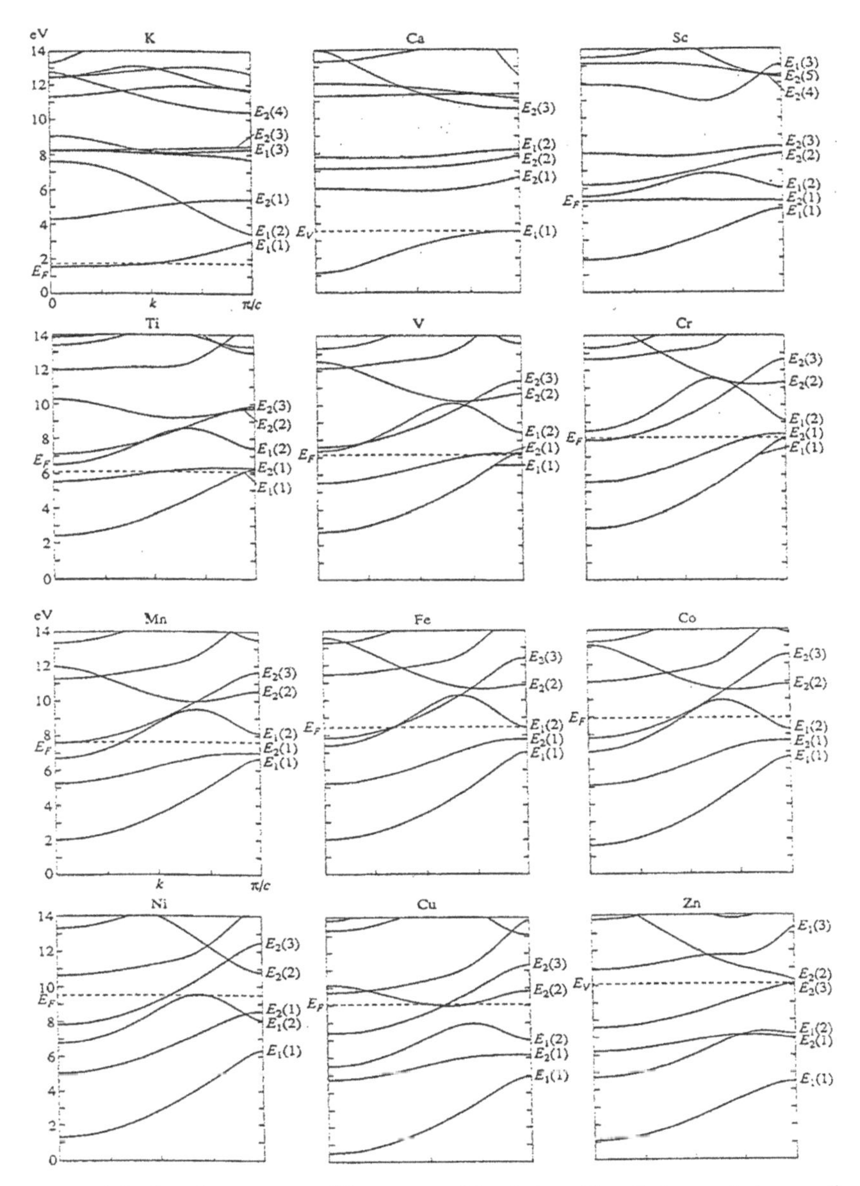

Fig. 1.28. The band structure of linear metallic chains. The distances between the nearest metal atoms were chosen equal to twice the metal radii of the elements. The radius a of the cylindrical potential barrier was equal to 3.5 Å.

3d metals have many common features. In all cases, the nondegenerate bands $E_1(1)$ and $E_1(2)$ are bonding and antibonding combinations of the 4s and 3$d(z^2)$ states, the $E_1(1)$ band being completely occupied with electrons.

The doubly degenerate narrow $E_2(1)$ band, which is located between the $E_1(1)$ and $E_1(2)$ bands, is formed mainly by $3d(x^2 - y^2)$ and $3d(xy)$ orbitals; an orientation of these functions shows that it is the δ bonding band. In chains from $[Sc]_\infty$ to $[V]_\infty$, the Fermi level crosses precisely this band and the electron conductivity in these chains is realized with the participation of δ-type states. The doubly degenerate $E_2(2)$ and $E_2(3)$ bands are located above. The lower of these bands can be correlated with the π-states formed by $3d(xz)$ and $3d(yz)$ functions. The origin of other high-energy bands can not be interpreted within the framework of a simple LCAO model; they can be associated with metallic conducting states with electron transfer through the interspherical region. Starting with the chain $[Cr]_\infty$, the electrons gradually occupy the $E_2(3)$ band. The band $E_2(2)$ lies entirely above the upper occupied levels in all metal chains except the copper. The maximum density of states at the Fermi level is observed in the $[Sc]_\infty$ and $[Zn]_\infty$ chains due to the presence of narrow partially occupied $E_2(1)$ band in $[Sc]_\infty$ and $E_2(3)$ band in $[Zn]_\infty$ that can result in a high conductivity of such nanowires.

The strength of the chemical bonding in the 3*d* metal chains is due to the bonding and antibonding character of bands and their electron occupation. In all cases, the bonding σ-band $E_1(1)$ is occupied, so its effect results in the same stabilization of all systems. In the series $[Sc]_\infty$, $[Ti]_\infty$, $[V]_\infty$, and $[Cr]_\infty$, the gradual broadening and occupation of the bonding δ-band $E_2(1)$ is observed. The wider this band, i.e. the stronger is the δ metal-metal interaction, and the greater is its electron population, the stronger becomes the chemical bonding between the atoms in chain. Thus, it is possible to foresee an increase in the stability of chains in the series from $[Sc]_\infty$ to $[Cr]_\infty$ with respect to the break of metal-metal bonds. Starting with the $[Mn]_\infty$ chain, the antibonding σ-band $E_1(2)$ becomes occupied, which should lead to a weakening of the chemical bonds. In the $[Ni]_\infty$, $[Cu]_\infty$, and $[Zn]_\infty$ chains, the σ-antibonding completely compensates for σ-binding and such chains must be less stable, since they are stabilized only due to the weak δ interactions. The occupation of the $E_2(3)$ band, which we associate with electron transport mainly along the interspherical region, should not lead to stabilization of the chains of the second half of the 3*d*-metal series. Thus, as the atomic number of the transition metal increases, the stability of the chains $[M]_\infty$ should increase by the middle of the period and then decrease (Kepp and Dyachkov 1999).

1.3.5.5. 3d-Metal Intercalated Nanotubes

Different transition metals can be introduced into the carbon nanotubes directly in the course of their arc discharge synthesis with the use of graphite electrodes doped with metal powders [Ajayan et al. 1994, Loiseau et al. 2000, Ugarte et al. 1996, Ugarte et al. 1998, Seltlur et al. 1996, Dai et al.

1996]. A special notation M@nanotube is suggested for such compounds, which means that a metal M is located inside a nanotube.

As a typical example, Fig. 1.29 shows the LACW band structures of copper intercalated armchair Cu@(10, 10) and zigzag Cu@(14, 0) nanotubes with one Cu atom located at the unit cell centers (D'yachkov and Kepp 2000).

The bands of pristine tubules change significantly due to intercalation. There is an optical gap in the (14, 0) tubule and only one crossing point of π bands in the Fermi level of the (10, 10) tubule. On the contrary, there are many dispersion curves in the Fermi energy region of both intercalants. The intercalation effect can be roughly described as a superposition and mixing of bands of the carbon nanotube and metal core.

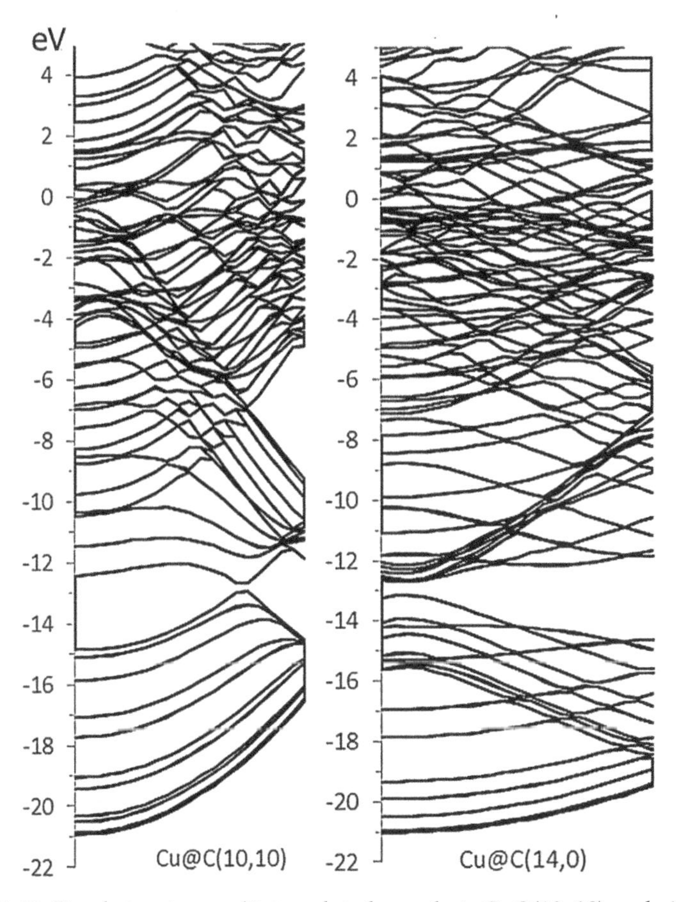

Fig. 1.29. Band structures of intercalated armchair Cu@(10, 10) and zigzag Cu@(14, 0) nanotubes. Fermi level corresponds to energy $E = 0$ eV. The radius of the external impenetrable cylindrical barrier a was taken to be the same as in pure undoped nanotubes.

Figure 1.30 shows the electron total DOS of the forty M_n@(5, 5) nanotubes as the functions of type and content of the 3*d* metals (D'yachkov et al. 2014). The Fermi level of the undoped carbon nanotube is located in the depression of the plot of DOS versus energy; as a result, the electrons DOS at the Fermi level is low (0.45 state/eV per unit cell), which, for example, can restrict the electrical conductivity of such a tube. The introduction of the transition metal atoms drastically changes the pattern: the depression at the Fermi level is filled, and the electron concentration at the Fermi level increases fourfold for V@(5, 5) to 40 fold for Ni_3@(5, 5), being as large as 17.6 state/eV per unit cell (Table 1.1). As expected, the DOS at the Fermi level increases generally with increasing metal content, but its maximum value is observed in compounds with three Ni and Cu atoms per cell, where this level coincides with position of the DOS peak.

Intercalation of the (10, 0) nanotube also results in strong metallization of system (Fig. 1.31).

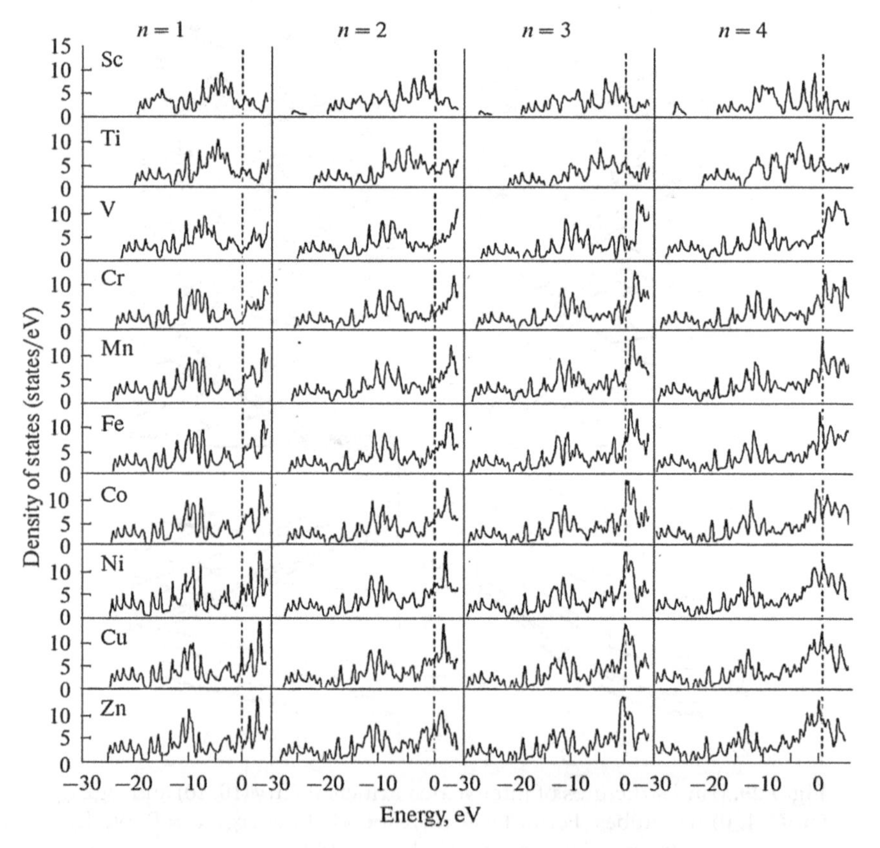

Fig. 1.30. Total DOS in M_n@(5, 5) nanotubes (n = 1–4).

Table 1.1. DOS at the Fermi level (states/eV/unit cell) in M_n@(5, 5) and M_n@(10, 0) nanotubes.

Metal	M_n@(5, 5)				M_n@(10, 0)			
n	1	2	3	4	1	2	3	4
Sc	2.863	6.263	6.837	3.491	17.288	13.397	14.460	16.082
Ti	4.771	4.028	6.788	6.023	12.426	9.319	15.014	19.633
V	1.811	5.900	3.906	6.123	12.137	19.077	4.974	4.593
Cr	2.462	3.386	4.572	6.710	12.324	6.115	10.158	13.068
Mn	2.66	5.289	5.573	13.56	14.653	7.176	15.186	11.691
Fe	3.554	5.013	8.496	10.751	13.945	7.927	12.499	12.683
Co	5.818	6.554	8.960	6.285	10.503	7.147	7.712	13.158
Ni	7.495	7.788	17.613	10.468	8.251	8.006	9.749	14.32
Cu	8.467	6.054	14.992	10.963	3.935	5.267	10.555	16.267
Zn	3.569	8.662	11.119	9.661	4.571	11.513	12.143	13.547

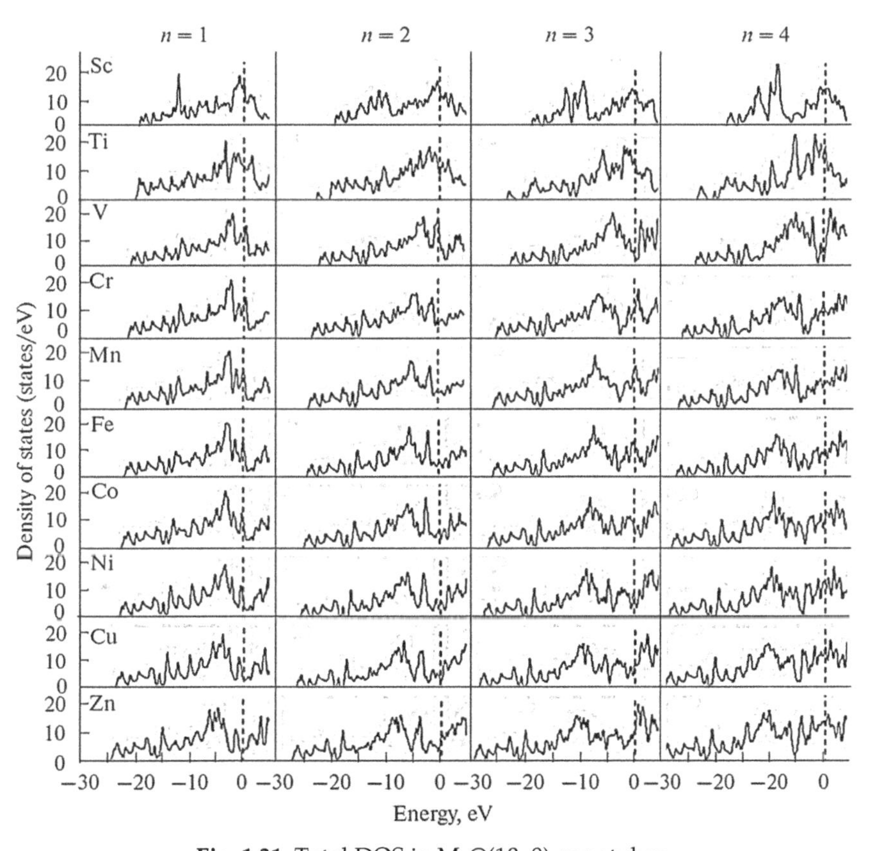

Fig. 1.31. Total DOS in M_n@(10, 0) nanotubes.

In the series of 40 M_n@(10, 0) compounds, the maximal DOS at Fermi level is observed in the Ti_4@(10, 0) nanowire. In Sc_n@(10, 0) nanowires, the density of states is high independently of the metal concentration. Thus, in order to create nanowires with the highest DOS at the Fermi level, either the metallic (5, 5) nanotube should be intercalated with nickel or copper or the semiconducting (10, 0) nanotube should be intercalated with scandium or titanium.

Figure 1.32 shows the partial *d*(Cu) and *p*(C) DOS for the particular case of Cu_n@(5, 5) nanowires (Bochkov and D'yachkov 2013).

One can see that both the metal *d* and carbon *p* states are almost equally involved in formation of electron states in the Fermi energy region. The Cu_4@(5, 5) nanowire is characterized by the highest *d*(Cu) and *p*(C) partial DOS at the Fermi level. The partial *d*(Cu) and *p*(C) DOS are roughly the same; therefore, the carbon shell and the copper core are equally involved in electron transport in intercalated wires.

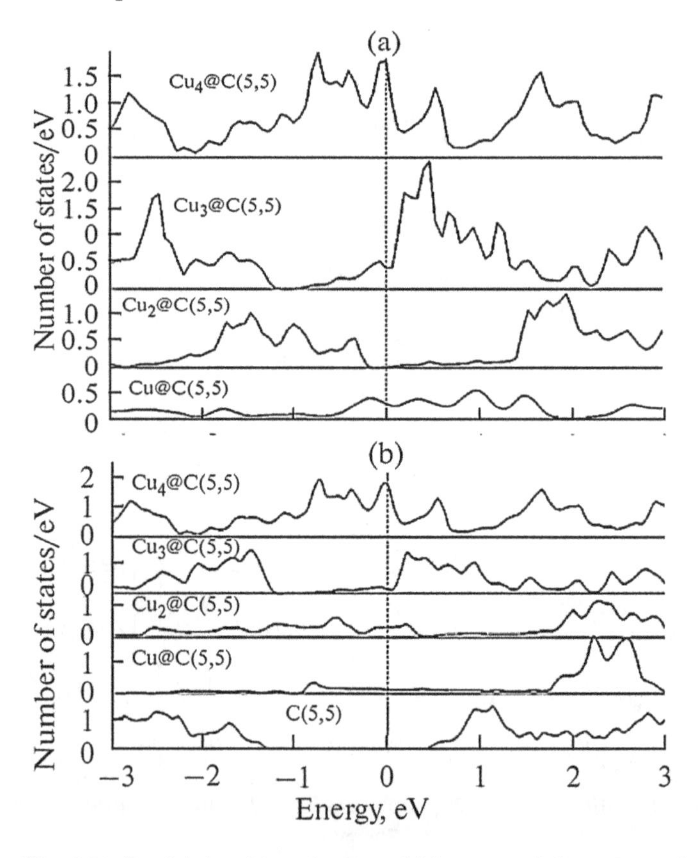

Fig. 1.32. Partial densities of states of (a) copper *d* electrons and (b) carbon *p* electrons for Cu_n@(5, 5) nanowires.

1.3.6. Beyond the MT Approximation

From the electronic structure theory of bulk solids, it is known that the MT approximation yields good results in the case of close-packed structures, in which the major part of the cell volume is occupied by MT spheres rather than by interspherical space, and hence, the potential in the compact interspherical space can be closely approximated by the constant V_{out}. For nanotubes, the interspherical space formally extends to infinity on the outer vacuum side and to the tube axis on the inner side, because the inner cavity is also free of atoms. In the cylindrical MT approximation by imposing impenetrable potential barriers, we reasonably restrict the interspherical space; however, equations in this case depend on the radii of the outer (a) and inner (b) potential barriers. In this section, in order to obtain some estimations of a validity of the cylindrical MT approach, we go beyond the MT approximation and consider the variation of the electron potential and electron density in the interspherical space. We still assume that the potential in the MT sphere region is spherically symmetric. The method is applied to calculating the band structures of carbyne and some nanotubes (Kirin and D'yachkov 2001).

1.3.6.1. Method of Calculation

To include the variation of the potential in the interspherical space, we represent the one-electron Hamiltonian H of the system in the form

$$H = H_{MT} + H_{out}. \tag{1.88}$$

Here, H_{MT} is the Hamiltonian of the system in the MT approximation with a constant potential V_{out}, in the interspherical space and H_{out} is the variable part of the potential in the interspherical space. In the local density approximation, the H_{out} is determined by the electron density ρ_{out} and represented as the sum of the Coulomb (V_c) and exchange-correlation ($V_{X\alpha}$) potentials:

$$H_{out} = V_c + V_{X\alpha}. \tag{1.89}$$

As above, the basic functions are the linearized augmented cylindrical waves $\Psi_{PMN}(\mathbf{k} \mid \mathbf{r})$, which are expanded in terms of spherical harmonics inside MT spheres and are the solutions of the Schrödinger equation for free motion of electron in the potential well confined by the cylindrical walls with radii a and b in the interspherical space. Unlike the MT approximation, where radii a and b were adjustable parameters, in this version of the LACW method, these values are only the convergence parameters. Convergence is achieved by increasing the radius of the outer potential barrier and decreasing the radius of the inner potential barrier. Thus, to include the variation of the potential in the interspherical space, it is necessary to add the values

$$\langle P_2M_2N_2|H_{out}|P_1M_1N_1\rangle|_{\Omega_{II}} = \langle P_2M_2N_2|V_c|P_1M_1N_1\rangle|_{\Omega_{II}} + \langle P_2M_2N_2|V_{X\alpha}|P_1M_1N_1\rangle|_{\Omega_{II}} \qquad (1.90)$$

to the matrix elements calculated in the cylindrical MT approximation $\langle P_2M_2N_2|H_{MT}|P_1M_1N_1\rangle$ and to solve the secular equation. In Eq. (1.90), the integration is performed over the interspherical space.

In the numerical calculations, the electron density distribution ρ_{out} was constructed as a superposition of atomic electron densities:

$$\rho(\mathbf{r}) = \sum_{\alpha} \rho_{\alpha}(|\mathbf{r} - \mathbf{r}_{\alpha}|), \qquad (1.91)$$

where $\mathbf{r}_{\alpha}$ is the position of the atom α and $\rho_{\alpha}(r)$ is the electron density distribution of the isolated atom.

The Coulomb potential at the point $\mathbf{r}$ in the interspherical space is represented by the sum of two components: $V_c = V_1 + V_2$. Here, V_1 is the potential produced by the interspherical electron density

$$V_1(\mathbf{r}) = 2\int_{\Omega_{II}} \frac{\rho(\mathbf{r} - \mathbf{r}')}{|\mathbf{r} - \mathbf{r}'|} d\mathbf{r}' \qquad (1.92)$$

and V_2 is the total potential produced by the atomic nuclei and the electron density located inside the MT spheres. The potential inside the MT spheres is assumed to be spherically symmetric; therefore,

$$V_2(\mathbf{r}) = \sum_{\alpha} \frac{Q_{\alpha}^{eff}}{|\mathbf{r} - \mathbf{r}_{\alpha}|}, \qquad (1.93)$$

where Q_{α}^{eff} is the effective charge of the MT spheres, $Q_{\alpha}^{eff} = z_{\alpha} + Q_{\alpha}^{MT}$, depending on the nuclear charge z_{α} and the total electronic charge inside the MT sphere

$$Q_{\alpha}^{MT} = -4\pi \int_0^{r_{\alpha}^{MT}} \rho_{\alpha}(r) r^2 dr\,. \qquad (1.94)$$

The exchange-correlation potential was calculated from Eq. (1.87) in the local Slater approximation.

For the calculation of H_{out} matrix elements, we used integration on a spatial grid with uniform steps in cylindrical coordinates Z, Φ and R. The grid was selected so that the calculated total electronic charge of the unit cell $O_{\Omega} = \int_{\Omega} \rho(\mathbf{r}) d\mathbf{r}$ was equal to the number of electrons in the unit cell Ω with the required accuracy.

1.3.6.2. Polyynic Carbyne

A linear chain of carbon atoms with alternating bond lengths 1.34 and 1.20 Å is the simplest nanowire. In this case, the inner cavity is obviously absent; therefore, we analyzed the convergence of the band structure as a function of the radius a of the outer potential barrier. The value a was varied from the covalent radius of carbon up to the a = 2.11 Å (the van der Waals radius of the C atom is equal to 1.7 Å). Physically reasonable band diagrams of carbyne were obtained at $a \geq 1.48$ Å. For $1.48 \leq a \leq 2.11$ Å, only about 3.5–0.2% of the electron density of the system is located beyond the potential cylinder and MT regions occupy 40–27% of the unit cell volume. Because of the rapid decrease in atomic electron density, for the calculation of the electron density distribution in the unit cell, it was sufficient to consider the contributions of atoms located in given and adjacent cells. The contribution of the other atoms of the chain to the Q_Ω is below 1%.

The grid on which the H_{out} potential and its matrix elements were calculated included 10–50 steps along radial coordinate R, 20 steps along angular coordinate Φ, and 20–50 steps along tube axis Z. On this grid, the error in the calculation of the Q_Ω is between 3 and 0.3%. For each value of a, we performed calculations with different numbers of basic functions. An increase in a value called for a growth of the numbers of basic functions and of points in the integration grid. For example, we monitored the shift of bands upon extending the basic set for a = 2.11 Å. On changing from 193 to 279 basic functions in the lower energy range of width 25 eV including the whole valence band and the π^* conduction band, the levels are shifted by less than 0.15 eV. On changing to 325–420 basic functions, the shift is below 0.05 eV. Thus, in this version of the LACW method, for calculating of carbyne with an accuracy ~0.1 eV, it is sufficient to take $a \geq 1.48$ Å and about 200 basic functions. Figure 1.33 shows the results of the calculation of the band structure of carbyne at a = 1.48, 1.9, and 2.11 Å.

It is seen that the main effect of increasing a consists in the shift of the whole band structure down along the energy axis and in only small variations of the mutual arrangement of valence band branches. The calculations are in close agreement with the experimental data for carbyne: the experimental band gap width is 1–2.2 eV, the gap between the σ^* and π bands is 2 eV, the width of the valence π band is 5–6 eV, and the width of the σ band is 6–7 eV (Kudryavtsev et al. 1993).

1.3.6.3. BC_2N Nanotubes

Layered materials of composition BC_2N can be obtained by saturating a mixture of boron and carbon black with nitrogen and by chemical vapor deposition from BCl_3 and CH_3CN (Liu et al. 1989). There can be different

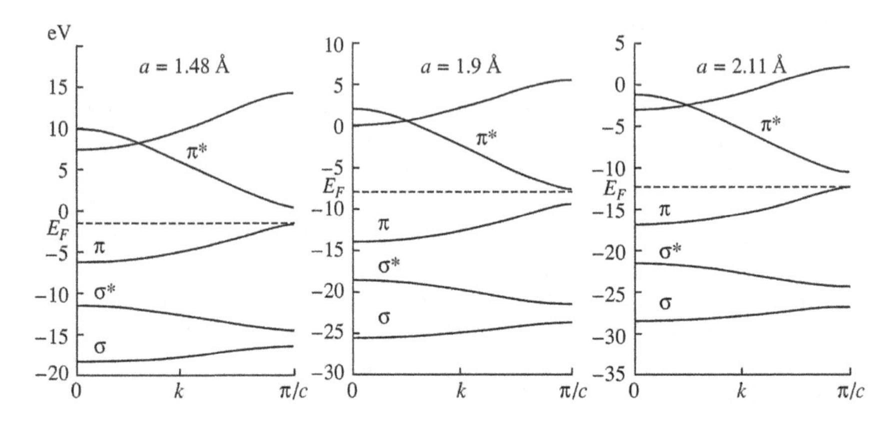

Fig. 1.33. Dependence of the band diagram of carbyne on the radius *a* of the cylindrical potential barrier.

arrangements of atoms in the graphite-like BC_2N layers, but the structures with the C–C and C-N bonds form more stable compounds. Figure 1.34 shows the three possible armchair tubules constructed by coiling such BC_2N layers, the atomic coordinates of which were determined by molecular mechanics techniques.

The band structure calculations were performed using the basic sets of 350–475 functions and potential integration grids containing the 20–30 nodes along *R*, 40–60 nodes along Φ, and 20-40 nodes along Z coordinates. The BC_2N-I nanotube is metallic with valence band width about 26 eV (Fig. 1.35).

The BC_2N-II nanotube is a semiconductor with $E_g = 1$ eV and valence band width of 21.5 eV. The BC_2N-III nanotube has a narrower optical gap (0–0.5 eV), and the valence band width equal to 27.5 eV.

1.3.6.4. GaAs Nanotubes

GaAs (5, 5) is an isoelectronic and isostructural analog of the (5, 5) boron nitride nanotube (Bolshakov et al. 2018). In calculations of the band structure of this nanotube, the Ga–As distance was assumed to be equal to 2.44 Å, the interatomic distance in sphalerite-like GaAs. The full-potential integration grid contained up to 30, 64, and 30 nodes along *R*, Φ, and Z, respectively. The full-potential distribution was constructed for two-unit cells in each direction, because the atomic electron densities of Ga and As atoms decrease more slowly than for C, B, and N atoms. The width of the valence band is 10.6 eV; the predominantly *s* band of width 2.7 eV is separated from the predominantly *p* part of the valence band by a gap of 4.5 eV (Fig. 1.36).

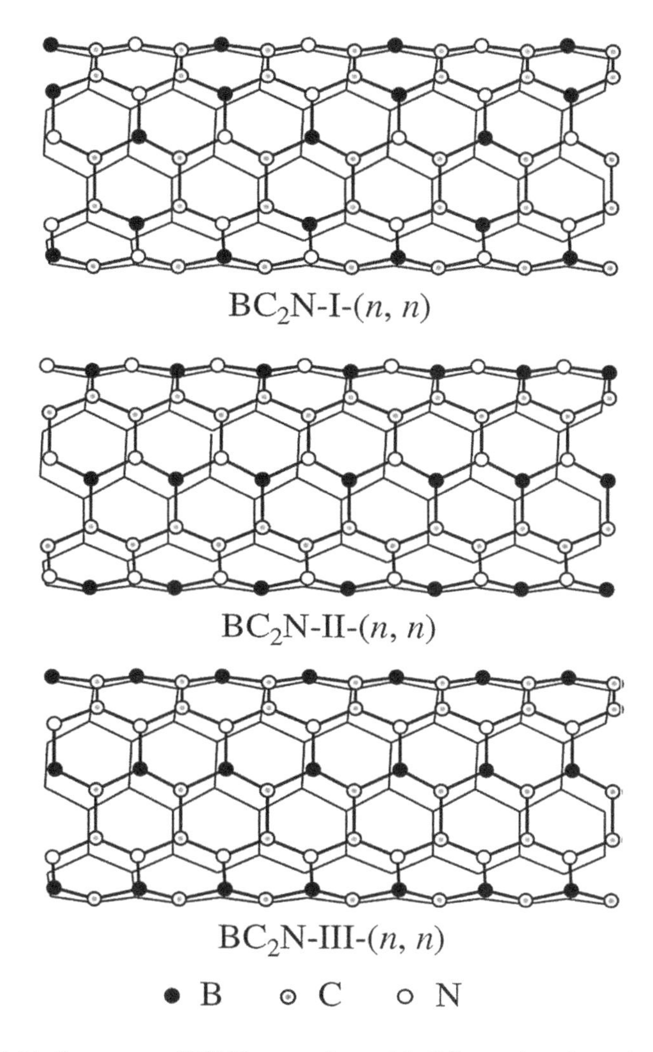

Fig. 1.34. Geometry of BC_2N nanotubes with different bond topologies.

Depending on the density of the full-potential integration grid, the resulting band diagram is either semimetallic with a small (~0.1 eV) band overlap or has a band gap of width 0.1–0.3 eV (the bulk cubic phase is a semiconductor with a gap of 1.428 eV). It is known (Rubio et al. 1994) that the presence of a wide band gap in boron nitride nanotubes is caused by the ionic component of chemical bonding. Chemical bonding in GaAs compound is less ionic than in BN system. Correspondingly, the change from BN to GaAs must lead to a decrease or disappearance of the band gap in nanotubes.

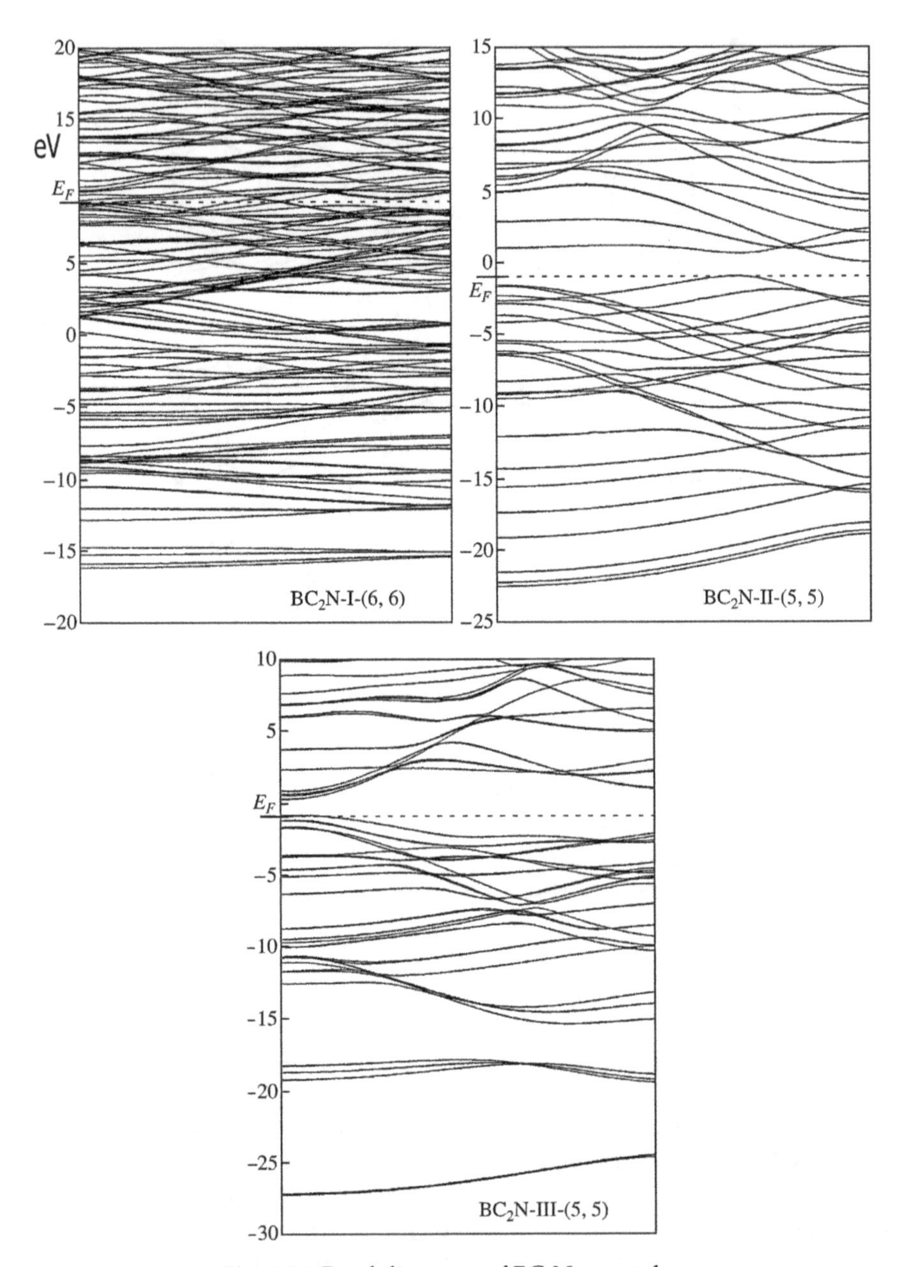

Fig. 1.35. Band diagrams of BC_2N nanotubes.

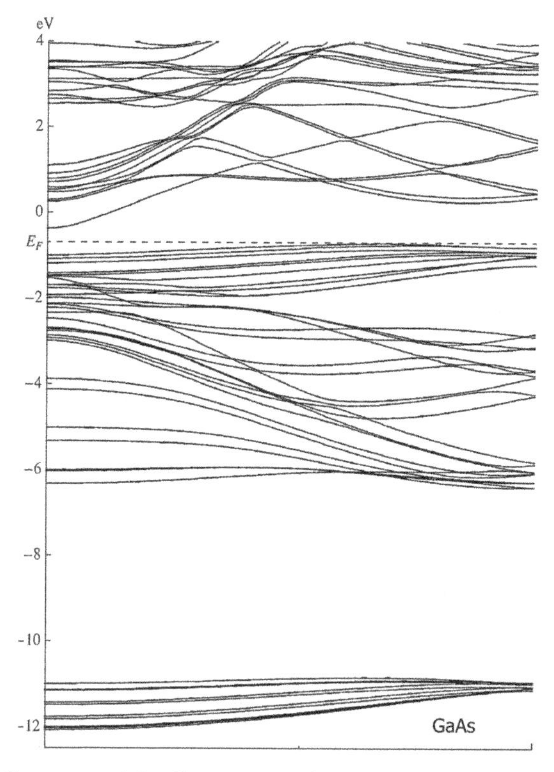

Fig. 1.36. Band diagrams of the GaAs nanotube (5, 5) (basic set of 458 functions).

CHAPTER

2

Augmented Cylindrical Waves for Embedded and Double-walled Nanotubes

In Chapter 1, the electronic structure of carbon nanotubes was considered for the individual single-walled nanotubes not interacting with their environment. However, the properties of nanotubes change on going to solid samples. In bundles of metallic or semiconducting carbon nanotubes and in multiwalled nanotubes, interaction between tubules leads to the opening or perturbation of the optical gaps (Ouyang et al. 2001, Kwon and Tomanek 1998). In films of semiconducting nanotubes under pressure, the gaps width decreases in comparison to the free tubules (Kazaoui et al. 2000). Nanotubes in electronic devices are in contact with crystals (Yao et al. 1999): the nanotubes are typically located on a crystalline substrate and interact with it, whereas the nanotube ends interact strongly with electrodes for good electric contacts. The hybrid electronic devices have been invented, in which nanotubes are embedded into bulk semiconductors (Gelin and Bondarev 2016, Grzelczak et al. 2006, Jensen et al. 2004). In multiwalled nanotubes, the intertube interaction results in perturbation of their electron bands. In this chapter, we study the two simple cases, where interaction of tubule with the surrounding cannot be neglected; namely the LACW techniques for the embedded and double-walled carbon nanotubes are developed and the electron properties of such tubules are discussed.

2.1. Embedded Nanotubes

There are attempts to integrate nanotubes in conventional two- or three-dimensional semiconductor systems. For example, small-diameter nanotubes in the channels of the zeolite crystals were fabricated by pyrolyzing tripropylamine (Tang et al. 2001). The encapsulated nanotubes in epitaxially grown semiconductor heterostructures show promise for realizing semiconductor hybrid devices such as nanotubes contacted by an electron gas (Gelin and Bondarev 2016, Jensen et al. 2004, Barraza-Lopez et al. 2009). In this section, we consider how the interaction of nanotubes with a surrounding crystal can change their band structure (D'yachkov and Makaev 2005a,b). To this end, we constructed, in terms of the LACW method, a model of the electronic structure of a single-walled carbon nanotube embedded into crystal matrix and study the effects of nanotube electrons delocalization into the bulk matrix on the nanotube's properties.

2.1.1. Theory

2.1.1.1. Potential for Embedded Nanotube

It should be noted that for an isolated nanotube, there are two vacuum regions Ω_v, on the inside and the outside of the nanotube. The nanotube and the vacuum regions are separated by impenetrable (in our model, infinite) cylindrical potential barriers. For an embedded nanotube, it is surrounded on the outside by a region of a crystal matrix Ω_m (Fig. 2.1).

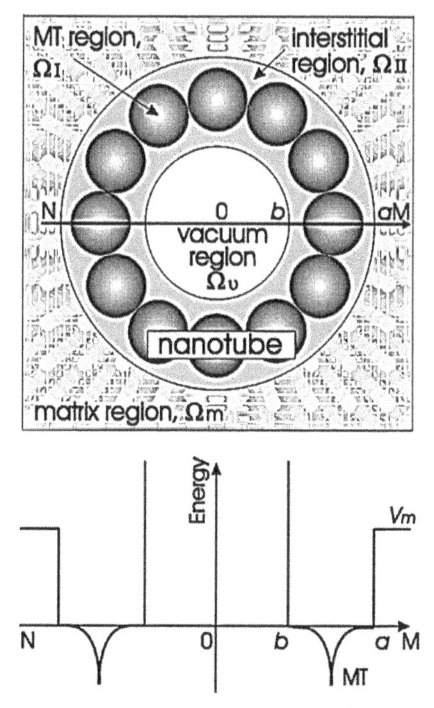

Fig. 2.1. Nanotube in a matrix (above) and cross section of the electronic potential along the N0M line (below).

We assume that the barrier V_m between the nanotube and the matrix is penetrable (in our model, finite), so that tunneling of electrons from the nanotube into the matrix is possible. Our task is to find the solutions of the Schrödinger equation for the orbitals and electronic energies of the nanotube in the matrix. The matrix is considered to be a homogeneous medium with a constant potential V_m; i.e., we ignore the atomic structure of matrix and complicated structure of the potential barrier of the transition region between the nanotube and the matrix, which corresponds to the model of a single-walled nanotube in contact with an electron gas. Let us consider the case where the barrier V_m is located noticeably above the Fermi level, so that the matrix has a relatively weak effect on the states of the valence and conduction bands of the nanotube, and calculate the eigenstates of the system located below V_m. These states are the nanotube's states, but modified by the effects of electron delocalization into the matrix region.

For the electron potential of the nanotube, the MT approximation is used again; i.e., we assume that the potential is spherically symmetric in the vicinity of MT spheres Ω_α, and constant, but different in the interatomic space Ω_{II} and matrix Ω_m regions. The constant potential in the space Ω_{II} is taken as the origin for measuring the energy.

2.1.1.2. Basic Functions

To construct the basic wave functions, the solutions of the wave equation for the matrix and the interspherical and MT regions should be sewn together so that the resulting functions are continuous and differentiable at the boundaries of MT spheres and interspherical and matrix regions. In the interspherical region of the nanotube and in the matrix region, the basic functions are the solutions of the Schrödinger equation for free electron movement with constant potentials. In the cylindrical coordinates Z, R and Φ, this equation takes the same form (1.8) as in Chapter 1, but with new potential $U(R)$

$$\left\{-\left[\frac{1}{R}\frac{\partial}{\partial R}\left(R\frac{\partial}{\partial R}\right)+\frac{1}{R^2}\frac{\partial^2}{\partial \Phi^2}+\frac{\partial^2}{\partial Z^2}\right]+U_m(R)\right\}\Psi^{II,m}(Z,\Phi,R) = E\Psi^{II,m}(Z,\Phi,R), \tag{2.1}$$

$$U_m(R)=\begin{cases} 0, & b\le R\le a \\ \infty, & R<b \\ V_m, & R>b. \end{cases} \tag{2.2}$$

In the region Ω_{II} plus Ω_m, due to cylindrical symmetry of the potential $U_m(R)$, the solutions of Eqs (2.1) and (2.2) are presented in the form $\Psi^{II,m}(Z,\Phi,R) = \Psi_P(Z,k)\Psi_M(\Phi)\,\Psi^{II,m}_{|M|,N}(R)$, where the function

$\Psi_P(Z,k)$ corresponds to the movement of an electron along the z axis in a one-dimensional system with periodic boundary conditions (1.10). The function $\Psi_M(\Phi)$ corresponds to the rotation of an electron about the z axis (1.11). The function $\Psi^{II,m}_{|M|,N}(R)$ is the solution of the equation

$$\left[-\frac{1}{R}\frac{d}{dR}R\frac{d}{dR}+\frac{M^2}{R^2}\right]\Psi^{II,m}_{|M|,N}(R)+U_m(R)\Psi^{II,m}_{|M|,N}(R)=E_{|M|,N}\Psi^{II,m}_{|M|,N}(R) \quad (2.3)$$

with the potential $U_m(R)$ (2.2). Now, the $\Psi^{II,m}_{|M|,N}(R)$ function describes the radial movement of an electron in the interspherical region Ω_{II} of the nanotube and in the matrix region Ω_m. In the nanotube region, $U_m(R) = 0$ and Eq. (2.3) takes the previous form of the Bessel equation (1.14), any solution of which is represented by a linear combination (1.15) of cylindrical Bessel functions of the first and second kinds

$$\Psi^{II,m}_{|M|,N}(R)=C^J_{MN}J_M\left(\kappa_{|M|,N}R\right)+C^Y_{MN}Y_M\left(\kappa_{|M|,N}R\right). \quad (2.4)$$

In the matrix region, $U_m(R) = V_m$ and $\Psi^m_{|M|,N}(R)$ must obey the equation

$$\left[\frac{d^2}{dR^2}+\frac{1}{R}\frac{d}{dR}-(V_m-\kappa^2_{|M|,N})-\frac{M^2}{R^2}\right]\Psi^m_{|M|,N}(R)=0. \quad (2.5)$$

Equation (2.5) at $V_m > 0$ is known as a modified Bessel equation. Its solutions (Fig. 2.2), which vanishes when R tends to infinity, are modified Bessel functions of the first kind K_M:

$$\Psi^m_{|M|,N}(R)=C^K_{MN}K_M\left(\kappa^K_{|M|,N}R\right), \quad (2.6)$$

where

$$\kappa^K_{|M|,N}=(V_m-\kappa^2_{|M|,N})^{1/2}. \quad (2.7)$$

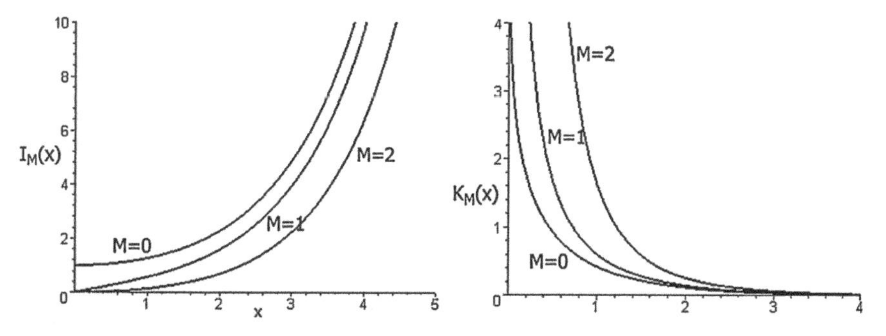

Fig. 2.2. Modified Bessel functions of the first K_M and second I_M kind.

The function $\Psi^{II,m}_{|M|,N}(R)$ should vanish at $R = b$, be continuous and differentiable at $R = a$, and be normalized. As a result, we obtain the set of equations for the coefficients $C^J_{MN}, C^Y_{MN}, C^K_{MN}$ and for the energy $\kappa^2_{|M|,N}$:

$$C^J_{MN} J_M\left(\kappa_{|M|,N} b\right) + C^Y_{MN} Y_M\left(\kappa_{|M|,N} b\right) = 0,$$

$$C^J_{MN} J_M\left(\kappa_{|M|,N} a\right) + C^Y_{MN} Y_M\left(\kappa_{|M|,N} a\right) = C^K_{MN} K_M\left(\kappa^K_{|M|,N} a\right),$$

$$\kappa_{|M|,N}\left[C^J_{MN} J'_M\left(\kappa_{|M|,N} a\right) + C^Y_{MN} Y'_M\left(\kappa_{|M|,N} a\right)\right] = \kappa^K_{|M|,N} C^K_{MN} K'_M\left(\kappa^K_{|M|,N} a\right),$$

$$\int_b^\infty \left|\Psi^{II,m}_{|M|,N}(R)\right|^2 R dR = 1. \tag{2.8}$$

From this system, we easily obtain the equation for $\kappa_{|M|,N}$

$$\kappa_{|M|,N}\left[Y_M\left(\kappa_{|M|,N} b\right) J'_M\left(\kappa_{|M|,N} a\right) - J_M\left(\kappa_{|M|,N} b\right) Y'_M\left(\kappa_{|M|,N} a\right)\right]$$
$$-\kappa^K_{|M|,N}\left[Y_M\left(\kappa_{|M|,N} b\right) J_M\left(\kappa_{|M|,N} a\right) - J_M\left(\kappa_{|M|,N} b\right) Y_M\left(\kappa_{|M|,N} a\right)\right]$$
$$\times \frac{K'_M\left(\kappa^K_{|M|,N} a\right)}{K_M\left(\kappa^K_{|M|,N} a\right)} = 0. \tag{2.9}$$

For each root $\kappa_{|M|,N}$ from the equation

$$\left(C^J_{MN}\right)^2 \left\{ \frac{a^2}{2}\left[Y_M\left(\kappa_{|M|,N} b\right) J'_M\left(\kappa_{|M|,N} a\right) - J_M\left(\kappa_{|M|,N} b\right) Y'_M\left(\kappa_{|M|,N} a\right)\right]^2 \right.$$
$$- \frac{b^2}{2}\left[Y_M\left(\kappa_{|M|,N} b\right) J'_M\left(\kappa_{|M|,N} b\right) - J_M\left(\kappa_{|M|,N} b\right) Y'_M\left(\kappa_{|M|,N} b\right)\right]^2$$
$$+ \left[Y_M\left(\kappa_{|M|,N} b\right) J_M\left(\kappa_{|M|,N} a\right) - J_M\left(\kappa_{|M|,N} b\right) Y_M\left(\kappa_{|M|,N} a\right)\right]^2$$
$$\left. \times \left[\left(K_M\left(\kappa^K_{|M|,N} a\right)\right)^{-2} \int_a^\infty K^2_M\left(\kappa^K_{|M|,N} R\right) R dR\right] \right\} = 1 \tag{2.10}$$

we obtain the C^J_{MN} values and calculate C^Y_{MN} and C^K_{MN} from equations:

$$C^Y_{MN} = -C^J_{MN} \frac{J_M\left(\kappa_{|M|,N} a\right)}{Y_M\left(\kappa_{|M|,N} a\right)},$$

$$C^K_{MN} = C^J_{MN} \frac{Y_M\left(\kappa_{|M|,N} b\right) J_M\left(\kappa_{|M|,N} a\right) - J_M\left(\kappa_{|M|,N} b\right) Y_M\left(\kappa_{|M|,N} a\right)}{Y_M\left(\kappa_{|M|,N} b\right) K_M\left(\kappa^K_{|M|,N} a\right)}. \tag{2.11}$$

Thus, the basic functions $\Psi^{II,m}_{M,N,P}(Z,\Phi,R)$ in the interspherical and in matrix regions are finally determined. It is worth noting that, in the Ω_{II} region of the embedded nanotube, the cylindrical wave $\Psi^{II}_{M,N,P}(Z,\Phi,R)$ has the same analytical form $\Psi_P(Z,k)\Psi_M(\Phi)\,\Psi^{II}_{|M|,N}(R)$ with $\Psi^{II}_{|M|,N}(R)$ written as the linear combination of the functions J_M and Y_M (2.4) as in the case of isolated nanotube; only Eqs. (2.8-2.11) for $\kappa_{|M|,N}$, C^J_{MN}, and C^Y_{MN} change for Eqs. (1.19), (1.20), and (1.25).

As in Chapter 1, inside the MT spheres α in the local spherical coordinate system (r, θ, φ), the basic functions are expanded in spherical harmonics $Y_{lm}(\theta,\phi)$ and the solutions of the radial Schrödinger equation $u_{l\alpha}(r,E_{l\alpha})$ and their radial derivatives $\dot{u}_{l\alpha}(r,E_{l\alpha})$

$$\Psi_{I\alpha}=\sum_{l=0}^{\infty}\sum_{m=-l}^{l}\left[A^{MNP}_{lm\alpha}u_{l\alpha}(r,E_{l\alpha})+B^{MNP}_{lm\alpha}\dot{u}_{l\alpha}(r,E_{l\alpha})\right]Y_{lm}(\theta,\phi). \tag{2.12}$$

The coefficients $A^{MNP}_{lm\alpha}$ and $B^{MNP}_{lm\alpha}$ are to be selected so that both the basic functions $\Psi_{k,PMN}$ and their radial derivative have no discontinuities at the boundaries of the MT spheres. However, the analytical form of the cylindrical wave $\Psi^{II}_{M,N,P}(Z,\Phi,R)$ near the MT spheres of the nanotube remains unaltered in going from the isolated nanotube to the tubule embedded into the matrix. Therefore, the analytical expressions (1.46)–(1.51) for coefficients $A_{lm\alpha}$ and $B_{lm\alpha}$ derived in the previous chapter remain valid in this case too.

Thus, at $b \le R \le a$, the LACWs $\Psi_{k,PMN}$ have the same analytical form as for the isolated nanotube, whereas, in the matrix region,

$$\Psi_{k,MNP}(Z,\Phi,R)=\frac{C^K_{MN}}{\sqrt{2\pi c}}e^{i(k+k_P)Z}e^{iM\Phi}K_M\left(\kappa^K_{|M|,N}R\right). \tag{2.13}$$

2.1.1.3. Overlap and Hamiltonian Integrals

The integral $\left\langle\Psi^*_{M_2N_2P_2}\middle|\Psi_{M_1N_1P_1}\right\rangle$ of the product of the functions $\Psi^*_{M_2N_2P_2}$ and $\Psi_{M_1N_1P_1}$ over the unit cell Ω is equal to the integral of the cylindrical waves $\Psi^{II,m}_{M,N,P}$ over the interspherical and matrix regions Ω_{II} + Ω_m plus the sum of the integrals of the spherical parts of the LACWs $\Psi_{I\alpha,MNP}$ over the MT regions:

$$\left\langle\Psi^*_{M_2N_2P_2}\middle|\Psi_{M_1N_1P_1}\right\rangle=\int_{\Omega_{II}+\Omega_{I\alpha}}\bar{\Psi}^{II}_{M_2N_2P_2}\Psi^{II}_{M_1N_1P_1}dV+$$
$$\sum_{\alpha}\int_{\Omega_\alpha}\bar{\Psi}_{I\alpha,M_2N_2P_2}\Psi_{I\alpha,M_1N_1P_1}dV. \tag{2.14}$$

The integral over $\Omega_{II} + \Omega_m$ is equal to the integral over Ω minus the sum of the integrals over the MT regions. Due to the fact that the cylindrical waves as solutions of the Schrödinger equation (2.1) are orthonormalized, the integral over Ω is equal to the product of the δ-functions. As a result, Eq. (2.14) takes the form

$$\left\langle \Psi_{M_2N_2P_2} \middle| \Psi_{M_1N_1P_1} \right\rangle = \delta_{M_1,M_2}\delta_{N_1,N_2}\delta_{P_1,P_2} - \sum_{\alpha} \int_{\Omega_\alpha} \Psi^*_{II\alpha,M_2N_2P_2} \Psi_{II\alpha,M_1N_1P_1} dV + \sum_{\alpha} \int_{\Omega_\alpha} \Psi^*_{I\alpha,M_2N_2P_2} \Psi_{I\alpha,M_1N_1P_1} dV. \quad (2.15)$$

Inasmuch as both sides of the LACWs $\Psi_{II\alpha}$ and $\Psi_{I\alpha}$ in the MT regions have the same form for both the embedded and isolated nanotubes, the expressions for overlap integrals (1.54) obtained for the separate nanotube are also valid for the more complex case of embedded nanotube.

Analogously, the Hamiltonian matrix elements for the embedded nanotube can be calculated using Eq. (1.62) obtained for the isolated nanotube. Indeed,

$$\left\langle \Psi^*_{M_2N_2P_2} \middle| H \middle| \Psi_{M_1N_1P_1} \right\rangle = \int_{\Omega_{II}+\Omega_m} \bar{\Psi}^{II,m}_{M_2N_2P_2} H \Psi^{II,m}_{M_1N_1P_1} dV + \sum_{\alpha} \int_{\Omega_\alpha} \bar{\Psi}_{I\alpha,M_2N_2P_2} H \Psi_{I\alpha,M_1N_1P_1} dV. \quad (2.16)$$

In the interspherical region, $\hat{H} = -\Delta$ and

$$\int_{\Omega_{II}} \bar{\Psi}^{II,m}_{M_2N_2P_2} (-\Delta) \Psi^{II,m}_{M_1N_1P_1} dV = \int_{\Omega} \bar{\Psi}^{II,m}_{M_2N_2P_2} (-\Delta) \Psi^{II,m}_{M_1N_1P_1} dV - \sum_{\alpha} \int_{\Omega_\alpha} \bar{\Psi}_{II\alpha,M_2N_2P_2} (-\Delta) \Psi_{II\alpha,M_1N_1P_1} dV. \quad (2.17)$$

The cylindrical waves Ψ_{II} are the eigenfunctions of the electron kinetic energy operator; therefore,

$$\int_{\Omega} \bar{\Psi}^{II,m}_{M_2N_2P_2} (-\Delta) \Psi^{II,m}_{M_1N_1P_1} dV = \left(K_{P_1} K_{P_2} + \kappa_{|M_1|,N_1} \kappa_{|M_2|,N_2} \right) \times \delta_{M_1,M_2}\delta_{N_1,N_2}\delta_{P_1,P_2} \quad (2.18)$$

while the analytical expressions for the integrals over the MT regions remain unaltered since the form of the basic functions in the nanotube region does not change.

Thus, to obtain dispersion curves for an embedded nanotube, integrals (1.54) and (1.62) should be substituted into the secular equations, while the $\kappa_{|M|,N}$, C^J_{MN}, and C^Y_{MN} values should be calculated by Eqs. (2.9)–(2.11).

2.1.2. Application

We have calculated the band structures and DOS of embedded metallic armchair (n, n) nanotubes with $4 \leq n \leq 12$ and semiconducting zigzag $(n, 0)$ nanotubes with $10 \leq n \leq 26$; in the case of $(n, 0)$ nanotubes, we have omitted the metallic structures having n evenly divisible by 3.

Figures 2.3 and 2.4 show the representative results for the embedded armchair nanotubes that can be compared with the analogous data for the nonembedded ones.

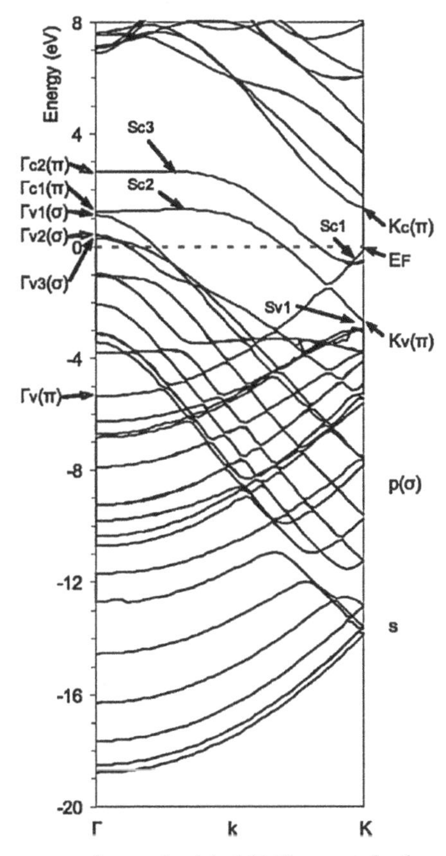

Fig. 2.3. Band structure of an embedded (5, 5) nanotube for a barrier parameter $\varepsilon_m = 2$. Here, S_{v1}, S_{c1}-S_{c3} are the van Hove singularities located between the Brillouin zone center Γ and boundary K; $\Gamma_{vi}(\pi/\sigma)$, $\Gamma_{ci}(\pi/\sigma)$, $K_v(\pi)$, and $K_c(\pi)$ stand for π/σ states at the Γ and K points. The subscripts v and c refer to the bands of a non-embedded nanotube; for example, the $\Gamma_{v1}(\sigma)$ state is located below the Fermi level of the non-embedded (5, 5) nanotube, but it is above the Fermi level of the embedded one. The dimensionless barrier parameter ε_m is defined as $\varepsilon_m = V_m/\delta$, where δ is energy gap between the Fermi level of the nanotube and the interatomic potential.

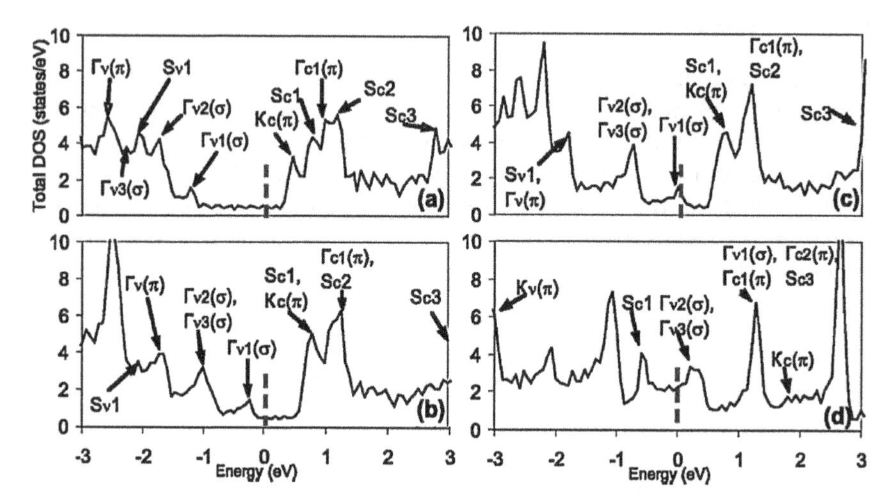

Fig. 2.4. DOS of a (5, 5) nanotube vs barrier parameter ε_m. Here, we use Gaussian broadening with a half-width of 0.05 eV. (a) $\varepsilon_m = \infty$ (nonembedded nanotube); (b–d) $\varepsilon_m = 6$, 4, and 2, respectively.

According to the LACW calculations, the armchair nanotubes have the metal-type electronic structure, with the Fermi level located at the intersection of the two π bands at about $k = (2/3)(\pi/c)$. In the Brillouin zone center, the highest occupied σ-state $\Gamma_{v1}(\sigma)$ is located above the highest occupied π-state $\Gamma_v(\pi)$ in all non-embedded armchair nanotubes and in graphite. The delocalization of electrons of the armchair nanotube into the matrix region results in a strong band structure perturbation. A high energy shift of the σ states at the Γ point relative to the π bands is the most significant effect of embedding. As a result, the former top of the valence σ-band $\Gamma_{v1}(\sigma)$ shifts to the nanotube conduction band region, and the σ electrons can take part in the nanotube charge transfer via the mechanism of electron tunneling into the matrix. The intersection point of the frontier π bands shifts in the direction of the Brillouin zone boundary, and the total valence band width decreases as one goes from the free nanotubes to the embedded ones. In the pristine nanotubes, the Fermi level is located in a dip of the electron DOS. Interaction with the matrix fills this dip, which results in a growth of the DOS at the Fermi level. The electrons tunneling into the matrix does not destroy the metallic character of the armchair nanotube's band structure.

In the case of semiconducting nanotubes, the band structure, DOS, and minimum energy gap E_{11} are very sensitive to the matrix effect (Figs. 2.5-2.7).

For example, the gap $E_{11} = 0.88$ eV of the (13, 0) nanotube grows slightly and then falls drastically with decreasing potential barrier. Finally, the gap at the Γ point is closed up, and the semiconducting behavior breaks

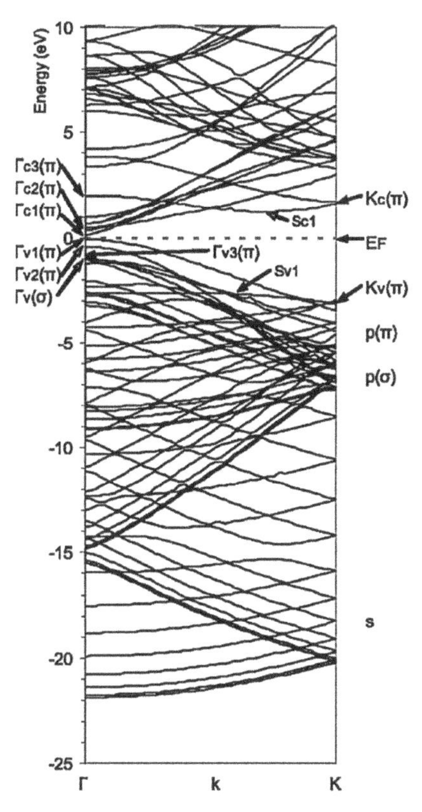

Fig. 2.5. Band structure of embedded (13, 0) nanotube for barrier parameter $\varepsilon_m = 2$.

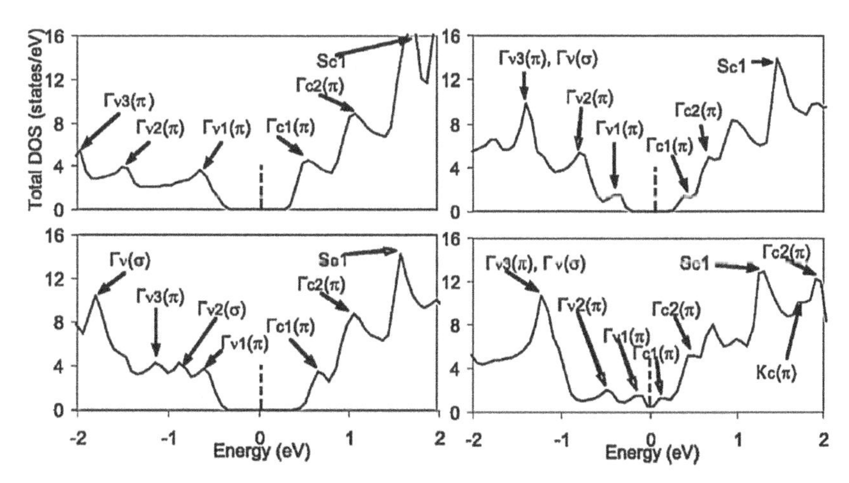

Fig. 2.6. DOS of a (13, 0) nanotube for various barrier parameters ε_m. (a) $\varepsilon_m = \infty$; (b–d) $\varepsilon_m = 6, 4$, and 2, respectively.

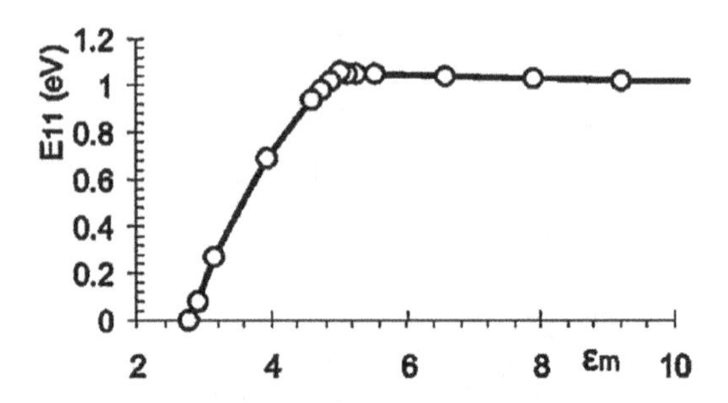

Fig. 2.7. Minimum energy gap E_{11} vs ε_m for (13, 0) nanotube.

down. In our one-parameter model, a large value of barrier parameter $\varepsilon = V_m/\delta > 5$ corresponds to an insulating crystal matrix with only a minor perturbation of the nanotube band structure. A small barrier $V_m/\delta < 5$ corresponds to semiconducting or metallic matrices.

The metallization of the embedded nanotubes is consistent with the measured transport properties of the tube-semiconductor devices with nanotubes encapsulated in a semiconductor crystal using molecular beam epitaxy. The results from a collection of 20 measured devices displayed no gate voltage dependence of the conductance at room temperature; i.e., the incorporated nanotubes were metallic (Jensen et al. 2004). A Luttinger liquid behavior with a conductance equal to 0.2 e^2/h confirmed that leak currents through the substrate can be neglected and that the electron transport is dominated by the properties of the embedded metallic nanotubes.

2.2. Double-Walled Carbon Nanotubes

The carbon double-walled nanotubes can be considered as a special case of multiwalled carbon nanotubes. They consist of two concentric graphene cylinders with extremely strong covalent bonding of atoms within the individual graphitic sheets, but very weak van der Waals type interaction between them (Shen et al. 2011, Ghedjatti et al. 2017, Levshov et al. 2017). The double-walled nanotubes are very important from the theoretical and experimental standpoint. They are essential for electronic device applications, because they are the molecular analogs to coaxial cables. Thus, a metallic@semiconducting or semiconducting@metallic nanotube can be, respectively, a molecular conductive wire covered by an insulator or a molecular capacitor in a memory device (Saito et al. 1993, Chen et al. 2003, Enyashin et al. 2004). It is believed that the double-shell carbon cylinders would exhibit enhanced field emission, mechanical, thermal, and filters properties when compared to the single-walled nanotubes (Kim et al. 2004).

The double-walled carbon nanotubes have been produced by several techniques such as the arc discharge method (Hutchison et al. 2001, Sugai et al. 2003), the catalytic chemical vapor deposition method (Bacsa et al. 2003), and a method utilizing fusion reactions of fullerenes in single-walled carbon nanotube (Bandow et al. 2001, 2004). The double-walled carbon nanotube yield can be greater than 95% (Endo et al. 2005). The two constituent tubules of such tubes can be characterized in detail by measuring the Raman spectra and high-resolution transmission electron microscopy (Ren et al. 2002, Li et al. 2003, Hashimoto et al. 2005). Particularly, the indices of the two coaxial layers of a double-walled system can be assigned based on the radial breathing mode frequencies, and the atomic correlations between two graphene layers in tubule can be obtained using electron microscopy. In the double-walled carbon compounds, the interlayer spacing is not a constant, ranging from 3.4 Å (the interlayer distance of graphite) to 4.1 Å (Li et al. 2003). Recently, the electronic structure of individual double-walled tubes suspended in water and in air over trenches was studied using optical absorption, emission, and time-resolved photoluminescence spectroscopy; it was shown experimentally that interaction of the two layers influences the optical transitions of the inner and outer tubules (Shen et al. 2011, Ghedjatti et al. 2017, Levshov et al. 2017, Hertel et al. 2005).

While different from the embedded tubes, the theoretical studies of the nanotube electronic structure change due to interwall interaction in double-walled systems have received much attention since 1993. The first calculation for a band structure of double-walled carbon tubules was done using the perturbative tight-binding π-electronic technique, which sensitively includes all symmetry constraints, but in a strongly simplified Hamiltonian (Saito et al. 1993). It was shown that the energy dispersion relations of single-walled counterparts are weakly perturbed by the interlayer interaction. The calculated energy band structures of the best matched commensurate metallic@metallic armchair (5, 5)@(10, 10) and zigzag (9, 0)@(18, 0) nanotubes with the number of carbon atoms ratio in the layers equal to 1:2 yield a metallic double-walled system when a weak interlayer coupling between the concentric nanotubes is introduced. The calculated coaxial incommensurate zigzag metallic@semiconducting (9, 0)@(17, 0) and semiconducting@metallic (10, 0)@(18, 0) nanotubes also retain their individual metallic and semiconducting identities when the weak interlayer interaction is turned on. Finally, two coaxial semiconducting zigzag tubules remain semiconducting when the weak interlayer coupling is introduced (Saito et al. 1993).

Closer examination of the interwall vibronic interaction effects was performed in terms of tight-binding calculations with parametrization of the Linear Combination of Atomic Orbitals (LCAO) matrix elements based on *ab initio* results for simpler structures (Lambin et al. 1994, Kwon and Tomanek 1998). The electronic structure was calculated for three nanotubes in the Fermi level E_F region. For the (5, 5)@(10, 10) nanotube, it was predicted that the weak intertube interaction periodically opens and closes four pseudogaps in the density of states due to symmetry lowering during the low-frequency librational motion about and vibrational motion normal to the double-tube axis. As the intertube interaction is switched on in the (9, 0)@(18, 0) nanotubes, the 30 meV gap opens. In the semiconducting (8, 0)@(17, 0) system, the intertube interaction reduces the gap of the noninteracting system by 0.1 eV. The band structure of the (5, 5)@(10, 10) nanotubes obtained using a pseudopotential method and a plane-wave basis (Miyamoto et al. 2001) resembles the tight-binding results. Using a scattering technique based on a LCAO Hamiltonian, the ballistic quantum conductance of (10, 10)@(15, 15) finite nanotube was calculated (Sanvito et al. 2000). It was found that the interwall interaction blocks certain conduction channels and redistributes the current non-uniformly across the walls providing an explanation for the unexpected integer and noninteger conductance values reported for multiwalled nanotubes (Frank et al. 1998).

Using the self-consistent plane-wave pseudopotential calculations, the work functions of the small-diameter double-wall nanotubes starting from the (4, 0)@(13, 0) and (3, 3)@(8, 8) nanotubes were studied (Shan and Cho 2005). In the case of nanotubes with ultra-small inner tubules, the calculations show that the electrostatic interwall charge transfer induced effects result in the few tenth of eV (up to 0.5 eV) band shift due to the large (up to 1.25 eV) difference between the Fermi energies of the inner and outer tubules. These effects decrease drastically as one goes to the double-walled systems with larger diameters, because the Fermi level energies are equal for the single-walled tubules with diameters larger than 1 nm. Particularly, in the largest zigzag (8, 0)@(17, 0) and armchair (5, 5)@(10, 10) tubules, the shifts are equal to 0.05 eV and 0.02 eV, respectively, that is negligible in comparison with the tunneling effects studied below.

Finally, with due account of the cylindrical geometry of the nanotubes, a numerical technique for a local-density functional calculation of the nanotube's electronic structure was presented and applied to double-walled carbon nanotubes (Östling et al. 1997), but for strongly oversimplified structural model of the nanotubes, where the point charges of the individual C^{4+} ions in the walls with graphitic honeycomb lattice were replaced by the two-dimensional infinitely thin structureless charged "sheets" of cylindrical symmetry with uniform surface-charge density.

The LACW method seems to be a more satisfying local-density-functional approach to determining the electronic states of double-walled nanotubes, in which one takes into account the cylindrical geometry of the nanotubes and considers a real approximately van der Waals width of the cylindrical layer and a real atomic structure of nanotubes (D'yachkov and

Makaev 2006, Makaev and D'yachkov 2006). It is the aim of this section to study the perturbations of the electronic band structures and DOS of the core and shell tubes due to the intertube coupling in double tubules in terms of this method. For this purpose, we calculate the complete band structure of the purely semiconducting zigzag $(n, 0)@(n', 0)$ and purely metallic armchair $(n, n)@(n', n')$ nanotubes and use the results to correlate the minimum direct energy gaps E_{11} between the conduction and valence band singularities with the nanotubes diameters. It is to be noted that the optical absorption and time-resolved photoluminescence measurements of nanotubes have shown that one can determine experimentally the partial DOS and interband optical transition energies associated with core and shell tubes (Hertel et al. 2005).

2.2.1. Theory

2.2.1.1. Electron Potential

As in Chapter 1, in which LACW technique for the single-walled nanotubes is developed, the one-electron and MT potentials and local density theory are used here to study the double-walled carbon systems, but the ideas of the cylindrical MT potential are to be somewhat refined here (Fig. 2.8).

The atoms of double-walled nanotube are considered to be enclosed between two essentially impenetrable infinite cylinder-shaped potential barriers Ω_{b1} and Ω_{a2}, because there are two vacuum regions Ω_v on the outside of the outer tubule 2 and on the inside of the inner tubule 1. On the other hand, the cylinder-shaped potential barriers Ω_{a1} and Ω_{b2} on the outside of the inner tubule 1 and on the inside of the outer tubule 2 are penetrable (finite) ones, so that electron tunneling exchange between the tubules 1 and 2 is possible. The radii a_1, b_1 and a_2, b_2 of these barriers are virtually the same as for the pristine single-wall systems. For zigzag nanotubes, we take $a_j = R_j + 2.3$ a.u., $b_j = R_j - 2.3$ a.u., and the cutoff energy $E_{cut} = 50$ eV (here, R_j is radius of the tubule j). For armchair nanotubes, we take $a_j = R_j + 2.4$ a.u., $b_j = R_j - 2.6$ a.u., and $E_{cut} = 100$ eV. This choice determines up to 600 basic functions for single-walled and up to 1100 basic functions for double-walled tubules and permits a description of all the valence and the most important low-lying conduction band states, but possibly not the nearly free electron states located about 2.3 Å away from the carbon layer (3 or 4 eV above E_F, 2 or 1 eV below vacuum) (Okada et al 2000).

For simplicity, we treat the classically forbidden region Ω_f between barriers Ω_{a1} and Ω_{b2} as a homogenous medium with a constant classically impenetrable potential V_f. This potential is treated as a parameter. To characterize the barrier V_f, we used the dimensionless barrier parameter ε defined as $\varepsilon = V_f/\delta$, where δ is the energy gap between the Fermi level of the single-walled nanotube and the interspherical potential. All the

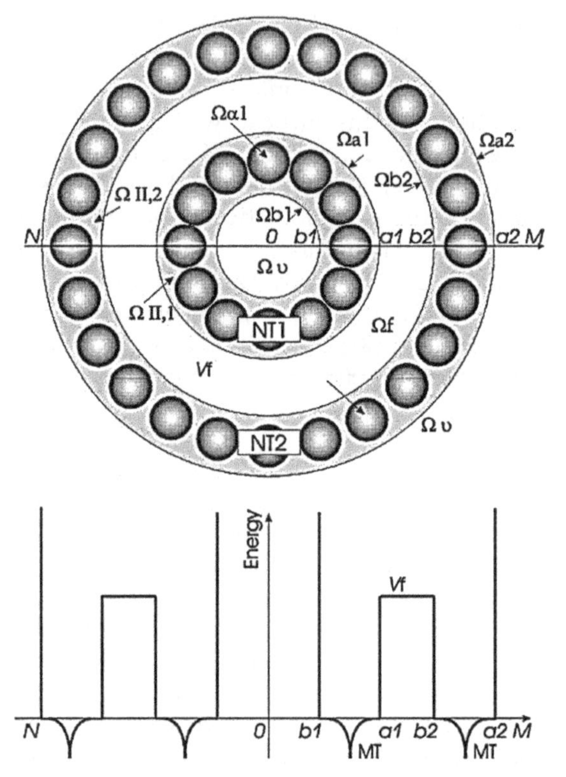

Fig. 2.8. Double-walled nanotube (above) and cross section of the electronic potential along N0M line (below). Here, NT1 and NT2 are the inner and outer tubes, respectively; $\Omega_{II,1}$ and $\Omega_{II,2}$ are the interstitial regions of these tubules; the $\Omega_{1\alpha}$ and $\Omega_{2\alpha}$ are the muffin-tin regions of the tubules 1 and 2, respectively; the Ω_{b1} and Ω_{a2} are the inner and outer impenetrable cylindrical potential barriers between the double-walled nanotube and vacuum regions Ω_v; V_f is a potential energy in Ω_f region between the finite cylindrical potential barriers Ω_{a1} and Ω_{b2}.

results presented here are obtained for ε = 7 determined with account of the graphite band structure. In graphite, the interlayer interactions introduce band splitting and shift of roughly 2–1 eV. For example, the three-dimensional graphite calculations using the first principle MT and full-potential Korringa-Kohn-Rostoker technique suggest that the perturbations of valence π and σ bands are equal to 4–2 and 1–0.5 eV, respectively (Tatar and Rabii 1982). In double-walled nanotube with the same interlayer separation, the band splitting and shift should be roughly less by half, because every graphene layer interacts with two neighbor layers in graphite, but there is only one interlayer coupling in the nanotube. Moreover, we would expect the interlayer interaction to decrease additionally for the nanotubes in comparison to graphite due to hybridization of π and σ states because of some curvature of the tubules.

Therefore, in this chapter, the potential V_f common for all double-walled carbon nanotubes studied is chosen so that the average band splitting and shifts due to the intertube interaction are about 0.5 eV in the (5, 5)@(10, 10) nanotubes; the (5, 5) and (10, 10) tubes show the graphitic interwall separation of 3.4 Å when nested.

Finally, the electronic potential is spherically symmetrical in the $\Omega_{I,1\alpha}$ and $\Omega_{I,2\alpha}$ regions of MT spheres of atoms of tubules 1 and 2, respectively. Inside these spheres, we calculate the electron potential by means of the local density approximation with the Slater exchange. In the interatomic regions $\Omega_{II,1}$ and $\Omega_{II,2}$, the one-electron potential is constant; this constant potential is taken as the origin for measurements of energy. Thus, in the MT LACW approach, the electronic spectrum of the double-walled nanotubes is governed by the free movement of the electron in the interatomic space of two cylindrical layers, by electron scattering on the MT spheres, and by electron tunneling through classically forbidden region Ω_f between the layers.

2.2.1.2. Basic Functions

To construct the basic wave functions for double-walled nanotubes, the solutions of the wave equation for the classically forbidden, interspherical, and MT regions of two tubules should be sewn together so that the resulting LACWs are continuous and differentiable anywhere in the system. In the interspherical region of the nanotubes and in the classically forbidden region between the tubules, the LACWs are the solutions of the Schrödinger equation (2.1) for a free electron movement written, but with more complex potential

$$\left\{-\left[\frac{1}{R}\frac{\partial}{\partial R}\left(R\frac{\partial}{\partial R}\right)+\frac{1}{R^2}\frac{\partial^2}{\partial \Phi^2}+\frac{\partial^2}{\partial Z^2}\right]+U_d(R)\right\}\Psi^{II,d}(Z,\Phi,R)$$
$$=E\Psi^{II,d}(Z,\Phi,R), \quad (2.19)$$

where

$$U_d(R)=\begin{cases}0, & b_1\le R\le a_1, b_2\le R\le a_2\\ \infty, & R<b_1, R>a_2\\ V_f, & a_1\le R\le b_2.\end{cases} \quad (2.20)$$

Similar to the case of free single-walled tubule, the potential $U_d(R)$ has cylindrical symmetry. Therefore, solutions of Eqs. (2.19) and (2.20) have the same form of the product $\Psi^{II,d}_{MNP}(Z,\Phi,R)=\Psi_P(Z,k)\Psi_M(\Phi)\Psi^{II,d}_{MN}(R)$, where the functions $\Psi_P(Z,k)$ and $\Psi_M(\Phi)$ do not change, but the $\Psi^{II,d}_{MN}(R)$ is to be calculated from equation

$$\left[-\frac{1}{R}\frac{d}{dR}R\frac{d}{dR}+\frac{M^2}{R^2}\right]\Psi^{II,d}_{|M|,N}(R)+U_d(R)\Psi^{II,d}_{|M|,N}(R)=E_{|M|,N}\Psi^{II,d}_{|M|,N}(R) \quad (2.21)$$

with potential (2.20) and corresponds to the radial movement of an electron in the interspherical regions $\Omega_{II,j}$ ($j = 1, 2$) of the two tubules and in the classically forbidden region Ω_f. In the regions $\Omega_{II,j}$, $U(R) = 0$ and Eq. (2.3) takes the form of the Bessel equation, any solution of which is represented by a linear combination of cylindrical Bessel functions of the first J_M and second Y_M kinds,

$$\Psi^{j}_{II,|M|,N}(R)=C^{J,j}_{M,N}J_M\left(\kappa_{|M|,N}R\right)+C^{Y,j}_{M,N}Y_M\left(\kappa_{|M|,N}R\right), j=1,2. \quad (2.22)$$

In the Ω_f region, $U_d(R) = V_f$ and $\Psi_{f,|M|,N}(R)$ must obey the equation

$$\left[\frac{d^2}{dR^2}+\frac{1}{R}\frac{d}{dR}-\left(V_f-\kappa^2_{|M|,N}\right)-\frac{M^2}{R^2}\right]\Psi_{f,|M|,N}(R)=0. \quad (2.23)$$

We calculate the nanotube's electron energy levels located below the potential V_f of the classically forbidden region. At $V_f > \kappa^2_{|M|,N}$, Eq. (2.23) is the modified Bessel equation (Watson 1966). Its general solutions are the linear combinations of modified Bessel functions of the first K_M and second I_M kinds,

$$\Psi_{f,|M|,N}(R)=C^{K}_{M,N}K_M\left(\kappa^f_{|M|,N}R\right)+C^{I}_{M,N}I_M\left(\kappa^f_{|M|,N}R\right), \quad (2.24)$$

where $\kappa^f_{|M|,N}=\left(V_f-\kappa^2_{|M|,N}\right)^{1/2}$. The functions (2.22) and (2.24) must vanish at $R = a_1$ and $R = b_2$

$$C^{J,1}_{M,N}J_M\left(\kappa_{|M|,N}b_1\right)+C^{Y,1}_{M,N}Y_M\left(\kappa_{|M|,N}b_1\right)=0, \quad (2.25)$$

$$C^{J,2}_{M,N}J_M\left(\kappa_{|M|,N}a_2\right)+C^{Y,2}_{M,N}Y_M\left(\kappa_{|M|,N}a_2\right)=0,$$

be continuous and differentiable at $R = a_1$ and $R = b_2$

$$\begin{aligned}&C^{J,1}_{M,N}J_M\left(\kappa_{|M|,N}a_1\right)+C^{Y,1}_{M,N}Y_M\left(\kappa_{|M|,N}a_1\right)\\&\quad=C^{K}_{M,N}K_M\left(\kappa^f_{|M|,N}a_1\right)+C^{I}_{M,N}I_M\left(\kappa^f_{|M|,N}a_1\right),\\&C^{J,2}_{M,N}J_M\left(\kappa_{|M|,N}b_2\right)+C^{Y,2}_{M,N}Y_M\left(\kappa_{|M|,N}b_2\right)\\&\quad=C^{K}_{M,N}K_M\left(\kappa^f_{|M|,N}b_2\right)+C^{I}_{M,N}I_M\left(\kappa^f_{|M|,N}b_2\right),\end{aligned} \quad (2.26)$$

$$\kappa_{|M|,N}\left[C^{J,1}_{M,N}J'_M\left(\kappa_{|M|,N}a_1\right)+C^{Y,1}_{M,N}Y'_M\left(\kappa_{|M|,N}a_1\right)\right]$$
$$=\kappa^f_{|M|,N}\left[C^{K}_{M,N}K'_M\left(\kappa^f_{|M|,N}a_1\right)+C^{I}_{M,N}I'_M\left(\kappa^f_{|M|,N}a_1\right)\right],$$
$$\kappa_{|M|,N}\left[C^{J,2}_{M,N}J'_M\left(\kappa_{|M|,N}b_2\right)+C^{Y,2}_{M,N}Y'_M\left(\kappa_{|M|,N}b_2\right)\right]$$
$$=\kappa^f_{|M|,N}\left[C^{K}_{M,N}K'_M\left(\kappa^f_{|M|,N}b_2\right)+C^{I}_{M,N}I'_M\left(\kappa^f_{|M|,N}b_2\right)\right],$$

and be normalized

$$\int_{b_1}^{a_2}|\Psi^{II,d}_{|M|N}(R)|^2\,RdR=1. \tag{2.27}$$

From seven Eqs. (2.25)–(2.27) we obtain the coefficients $C^{J,j}_{M,N}$ and $C^{Y,j}_{M,N}$ $(j=1,2)$, $C^{K}_{M,N}$, $C^{I}_{M,N}$ and $\kappa_{|M|,N}$. Thus, in regions $\Omega_{II,1}$, $\Omega_{II,2}$, and Ω_f, the form of the basic function $\Psi^{II,d}_{|M|N}$ for the double-walled nanotubes is finally determined. It is a cylindrical wave and

$$\hat{H}\Psi^{II,d}_{MNP}=\left(K_P^2+\kappa^2_{|M|,N}\right)\Psi^{II,d}_{MNP}. \tag{2.28}$$

It is worth noting that, in the interspherical regions $\Omega_{II,1}$ and $\Omega_{II,2}$ of double-walled system, the cylindrical wave has the same analytical form of the linear combination of the functions J_M and Y_M as in the case of an isolated single-walled nanotube; only the equations for $\kappa_{|M|,N}$, $C^{J,j}_{M,N}$ and $C^{Y,j}_{M,N}$, as well as their numerical values change.

As in the LACW model of single-walled or embedded nanotubes, inside the MT sphere of the inner and outer tubules in the local spherical coordinate system (r,θ,ϕ), the LACW of the double tube is expanded in spherical harmonics $Y_{lm}(\theta,\phi)$

$$\Psi_{1,j\alpha}=\sum_{l=0}^{\infty}\sum_{m=-l}^{l}\left[A^{MNP}_{lm,j\alpha}u_{l,j\alpha}\left(\rho,E_{l,j\alpha}\right)+B^{MNP}_{lm,j\alpha}\dot{u}_{l,j\alpha}\left(\rho,E_{l,j\alpha}\right)\right]Y_{lm}(\theta,\phi), j=1,2 \tag{2.29}$$

Here, $u_{l,j\alpha}$ is the solution of the radial Schrödinger equation for MT sphere $j\alpha$ and $\dot{u}_{l,j\alpha}$ is its energy derivative. Coefficients $A^{MNP}_{lm,j\alpha}$ and $B^{MNP}_{lm,j\alpha}$ are selected so that both the LACW $\Psi_{k,PMN}(\mathbf{r})$ and its derivative have no discontinuities at the boundaries of the MT spheres. However, the analytical form of the cylindrical wave $\Psi^{II,d}_{MNP}$ near the MT spheres of the core and shell nanotubes remains unaltered in going from single-walled to the double-walled nanotubes. Therefore, the analytical Eqs. (1.46)-(1.52) for coefficients $A_{lm,j\alpha}^{MNP}$ and $B_{lm,j\alpha}^{MNP}$ presented for every isolated single-wall remain valid in the case of double-walled nanotubes. Thus, at $b_1 < R < a_1$ and $b_2 < R < a_2$, the basic functions $\Psi_{k,PMN}$ of double-walled nanotubes have the same analytical form as for the noninteracting tubules, whereas, in the Ω_f region of the double-walled system

$$\Psi_{k,PMN}(Z,\Phi,R) = e^{i(k+k_P)Z}e^{iM\Phi}\left[C_{M,N}^{K}K_M\left(\kappa_{|M|,N}^{f}R\right) + C_{M,N}^{I}I_M\left(\kappa_{|M|,N}^{f}R\right)\right]. \quad (2.30)$$

2.2.1.3. Overlap Integrals and Hamiltonian Matrix Elements

The integral $\left\langle \Psi_{M_2N_2P_2} \middle| \Psi_{M_1N_1P_1} \right\rangle$ of the product of the LACWs $\Psi_{P_2M_2N_2}^{*}$ and $\Psi_{P_1M_1N_1}$ over the unit cell Ω is equal to the integral of the product of cylindrical waves over the interspherical regions $\Omega_{II,1}$ and $\Omega_{II,2}$ and classically forbidden region Ω_f plus the sum of the integrals of the product of spherical parts of the LACWs $\Psi_{I,j\alpha,P_2,M_2,N_2}^{*}$ and $\Psi_{I,j\alpha,P_1,M_1,N_1}$ over the MT regions,

$$\left\langle \Psi_{M_2N_2P_2} \middle| \Psi_{M_1N_1P_1} \right\rangle = \int\limits_{\Omega_{II,1}+\Omega_{II,2}+\Omega_f} \bar{\Psi}_{P_2M_2N_2}^{IId}\Psi_{P_1M_1N_1}^{IId}dV + \sum_{j=1,2}\sum_{\alpha}\int\limits_{\Omega_{j\alpha}} \Psi_{I,j\alpha,P_2M_2N_2}^{*}\Psi_{I,j\alpha,P_1M_1N_1}dV. \quad (2.31)$$

The integral over $\Omega_{II,1} + \Omega_{II,2} + \Omega_f$ is equal to the integral over Ω minus the sum of the integrals over the MT regions. Due to the fact that the cylindrical waves as solutions of the Schrödinger equation (2.20) are orthonormalized, the integral over Ω is equal to the product of the δ functions. As a result, Eq. (2.30) takes the form

$$\left\langle \Psi_{M_2N_2P_2} \middle| \Psi_{M_1N_1P_1} \right\rangle = \delta_{P_2M_2N_2,P_1M_1N_1} - \sum_{j=1,2}\sum_{\alpha}\int\limits_{\Omega_{j\alpha}} \bar{\Psi}_{IIj\alpha,P_2M_2N_2}\Psi_{IIj\alpha,P_1M_1N_1}dV + \sum_{j=1,2}\sum_{\alpha}\int\limits_{\Omega_{j\alpha}} \bar{\Psi}_{I,j\alpha,P_2M_2N_2}\Psi_{I,j\alpha,P_1M_1N_1}dV. \quad (2.32)$$

Inasmuch as both the cylindrical wave $\Psi_{II,PMN}$ and spherically symmetrical part $\Psi_{I,j\alpha,PMN}$ of LACW in the MT regions have the same form for double-walled nanotube and constituent tubules, the expression for overlap integrals obtained for the single-walled nanotubes can be easily rewritten for the double-walled case:

$$\left\langle \Psi_{P_2M_2N_2} \mid \Psi_{P_1M_1N_1} \right\rangle = \delta_{M_2,M_1}\delta_{N_2,N_1}\delta_{P_2,P_1}$$
$$-\frac{1}{c}(-1)^{M_2+M_1}\sum_{j=1,2}\sum_{\alpha}\exp\left\{i\left[\left(k_{P_1}-k_{P_2}\right)Z_{j\alpha}+\left(M_1-M_2\right)\Phi_{j\alpha}\right]\right\}$$
$$\times\sum_{m=-\infty}^{\infty}\left[C^{J,j}_{M_2,N_2}J_{m-M_2}\left(\kappa_{|M_2|,N_2}R_{j\alpha}\right)+C^{Y,j}_{M_2,N_2}Y_{m-M_2}\left(\kappa_{|M_2|,N_2}R_{j\alpha}\right)\right]$$
$$\times\left[C^{J,j}_{M_1,N_1}J_{m-M_1}\left(\kappa_{|M_1|,N_1}R_{j\alpha}\right)+C^{Y,j}_{M_1,N_1}Y_{m-M_1}\left(\kappa_{|M_1|,N_1}R_{j\alpha}\right)\right]$$
$$\times\left\{I^{P_2M_2N_2,P_1M_1N_1}_{3,j\alpha}-r^4_{j\alpha}\sum_{l=|m|}^{\infty}\frac{(2l+1)\left[\left(l-|m|\right)!\right]}{2\left[\left(l+|m|\right)!\right]}S^{P_2M_2N_2,P_1M_1N_1}_{lm,j\alpha}\left(r_{j\alpha}\right)\right\}. \quad (2.33)$$

Analogously, for the Hamiltonian matrix elements of double-walled system we have

$$\left\langle \Psi_{M_2N_2P_2}\left|H\right|\Psi_{M_1N_1P_1}\right\rangle = \int_{\Omega_{\text{II},1}+\Omega_{\text{II},2}+\Omega_f}\bar{\Psi}^{\text{II}d}_{P_2M_2N_2}\,\hat{H}\,\Psi^{\text{II}d}_{P_1M_1N_1}dV$$
$$+\sum_{j=1,2}\sum_{\alpha}\int_{\Omega_{j\alpha}}\bar{\Psi}_{\text{I},j\alpha,P_2M_2N_2}\,\hat{H}\,\Psi_{\text{I},j\alpha,P_1M_1N_1}dV. \quad (2.34)$$

Again, the integral over the interspherical regions $\Omega_{\text{II},1}+\Omega_{\text{II},2}+\Omega_f$ is equal to the integral over Ω minus the sum of the integrals over the MT regions. In Ω, cylindrical wave $\Psi_{\text{II},PMN}$ is the solution of Schrödinger equation (2.1), (2.19) with energy $K_P^2+\kappa^2_{|M|,N}$. As a result, Eq. (2.32) takes the form

$$\left\langle P_2M_2N_2\left|\hat{H}\right|P_1M_1N_1\right\rangle = \left(K_{P_2}K_{P_1}+\kappa_{|M_2|,N_2}\kappa_{|M_1|,N_1}\right)\delta_{P_2M_2N_2,P_1M_1N_1}$$
$$-\sum_{j=1,2}\sum_{\alpha}\int_{\Omega_{j\alpha}}\bar{\Psi}_{\text{II},P_2M_2N_2}\left(-\Delta\right)\Psi_{\text{II},P_1M_1N_1}dV$$
$$+\sum_{j=1,2}\sum_{\alpha}\int_{\Omega_{j\alpha}}\Psi^*_{\text{I},j\alpha,P_2M_2N_2}\left|\hat{H}_{MT}\right|\Psi_{\text{I},j\alpha,P_1M_1N_1}dV. \quad (2.35)$$

In the MT regions, the functions $\Psi_{\text{II},j\alpha,P_1M_1N_1}$ and $\Psi_{\text{I},j\alpha,PMN}$ have the same analytical form in the case of double-tubule and noninteracting constituent tubules. Thus, the equation for Hamiltonian matrix elements of the double-walled tubule can be easily obtained from that of single-walled tubule

$$\left\langle \Psi_{P_2M_2N_2}\mid\hat{H}\mid\Psi_{P_1M_1N_1}\right\rangle = \left(K_{P_2}K_{P_1}+\kappa_{|M_2|,N_2}\kappa_{|M_1|,N_1}\right)\delta_{M_2,M_1}\delta_{N_2,N_1}\delta_{P_2,P_1}$$
$$-\frac{1}{c}(-1)^{M_2+M_1}\sum_{j=1,2}\sum_{\alpha}\exp\left\{i\left[\left(k_{P_1}-k_{P_2}\right)Z_{j\alpha}+\left(M_1-M_2\right)\Phi_{j\alpha}\right]\right\}\times$$

$$\times \sum_{m=-\infty}^{\infty} \left[C_{M_2,N_2}^{J,j} J_{m-M_2}\left(\kappa_{|M_2|,N_2} R_{j\alpha}\right) + C_{M_2,N_2}^{Y,j} Y_{m-M_2}\left(\kappa_{|M_2|,N_2} R_{j\alpha}\right)\right]$$

$$\times\left[C_{M_1,N_1}^{J,j} J_{m-M_1}\left(\kappa_{|M_1|,N_1} R_{j\alpha}\right) + C_{M_1,N_1}^{Y,j} Y_{m-M_1}\left(\kappa_{|M_1|,N_1} R_{j\alpha}\right)\right]$$

$$\times\left\{ K_{P_2} K_{P_1} I_{3,j\alpha}^{P_2M_2N_2,P_1M_1N_1} + \kappa_{|M_2|,N_2}\kappa_{|M_1|,N_1} I_{3,j\alpha}^{'P_2M_2N_2,P_1M_1N_1}\right.$$

$$+m_4^2 I_{4,j\alpha}^{P_2M_2N_2,P_1M_1N_1} - r_{j\alpha}^4 \sum_{l=|m|}^{\infty} \frac{(2l+1)\left[(l-|m|)!\right]}{2\left[(l+|m|)!\right]}$$

$$\left.\times\left(E_{l,j\alpha} S_{lm,j\alpha}^{P_2M_2N_2,P_1M_1N_1}(r_{j\alpha}) + \gamma_{lm,j\alpha}^{P_2M_2N_2,P_1M_1N_1}(r_{j\alpha})\right)\right\}. \tag{2.36}$$

Finally, using the secular equation we calculate the dispersion curves of double-wall tubule and corresponding electronic wave functions.

In order to better appreciate how the electronic structure of double-walled nanotube evolves from that of the two constituent tubules, for the eigenstate $\psi_{nk}(\mathbf{r}) = \sum_{PMN} a_{PMN}^{nk} \Psi_{k,PMN}(\mathbf{r})$, it is instructive to calculate the probabilities $w_{j,nk}$ and $w_{f,nk}$ that the electron is located on the tubule j and in the classically forbidden region

$$w_{j,nk} = \sum_{PMN} \left|a_{PMN}^{nk}\right|^2 \frac{\left(C_{M,N}^{J,j}\right)^2 + \left(C_{M,N}^{Y,j}\right)^2}{\sum_{i=1,2}\left[\left(C_{M,N}^{J,j}\right)^2 + \left(C_{M,N}^{Y,j}\right)^2\right] + \left(C_{M,N}^{I}\right)^2 + \left(C_{M,N}^{K}\right)^2}; j = 1,\ 2. \tag{2.37}$$

$$w_{f,nk} = \sum_{PMN} \left|a_{PMN}^{nk}\right|^2 \frac{\left(C_{M,N}^{I}\right)^2 + \left(C_{M,N}^{K}\right)^2}{\sum_{i=1,2}\left[\left(C_{M,N}^{J,j}\right)^2 + \left(C_{M,N}^{Y,j}\right)^2\right] + \left(C_{M,N}^{I}\right)^2 + \left(C_{M,N}^{K}\right)^2}. \tag{2.38}$$

In our case of high barrier V_f and weak interlayer coupling, the probabilities $w_{j,nk}$ are about 1 or 0 for different dispersion curves and virtually independent on momentum k. In all the nanotubes considered below, the probabilities $w_{j,nk}$ were either larger than 0.99, or smaller than 0.01. Therefore, the energy dispersion relations $E_{j,n}(k)$ of double-walled nanotube can be characterized by the number of tubule j, on which the electron is basically located, and one can present two band structures for the double-walled nanotube corresponding to the states of the inner and outer tubules, the total band structure of the double-walled system being just a superposition of band structures of the core and shell systems.

2.2.2. Applications

We first consider the semiconducting zigzag $(n, 0)@(n', 0)$ nanotube. We have calculated the complete band structures and densities of states in the Fermi level region of the 20 purely semiconducting nanotubes $(n,0)@(n',0)$ with $10 \le n \le 23$ and $19 \le n' \le 32$. We omit the metallic structures having n or n' evenly divisible by 3 and consider the nanotubes with interlayer distance 3.2 Å $\le d \le$ 3.7 Å.

The experimental data and LACW band structure calculations of semiconducting zigzag $(n, 0)$ single-walled nanotubes with the values of diameter d from 4 to 20 Å testify that the minimum direct energy difference $E_{11} = E[\Gamma_{c1}(\pi)] - E[\Gamma_{v1}(\pi)]$ between the singularities of the conduction and valence bands depends on whether dividing n by three leaves a remainder of 1 or 2 (n mod 3 = 1 or n mod 3 = 2). The gap energy $E_{11}(d^{-1})$ is an oscillating function that gradually decays to zero as d^{-1} goes to zero, reaches a maximum at d^{-1} between 0.08 and 0.1 Å^{-1} ($13 \le n \le 16$), and decreases abruptly at $d^{-1} > 0.1$ Å^{-1} ($n \le 11$). The curve $E_{11}(d^{-1})$ for n mod 3 = 1 is located totally above analogous curve for n mod 3 = 2. For zigzag single-walled nanotubes with about the same diameters, these gaps are

Table 2.1. Minimum energy gaps E_{11} of the core and shell nanotubes in double-walled nanotube and the shifts of the gaps ΔE_{11} due to formation of double-walled system from pairs of single-walled tubules.

Nanotube	E_{11}, eV		ΔE_{11}, eV	
	Core	Shell	Core	Shell
(10,0)@(19,0)	0.64	0.65	0.32	-0.15
(10,0)@(20,0)	0.63	0.53	0.32	0.07
(10,0)@(19,0)	0.64	0.65	0.32	-0.15
(10,0)@(20,0)	0.63	0.53	0.32	0.07
(11,0)@(19,0)	0.71	0.65	0.39	-0.16
(11,0)@(20,0)	0.71	0.53	0.39	0.07
(13, 0)@(22,0)	1.02	0.55	0.19	-0.19
(13, 0)@(23,0)	1.02	0.50	0.19	0.15
(14, 0)@(22,0)	0.70	0.56	0.14	-0.19
(14, 0)@(23,0)	0.70	0.50	0.14	0.15
(16,0)@(25,0)	0.94	0.52	0.04	-0.18
(16,0)@(26,0)	0.93	0.48	0.04	0.07
(17,0)@(25,0)	0.45	0.52	-0.05	-0.18
(17,0)@(26,0)	0.45	0.48	-0.05	0.07
(19,0)@(28,0)	0.76	0.46	-0.05	-0.20
(19,0)@(29,0)	0.76	0.46	-0.05	0.07
(20,0)@(28,0)	0.42	0.46	-0.05	-0.20
(20,0)@(29,0)	0.42	0.46	-0.05	0.07
(22, 0)@(31,0)	0.75	0.40	0.00	-0.22
(23, 0)@(31,0)	0.40	0.40	0.06	-0.22

approximately less by half for tubules with n mod 3 = 2 than for those with n mod 3 = 1. Thus, the $(n, 0)$ single-walled nanotubes with n mod 3 = 1 and 2 can be thought of as the wide and low gap semiconductors, respectively.

Table 2.1 shows the minimum gaps E_{11} in the double-walled nanotubes and the shifts ΔE_{11} of these gaps due to the interwall perturbation.

Figures 2.9, 2.10, and 2.11 show representative results of the band structure calculations of double-walled nanotubes with semiconducting tubules.

The complete band structures and DOS in the Fermi level region of the single-walled nanotubes (13, 0) and (22, 0) can be compared with the analogous data for the core (13, 0) and shell (22, 0) tubules of the double tube. In the (13, 0)@(22, 0) nanotube, both inner and outer nanotubes belong to the n mod 3 = 1 series, the minimum energy gap (0.83 eV) of the small-diameter (13, 0) tubule being larger than that of the large-diameter

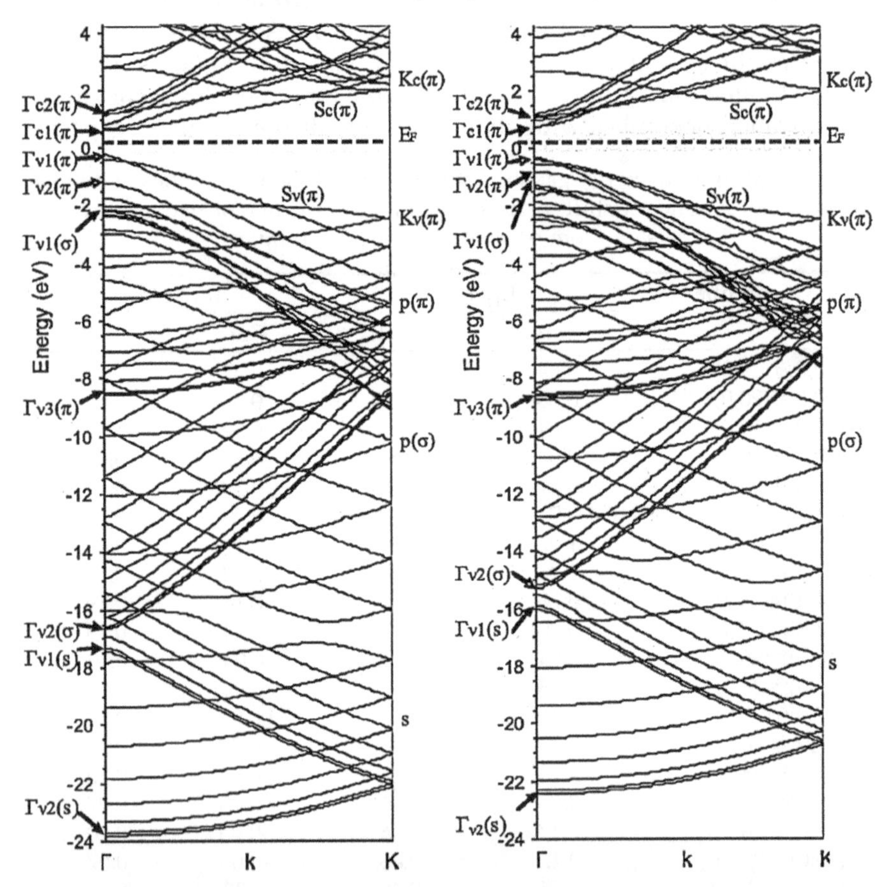

Fig. 2.9. Band structure of (13, 0) single-walled nanotubes (left) and of core (13, 0) tubule (right) located inside (22, 0) tubule.

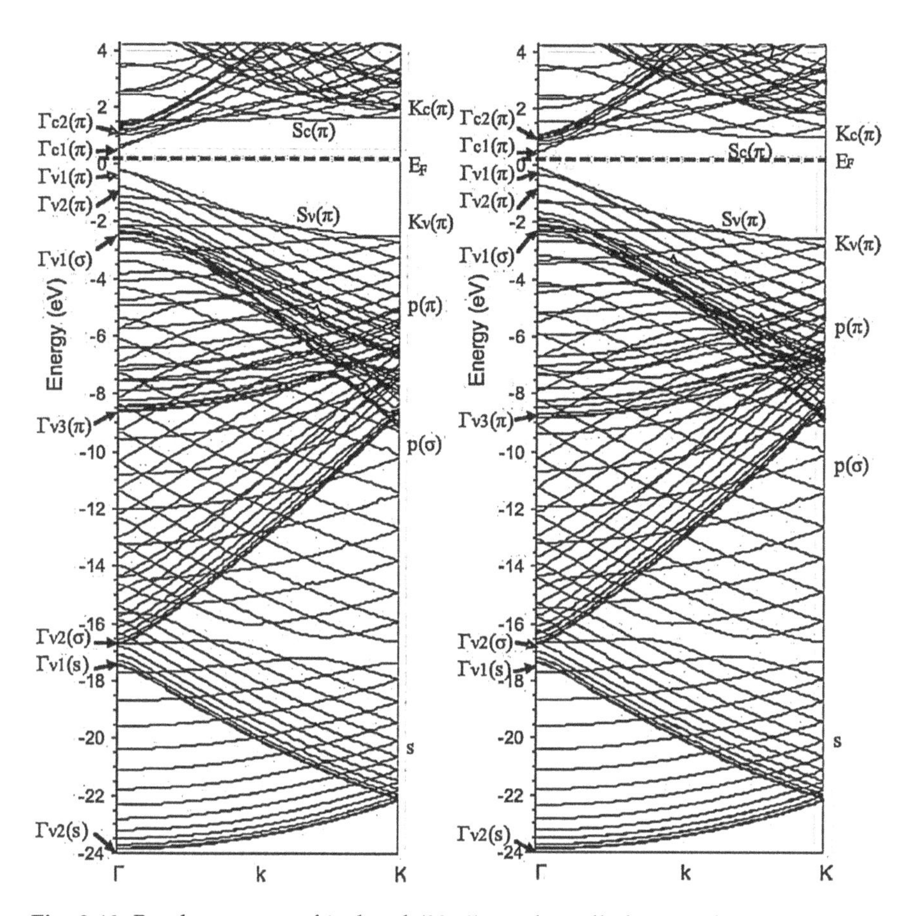

Fig. 2.10. Band structure of isolated (22, 0) single-walled nanotubes (left) and of outer (22, 0) tubule (right) interacting with inner (13, 0) tubule.

tubule (22, 0) (0.76 eV) in agreement with simple approximate equation $E_{11} = d^{-1}$ known from π-electronic band structure calculations of the single-walled nanotubes. Our calculations indicate that the minimum gap E_{11} of (13, 0) tube increases by 0.19 eV and that of (22, 0) tube decreases by 0.19 eV due to the formation of the double system, a significant value in device physics. The DOS curves in the Fermi level region show that there are analogous low energy shifts of the second gap $E_{22} = E_{c2}(\pi) - E_{v2}(\pi)$ equal to 0.3 and 0.4 eV in the case of (13, 0) and (22, 0) tubes, respectively. The interwall coupling results in a distinctly stronger perturbation of the band structure of inner tube as compared to that of the outer one. The reason is that an additional space located between the Ω_{b2} and Ω_{a2} barriers that become accessible to electrons of the small-diameter (13, 0) tube due to the formation of double-walled nanotube is about two times greater in

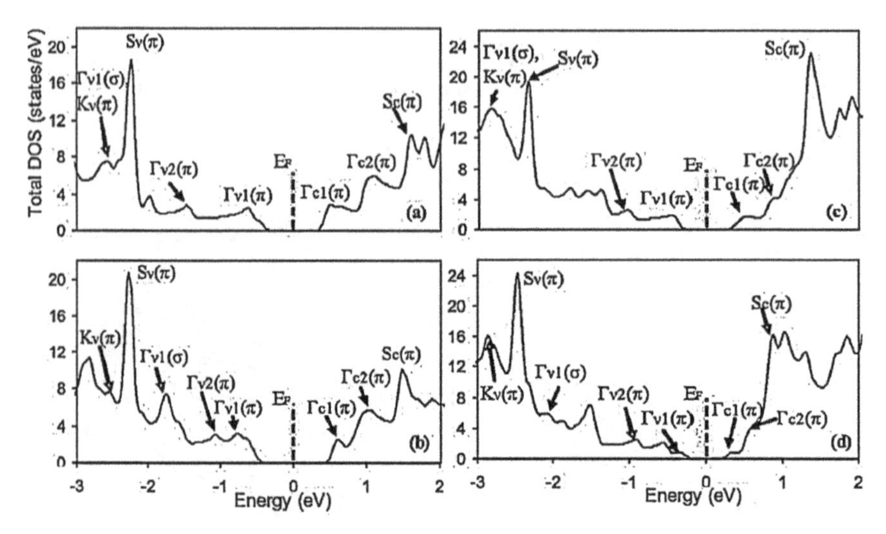

Fig. 2.11. LACW DOS in the Fermi level region. (a) (13, 0) single-walled nanotube; (b) core (13, 0) tubule nested in (22, 0) tubule; (c) (22, 0) single-walled nanotube; and (d) shell (22, 0) tubule with nested (13, 0) tubule.

comparison with a new accessible region between the Ω_{b1} and Ω_{a1} barriers in the case of electrons of the large-diameter (22, 0) tube. For example, as one goes to the double-walled nanotube, the total valence band width $E_v = E_F - E[\Gamma_{2v}(s)]$ of (13, 0) tube decreases by 1.40 eV and that of (22, 0) tube increases by 0.04 eV only (Table 2.2).

In the (14, 0)@(22, 0) nanotube, the inner tube belongs to the n mod 3 = 2 small gap series. Here, the gap of the inner (14, 0) tubule equal to 0.42 eV is smaller than that of the outer n mod 3 = 1 (22, 0) tubule in conflict with the $E_{11} - d^{-1}$ equation. As the intertube interaction is switched on, the gap of the inner tubule grows by 0.14 eV and that of the outer tubule decreases by 0.19 eV, the relative values of the gaps E_{11} of the tubules being reversed in the single-walled and double-walled nanotubes. For the core and shell nanotubes, the gap shifts ΔE_{11} induced by the intertube coupling are oppositely directed both in the (13, 0)@(22, 0) and (14, 0)@(22, 0) nanotubes. The ΔE_{11} values are positive and negative for the inner and outer tubules, respectively. In the (13, 0)@(23, 0) nanotube with the wide gap inner and low gap outer nanotubes, the gap shifts of the two tubules ΔE_{11} = 0.19 and 0.15 eV are almost equal and have the same positive sign. This is also true for the (14, 0)@(23, 0) nanotube, where both tubules belong to the low gap n mod 3 = 2 series; here, the ΔE_{11} = 0.14 and 0.15 eV for the core and shell systems. Table 2.1 shows that, independent on the type of core tubule, the gap E_{11} of the shell nanotube decreases by 0.15–0.22 eV, if this tube belongs to the n mod 3 = 1 series. On the other hand, for the shell tubes with n mod 3 = 2, the gap shift ΔE_{11} is always positive: $0.7 \leq \Delta E_{11} \leq$

Table 2.2. Energy levels of the (13, 0) and (22, 0) single-walled nanotube and of these tubules in the (13, 0)@(22, 0) nanotube.

	Energy (eV)			
Level	**(13, 0) Single-walled nanotube**	**(13, 0) Core of double-walled nanotube**	**(22, 0) Single-walled nanotube**	**(22, 0) Shell of double-walled nanotube**
$\Gamma_{C2}(\pi)$	0.95	0.79	0.81	0.53
$\Gamma_{C1}(\pi)$	0.41	0.51	0.37	0.29
$\Gamma_{v1}(\pi)$	-0.42	-0.50	-0.38	-0.28
$\Gamma_{v2}(\pi)$	-1.42	-0.75	-0.99	-0.89
$\Gamma_{v1}(\sigma)$	-2.35	-2.52	-2.53	-2.27
$\Gamma_{v2}(\pi)$	-8.74	-8.89	-8.79	-9.05
$\Gamma_{v2}(\sigma)$	-17.12	-15.48	-16.89	-16.88
$\Gamma_{v1}(s)$	-17.50	-16.10	-17.29	-17.19
$\Gamma_{v2}(s)$	-24.01	-22.63	-24.05	-24.10
$S_c(\pi)$	1.55	1.45	1.36	0.83
$S_v(\pi)$	-2.22	-2.25	-2.32	-2.43
$K_c(\pi)$	1.79	1.77	1.46	1.83
$K_v(\pi)$	-2.65	-2.67	-2.68	-2.74

0.15 eV. In both cases, the ΔE_{11} shifts do not decay, but slightly oscillates as one goes to the tubules with larger diameters d. For inner tubules, the ΔE_{11} shift depends strongly on the d. For n mod 3 = 1 and n mod 3 = 2 series with $10 \leq n \leq 16$, the shifts ΔE_{11} are positive, the maximum values of ΔE_{11} being equal to 0.32 and 0.39 eV, respectively. As one goes to the inner tubules with larger diameters, the shift ΔE_{11} quickly decays and thereupon varies between 0.06 and –0.05 eV.

Now let us consider the coaxial best matched metallic (5, 5)@(10, 10) nanotube consisting of a D_{5d} (5, 5) nanotube nested inside the D_{10h} (10, 10) nanotube with the diameter ratio 1:2 (Figs. 2.12 and 2.13).

In this most commensurate structure, there are two carbon atoms of the outer tubule for each carbon atom of the inner tubule; this geometry has many similarities to the AB stacking of graphite. Figure 2.12 shows all the occupied and unoccupied electronic states of the single-walled (5, 5) nanotube and the states of the core tubule (5, 5) nested inside of the (10,

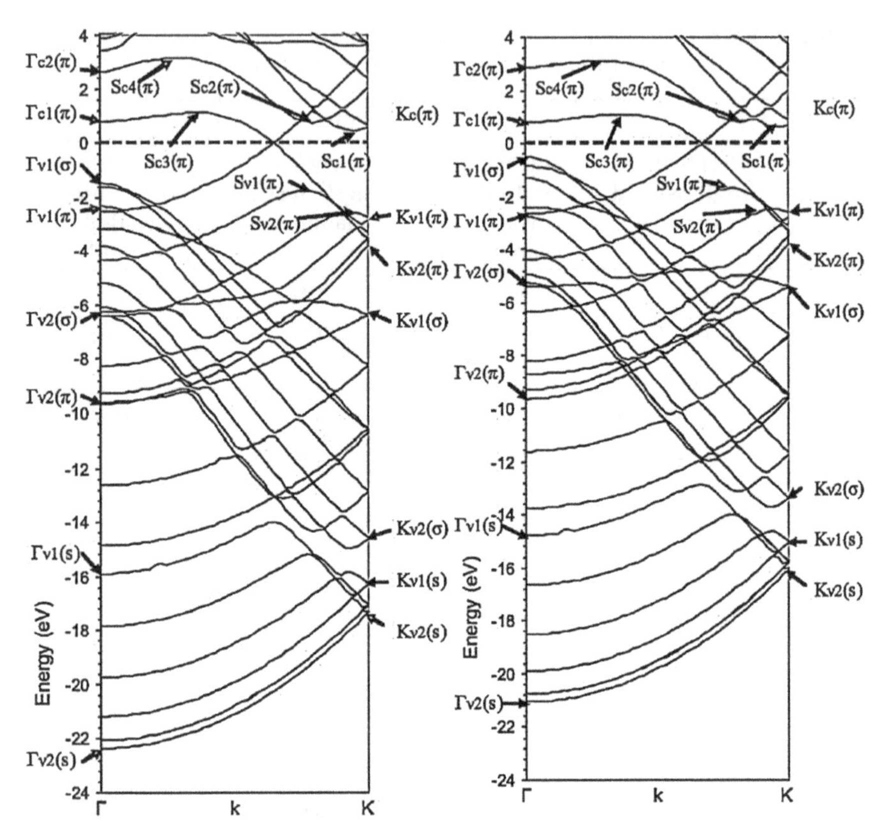

Fig. 2.12. Band structure of (5, 5) single-walled nanotube (left) and of the core (5, 5) nanotube (right) located inside (10, 10) tubule.

10) tubule. The band structure of the (10, 10) single-walled tubule can be compared with that of the (10, 10) outer tubule as the latter interacts with the inner (5, 5) tubule. Figure 2.14 shows the influence of the interlayer interaction on the DOS of (5, 5)@(10, 10) nanotube.

The geometry was chosen to give rise to the most commensurate interlayer stacking, and the energy dispersion relations are seen to be strongly perturbed by the interlayer coupling; however, the interlayer coupling does not break down the metal-type character of the band structures of the (5, 5) and (10, 10) tubules. The Fermi level is located at the intersection of the π bands at about $k = (2/3)(\pi/c)$ both in the single-walled tubules and in the double-walled pair. Formation of the (5, 5)@(10, 10) nanotube results in increase of the valence band width E_v of the (5, 5) tubule by 1.3 eV, the increase of the E_v of the (10, 10) tubule being only 0.15 eV; again, perturbation of the bands of core tubule is stronger in comparison with that of the shell tubule (Table 2.3).

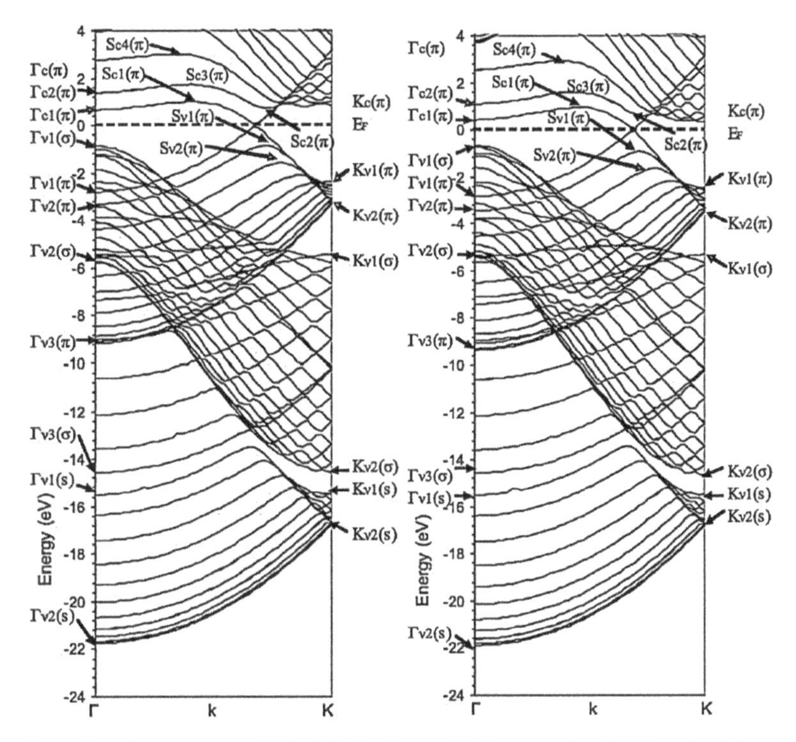

Fig. 2.13. Band structure of (10, 10) single-walled nanotube (left) and of the outer (10, 10) tubule (right) interacting with inner (5, 5) tubule.

A high-energy shift of the σ states relative to the occupied π states is seen to be the most significant effect of the interlayer interaction in the armchair double-walled pair. This shift results in the significant perturbation of the most important direct optical σ-π* and π-π* transitions. In the Brillouin zone center, the highest occupied σ state $\Gamma_{v1}(\sigma)$ is located above the highest occupied π state $\Gamma_{v1}(\pi)$ in all armchair single-walled nanotubes. As the core tubule is nested in the shell tubule, the direct gap in the Γ point $E_{11}(\sigma,\pi^*) = E[\Gamma_{c1}(\pi)] - E[\Gamma_{v1}(\sigma)]$ of the (5, 5) and (10, 10) tubules decreases by 0.96 and 0.40 eV, respectively. The energy gap $E_{11}(\pi,\pi^*) = E[S_{c1}(\pi^*)] - E[S_{v1}(\pi)]$ between the conduction and valence π-band singularities of the (5, 5) tube increases and that of the (10, 10) tubule decreases by 0.1 eV. Formation of the (4, 4)@(9, 9) nanotube results in the similar perturbation of the electronic structure of the core and shell tubules.

In conclusion, the large shifts of optical gaps of the tubules due to formation of the double-walled nanotubes complicate the determination of the structure of double-walled nanotubes on the basis of optical data. On the other hand, the results obtained open the opportunity to classify experimental data on the double-walled nanotubes more specifically.

Table 2.3. Energy levels of the (5, 5) and (10, 10) single-wall tubules and of these tubules in the (5, 5)@(10, 10) nanotube.

Level	Energy, eV			
	(5, 5) Single-walled nanotube	**(5, 5) Core of double-walled nanotube**	**(10, 10) Single-walled nanotube**	**(10, 10) Shell of double-walled nanotube**
$\Gamma_{c2}(\pi)$	2.66	2.87	1.37	1.10
$\Gamma_{c1}(\pi)$	0.78	0.78	0.65	0.43
$\Gamma_{v1}(\sigma)$	-1.47	-0.51	-0.85	-0.67
$\Gamma_{v1}(\pi)$	-2.54	-2.66	-2.64	-2.82
$\Gamma_{v2}(\sigma)$	-6.38	-6.25	-5.72	-5.34
$\Gamma_{v2}(\pi)$	-9.59	-9.64	-9.17	-9.30
$\Gamma_{v1}(s)$	-15.89	-14.78	-15.49	-15.50
$\Gamma_{v2}(s)$	-22.35	-21.04	-21.75	-21.90
S_{c1}	0.76	0.83	0.48	0.48
S_{c2}	1.13	1.08	0.98	0.96
S_{c3}	3.16	3.10	1.73	1.59
S_{v1}	-1.76	-1.69	-0.84	-0.90
S_{v2}	-2.57	-2.48	-1.58	-1.64
$K_c(\pi)$	0.66	0.71	1.02	0.68
$K_v(\pi)$	-2.74	-2.59	-2.33	-2.45
$K_{v1}(\sigma)$	-6.34	-5.39	-5.44	-5.30
$K_{v2}(\sigma)$	-14.58	-13.35	-14.56	-14.65
K_s	-17.31	-16.01	-16.79	-16.75

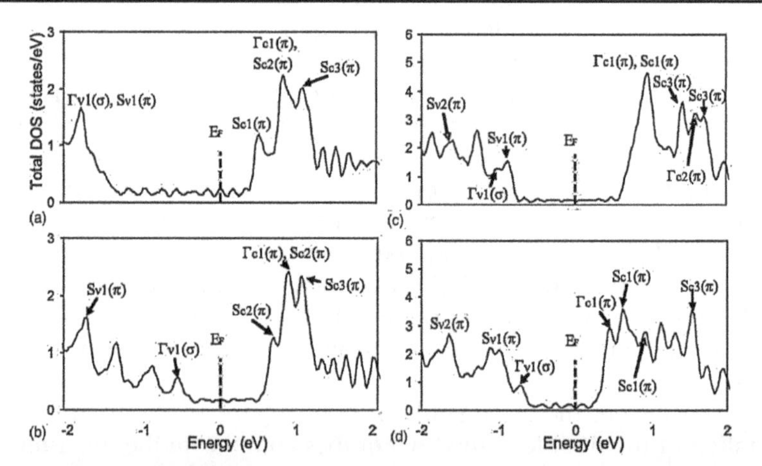

Fig. 2.14. LACW DOSs in the Fermi level region. (a) (5, 5) single-walled tubule; (b) core (5, 5) nested in (10, 10) tube; (c) (10, 10) single-walled tubule; (d) shell (10, 10) with nested (5, 5) tube.

CHAPTER

3

Symmetrized Augmented Cylindrical Waves for Chiral Nanotubes

Here our goal is to develop a method for calculating any nanotubes, regardless of their chirality and the number of atoms in the translation cells. As noted earlier, all the carbon single-walled nanotubes can be constructed by rolling up a single graphite sheet and the structures of tubules can be visualized as a conformal mapping of a two-dimensional graphitic lattice on the surface of a cylinder. One can make such a seamless tubule without any special distortion of their bonding angles other than the introduction of curvature to the carbon hexagons through the rolling process. Each tubule can be labeled by the pair of integers (n_1, n_2) where $n_1 \geq n_2 \geq 0$, which, together with C-C bond length determine the nanotubes geometry (Fig. 1.1).

The nanotubes generated by the mapping are translationally periodic along the tubule axis, and application of the Bloch's theorem facilitates the theoretical studies of the single- and double-walled nanotubes band structures performed in Chapters 1 and 2. However, all previous LACW calculations were limited by the achiral (n, n) armchair and $(n, 0)$ zigzag tubules with a relatively small number of atoms in the minimum translational unit cell $N_{tr} = 4n$. In the case of the chiral tubules even with small diameters, the minimum number of atoms per unit cell can be a very large one. For example, the translational unit cells of the achiral (10, 10) and chiral (10, 9) nanotubes with virtually equal diameters contain 40 and 1084 carbon atoms, respectively; the (100, 99) nanotube contains a total of 118,804 atoms per translational unit cell, but this version of the LACW theory must be applicable to this giant system too. The large number of atoms that can occur in the minimum translational unit cell makes

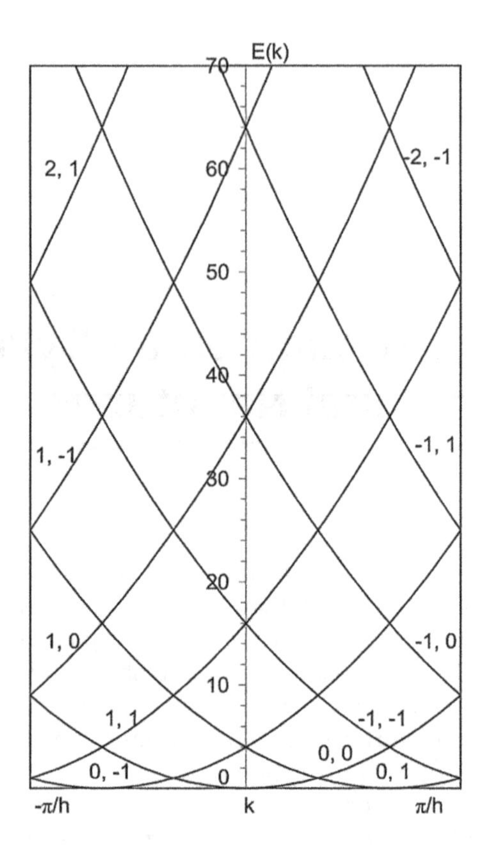

Fig. 3.1. Dispersion diagram for an empty 1D lattice with screw axis for $\omega = 2\pi/3$. The indices label the values of P and k_{Φ}, e.g., 2, 1 corresponds to dispersion curves with $P = 2$ and $k_{\Phi} = 1$. Energy $E(k)$ is in units of π/h.

recourse to the other point-group symmetries of these tubules practically mandatory.

Fortunately, every carbon single-walled nanotube can be generated by first mapping only two nearest-neighbor C atoms onto the surface of a cylinder and then using the rotational and helical symmetry operators to determine the remainder of the tubule (White et al. 1993). Herein, we use these symmetries to develop the symmetry-adapted version of the LACW method (D'yachkov and Makaev 2007). In this case, the cells contain only two carbon atoms, and the theory becomes applicable to any tubule, including those with screw axis independent of the number of atoms in the translational unit cell. It is very important that the band structure of any tubule can be easily calculated and the results can be presented in the standard form with four curves in the valence band plus one curve for the low-energy states of the conduction band.

3.1. Theory

3.1.1. Helical and Rotational Symmetries of Tubules Geometry

It should be noted how all the graphitic tubules defined by the integers (n_1, n_2) can also be defined in terms of their helical and rotational symmetries (White et al. 1993). In order to construct the carbon nanotube, one has to map the first atom to an arbitrary point T_1 on the cylinder surface, which requires that the position of the second one T_2 be found by rotating this point

$$\Phi_{T_2} = \pi \frac{n_1 + n_2}{n_1^2 + n_2^2 + n_1 n_2} \tag{3.1}$$

radian about the cylinder axis in conjunction with a translation

$$\delta_{T_2} = \frac{d_{C-C}}{2} \frac{n_1 - n_2}{\left(n_1^2 + n_2^2 + n_1 n_2\right)^{1/2}} \tag{3.2}$$

along this axis.

Let us map the first C atom to the point with cylindrical coordinates $Z_1 = 0$, $\Phi_1 = 0$, and $R_1 = R_{NT}$. In this case, the cylindrical coordinates of the second C atom are $Z_2 = \delta_{T_2}$, $\Phi_2 = \Phi_{T_2}$, and $R_2 = R_{NT}$, where

$$R_{NT} = \frac{d_{C-C}\sqrt{3}}{2\pi} \left(n_1^2 + n_2^2 + n_1 n_2\right)^{1/2} \tag{3.3}$$

is the nanotube radius.

The cylinder axis coincides with a C_n rotational axis of the tubule, where n is the largest common divisor of n_1 and n_2. Thus, the positions of these first two atoms can be used to locate $2(n-1)$ additional atoms on the cylinder surface by $n-1$ successive $2\pi/n$ rotations about the cylinder axis. Altogether, these $2n$ atoms complete the specification of the helical motif that can then be used to tile the remainder of the tubule by repeated operation of a single screw operation $S(h, \omega)$ representing a translation

$$h = \frac{3d_{C-C}}{2} \frac{n}{\left(n_1^2 + n_2^2 + n_1 n_2\right)^{1/2}} = \frac{3\sqrt{3}d_{C-C}^2}{4\pi} \frac{n}{R_{NT}} \tag{3.4}$$

along the cylinder axis in conjunction with a rotation

$$\omega = 2\pi \frac{n_1 p_1 + n_2 p_2 + \left(n_2 p_1 + n_1 p_2\right)/2}{n_1^2 + n_2^2 + n_1 n_2} \tag{3.5}$$

radian about this axis. The angle ω is defined mod 2π, the integer $p_1 \geq 0$, and the positive values of integers p_2 are obtained from the equation

$$p_2 n_1 - p_1 n_2 = n. \tag{3.6}$$

The different sets of pairs p_2 and p_1 correspond to different numberings of the C atoms of the nanotube; for uniqueness, we choose the minimum value of the p_2 and p_1.

The nanotube chiral angle θ is calculated from the equation

$$\arccos\theta = \frac{n_1 + (1/2)n_2}{\left(n_1^2 + n_2^2 + n_1 n_2\right)^{1/2}} \tag{3.7}$$

Finally, the number of atoms N_{tr} in the smallest translational unit cell can be expressed by

$$N_{tr} = 4\left(n_1^2 + n_2^2 + n_1 n_2\right)/L_{tr}, \tag{3.8}$$

where L_{tr} is the largest common factor of $(2n_1 + n_2)$ and $(2n_2 + n_1)$.

3.1.2. Helical and Rotational Symmetries of Eigenstates

In the perfect nanotube with the C_n symmetry, the nuclei are arranged in a regular array described by a set of rotations by the angles $\omega_n t = (2\pi/n)$ t with arbitrary integer t. Therefore, we can introduce the discrete values of a wave vector k_Φ corresponding to the periodic rotation operator and write, using a cylindrical coordinate system (Z, Φ, R),

$$\Psi(Z, \Phi + t\omega_n, R) = e^{ik_\Phi t\omega_n}\Psi(Z, \Phi, R). \tag{3.9}$$

Substituting $t = n$ and taking into account an equation $\Psi(Z, \Phi + 2\pi, R) = \Psi(Z, \Phi, R)$, we obtain that the values of k_Φ are integers and can be written as

$$k_\Phi = L + nM, \tag{3.10}$$

where $M = 0, \pm1, ...,$ and $L = 0, 1, ..., n - 1$.

The perfect nanotube, being infinite in the Z direction, is also invariant under the screw $S(h, \omega)$ operation representing a translation h along this axis in conjunction with a rotation ω about it. The screw transformations $S(h, \omega)$ form an Abelian group isomorphous with the usual translation group $T(h)$. Thus, according to Bloch's theorem, the wave function $\Psi(Z,\Phi,R)$ can be characterized by a continuous wave vector K_P,

$$\Psi(Z + th, \Phi + t\omega, R) = e^{iK_P th}\Psi(Z, \Phi, R), \tag{3.11}$$

where

$$K_P = k + k_P, k_P = \frac{2\pi}{h}P, P = 0, \pm1, ... \tag{3.12}$$

The vector k belongs to the first one-dimensional Brillouin zone $-(\pi/h) \leq k \leq (\pi/h)$. Note that if $\omega = 2\pi/\mu$ and $h = c/\mu$ with integer μ, where c is the translational lattice constant, then $\mathrm{k} = \mu k$, where k is the traditional One-Dimensional (1D) wave vector.

3.1.3. Cylindrical Waves with Helical and Rotational Symmetries

Now we have to construct the basic wave functions with account of the helical and rotational symmetries of nanotubes.

3.1.3.1. Interspherical Region

Similar to the approach developed in Chapter 1, in the interspherical region Ω_{II}, the basic functions are the solutions of the Schrödinger equation corresponding to an electron in the tubular potential well. For the sake of convenience, we give here again the original equation, written with a cylindrical coordinate system

$$\left\{-\left[\frac{1}{R}\frac{\partial}{\partial R}\left(R\frac{\partial}{\partial R}\right)+\frac{1}{R^2}\frac{\partial^2}{\partial \Phi^2}+\frac{\partial^2}{\partial Z^2}\right]+U(R)\right\}\Psi(Z,\Phi,R)=E\Psi(Z,\Phi,R),$$

$$U(R)=\begin{cases}0, & b\le R\le a\\ \infty, & R<b,\ R>a.\end{cases} \tag{3.13}$$

Again, the solutions of these equations is the product $\Psi(Z,\Phi,R)$ = $\Psi(Z)\Psi(\Phi)\Psi(R)$ of functions depending on Z, Φ, and R coordinates, but the functions $\Psi(Z)$, $\Psi(\Phi)$, and $\Psi(R)$ have the new form. Now the new function

$$\Psi_M(\Phi\,|\,L)=\frac{1}{\sqrt{2\pi/n}}\exp i\left[(L+nM)\Phi\right] \tag{3.14}$$

corresponds to a free rotation of an electron and the function

$$\Psi_{PM}(Z\,|\,\mathrm{k},L)=\frac{1}{\sqrt{h}}\exp\left\{i\left[\mathrm{k}+\mathrm{k}_P-(L+nM)\frac{\omega}{h}\right]Z\right\} \tag{3.15}$$

corresponds to the free movement of electron along axis Z in an infinite chiral one-dimensional system with the C_n symmetry and $S(h,\omega)$ screw axis. The coefficients $(2\pi/n)^{-1/2}$ and $h^{-1/2}$ in Eqs. (3.14) and (3.15) are determined from normalization conditions

$$\int_0^{2\pi/n}\left|\Psi_M(\Phi\,|\,L)\right|^2 d\Phi=1,\ \int_0^{h}\left|\Psi_{P,M}(Z\,|\,k,L)\right|^2 dZ=1\,. \tag{3.16}$$

The function $\Psi_{|M|,N}(R\,|\,L)$ corresponds to the radial movement of an electron in the interspherical regions Ω_{II} of the tubule. Substituting the explicit form of the functions $\Psi_M(\Phi\,|\,L)$ and $\Psi_{PM}(Z\,|\,\mathrm{k},L)$ into the equation (3.13), we find the equation for calculating the $\Psi_{|M|,N}(R\,|\,L)$

$$\left(-\frac{1}{R}\frac{d}{dR}R\frac{d}{dR}+\frac{(L+nM)^2}{R^2}\right)\Psi_{M,N}(R\,|\,L)+U(R)\Psi_{M,N}(R\,|\,L)$$

$$=E_{M,N}(L)\Psi_{M,N}(R\,|\,L). \tag{3.17}$$

Here, $E_{|M|,N}(L)$ is the energy spectrum, and N is the radial quantum number. The energy

$$E_{PMN}(\mathrm{k}, L) = \left[\mathrm{k} + \mathrm{k}_P - (L + nM)\frac{\omega}{h}\right]^2 + E_{M,N}(L) \tag{3.18}$$

corresponds to the cylindrical wave function $\Psi_{PMN}(Z,\Phi,R\,|\,\mathrm{k}, L)$. Note that the product $\Psi_M(\Phi\,|\,L)\Psi_{PM}(Z\,|\,\mathrm{k},L)$ and, consequently, $\Psi_{PMN}(Z,\Phi,R\,|\,\mathrm{k},L)$ satisfy the symmetry conditions (3.9) and (3.11). In Eq. (3.18), the term

$$E(\mathrm{k}) = \left[\mathrm{k} + \mathrm{k}_P - (L + nM)\frac{\omega}{h}\right]^2 = \left(\mathrm{k} + \frac{2\pi}{h}P - \frac{\omega}{h}\mathrm{k}_\Phi\right)^2 \tag{3.19}$$

determines the energy of free electrons in the 1D empty lattice with screw axis.

Figure 3.1 shows a dispersion diagram for the special case of the empty lattice with $\omega = 2\pi/3$. Note that the curves are symmetric with respect to a change k for –k, and one can calculate the diagram for the region $0 \le \mathrm{k} \le \pi/h$ only. It is also instructive to compare this diagram with the dispersion curves of the purely translational 1D empty lattice with $c = 3h$ (Fig. 3.2).

In the regions Ω_{II}, $U(R) = 0$ and Eq. (3.17) takes the form of the following Bessel equation:

$$\left(\frac{d^2}{dR^2} + \frac{1}{R}\frac{d}{dR} + \left(\kappa_{|L+nM|,N}\right)^2 - \frac{(L+nM)^2}{R^2}\right)\Psi_{M,N}(R\,|\,L) = 0\,, \tag{3.20}$$

where $\kappa_{|L+nM|,N} = \left[E_{M,N}(L)\right]^{1/2}$. Any solution of this equation can be represented by the following linear combination of cylindrical Bessel functions of the first J_M and second Y_M kinds,

$$\Psi_{M,N}(R\,|\,L) = C^{J,L}_{M,N} J_{L+nM}\left(\kappa_{|L+nM|,N}R\right) + C^{Y,L}_{M,N} Y_{L+nM}\left(\kappa_{|L+nM|,N}R\right). \tag{3.21}$$

The function $\Psi_{M,N}(R\,|\,L)$ should vanish at $R = a$ and $R = b$

$$\begin{aligned} C^{J,L}_{M,N} J_{L+nM}\left(\kappa_{|L+nM|,N}a\right) + C^{Y,L}_{M,N} Y_{L+nM}\left(\kappa_{|L+nM|,N}a\right) &= 0, \\ C^{J,L}_{M,N} J_{L+nM}\left(\kappa_{|L+nM|,N}b\right) + C^{Y,L}_{M,N} Y_{L+nM}\left(\kappa_{|L+nM|,N}b\right) &= 0 \end{aligned} \tag{3.22}$$

and be normalized

$$\int_a^b \left|\Psi_{M,N}(R\,|\,L)\right|^2 R\,dR = 1. \tag{3.23}$$

From Eqs. (3.22) and (3.23), we obtain the algebra equation for $\kappa_{|L+nM|,N}$

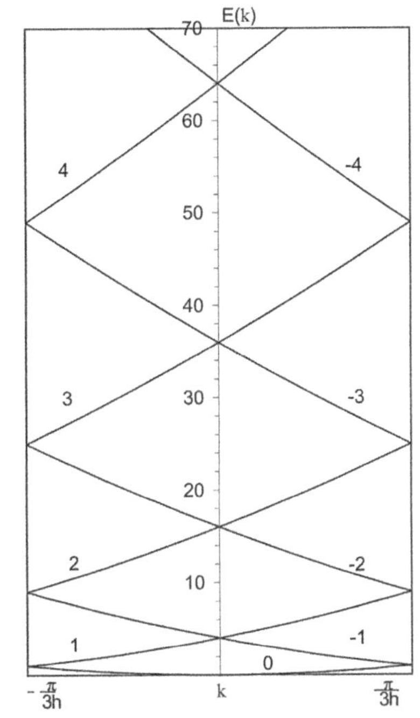

Fig. 3.2. Triply degenerate dispersion curves for the same system as in Fig. 3.1, but calculated taking into account only translational symmetry. Account of the screw symmetry results in a shift of some curves. The indices label the values of P. Energy $E(k)$ is in units of $\pi/3h$.

$$J_{L+nM}\left(\kappa_{|L+nM|,N}a\right)Y_{L+nM}\left(\kappa_{|L+nM|,N}b\right)$$
$$= J_{L+nM}\left(\kappa_{|L+nM|,N}b\right)Y_{L+nM}\left(\kappa_{|L+nM|,N}a\right) \qquad (3.24)$$

and the relationship between $C_{M,N}^{Y,L}$ and $C_{M,N}^{J,L}$

$$C_{M,N}^{Y,L} = -C_{M,N}^{J,L}\frac{J_{L+nM}\left(\kappa_{|L+nM|,N}a\right)}{Y_{L+nM}\left(\kappa_{|L+nM|,N}a\right)}. \qquad (3.25)$$

To calculate integral (3.23), let us again use Eq. (1.21) for an indefinite integral and recurrence formulas (1.22) and (1.23) for the cylindrical functions. Then,

$$\frac{R^2}{2}\left[C_{M,N}^{J,L}J'_{L+nM}\left(\kappa_{|L+nM|,N}R\right)+C_{M,N}^{Y,L}Y'_{L+nM}\left(\kappa_{|L+nM|,N}R\right)\right]^2\Bigg|_b^a = 1. \qquad (3.26)$$

Finally, the basic function in the Ω_{II} region in the general cylindrical coordinate system takes the form

$$\Psi_{II,PMN}(Z,\Phi,R\,|\,\mathrm{k},L)=\frac{1}{\sqrt{2\pi h/n}}\exp i\left\{\left[\mathrm{k}+\mathrm{k}_P-(L+nM)\frac{\omega}{h}\right]Z+(L+nM)\Phi\right\}$$
$$\times\left[C^{J,L}_{M,N}J_{L+nM}\left(\kappa_{|L+nM|,N}R\right)+C^{Y,L}_{M,N}Y_{L+nM}\left(\kappa_{|L+nM|,N}R\right)\right].$$
$$(3.27)$$

Here, the $\kappa_{|L+nM|,N}$ values are found from Eq. (3.24) and $C^{J,L}_{M,N}, C^{Y,L}_{M,N}$ are derived from the set of Eqs. (3.25) and (3.26). Thus, in regions Ω_{II}, the form of the basic function $\Psi_{II,PMN}(Z,\Phi,R\,|\,\mathrm{k},L)$ is finally determined. It is the symmetrized cylindrical wave for the systems with rotational and helical axes C_n and $S(h,\omega)$ and

$$\hat{H}\Psi_{II,PMN}(Z,\Phi,R\,|\,\mathrm{k},L)=E_{PMN}(\mathrm{k},L)\Psi_{II,PMN}(Z,\Phi,R\,|\,\mathrm{k},L). \quad (3.28)$$

3.1.3.2. Muffin-Tin Region and Sewing Conditions

As in Chapter 1, inside the MT spheres, the Hamiltonian is spherically symmetric, and, in terms of the local spherical coordinate system, the basic function $\Psi_{I\alpha,PMN}(\rho,\theta,\varphi\,|\,\mathrm{k},L)$ are expanded in spherical solutions $Y_{lm}(\theta,\varphi)$

$$\Psi_{I\alpha,PMN}(\rho,\theta,\varphi\,|\,\mathrm{k},L)=\sum_{l=0}^{\infty}\sum_{m=-l}^{l}\left[A^{PMN,\mathrm{k}L}_{lm\alpha}u_{l\alpha}\left(\rho,E_{l\alpha}\right)\right.$$
$$\left.+B^{PMN,\mathrm{k}L}_{lm\alpha}\dot{u}_{l\alpha}\left(\rho,E_{l\alpha}\right)\right]Y_{lm}(\theta,\varphi), \quad (3.29)$$

where $u_{l\alpha}(\rho,E_{l\alpha})$ and $\dot{u}_{l\alpha}(\rho,E_{l\alpha})$ are the radial wave functions and their derivatives which are the solutions of the radial Eqs (1.27)-(1.36), but the coefficients $A^{PMN,\mathrm{k}L}_{lm\alpha}$ and $B^{PMN,\mathrm{k}L}_{lm\alpha}$ are to be selected so that both the two parts of symmetrized basic functions $\Psi_{II,PMN}(Z, \Phi, R\,|\,\mathrm{k}, L)$ and $\Psi_{I\alpha,PMN}(\rho, \theta, \varphi\,|\,\mathrm{k}, L)$ as well as their radial derivatives have no discontinuities at the boundaries of the MT spheres. To do this, using the approach developed in Chapter 1, we can express the symmetrized cylindrical waves $\Psi_{II,PMN}(Z,\Phi,R\,|\,\mathrm{k},L)$ through the cylindrical coordinates Z_α, R_α, and Φ_α of the center of the sphere α and through the local spherical coordinate system and apply the theorem of addition for cylindrical functions. As a result, similar to Eq. (1.42) the symmetrized cylindrical wave (3.27) can be written in terms of local spherical coordinates ρ, ϕ, and θ as

$$\Psi_{II\alpha,PMN}(\rho,\varphi,\theta,|\,\mathrm{k},L)=\frac{1}{\sqrt{2\pi h/n}}$$
$$\times\exp\left\{i\left[\left(\mathrm{k}+\mathrm{k}_P-(L+nM)\frac{\omega}{h}\right)Z_\alpha+(L+nM)\Phi_\alpha\right]\right\}\times$$

$$\exp\left\{i\left[\mathrm{k}+\mathrm{k}_P-(L+nM)\frac{\omega}{h}\right]\rho\cos\theta\right\}(-1)^{L+nM}\times$$

$$\sum_{m=-\infty}^{+\infty}\left[C_{MN}^{J,L}J_{m-(L+nM)}\left(\kappa_{|L+nM|,N}R_\alpha\right)+\right.$$

$$\left.C_{MN}^{Y}Y_{m-(L+nM)}\left(\kappa_{|L+nM|,N}R_\alpha\right)\right]J_m\left(\kappa_{|L+nM|,N}\rho\sin\theta\right)e^{im\varphi}. \tag{3.30}$$

From the equality of $\Psi_{II\alpha,PMN}(\mathrm{k},L)$ and $\Psi_{I\alpha,PMN}(\mathrm{k},L)$ functions and of their radial derivatives at the MT spheres boundary, similar to Eqs. (1.45) – (1.51), we obtain

$$A_{lm,\alpha}^{PMN,\mathrm{k}L}=r_\alpha^2D(\mathrm{k},L)_{lm,\alpha}^{PMN}\,a_{lm,\alpha}^{PMN,L}(r_\alpha). \tag{3.31}$$

$$B_{lm,\alpha}^{PMN,\mathrm{k}L}=r_\alpha^2D(\mathrm{k},L)_{lm,\alpha}^{PMN}\,b_{lm,\alpha}^{PMN,L}(r_\alpha). \tag{3.32}$$

Here

$$D(\mathrm{k},L)_{lm,\alpha}^{PMN}=\frac{1}{\sqrt{2h/n}}(-1)^{\frac{m+|m|}{2}+l}i^l\left[(2l+1)\frac{(l-|m|)!}{(l+|m|)!}\right]^{1/2}(-1)^{L+nM}\times$$

$$\exp\left\{i\left(\mathrm{k}+\mathrm{k}_P-(L+nM)\frac{\omega}{h}\right)Z_\alpha+(L+nM)\Phi_\alpha\right\}\times$$

$$\left[C_{M,N}^{J,L}J_{m-(L+nM)}\left(\kappa_{|L+nM|,N}R_\alpha\right)\right.$$

$$\left.+C_{M,N}^{Y,L}Y_{m-(L+nM)}\left(\kappa_{|L+nM|,N}R_\alpha\right)\right], \tag{3.33}$$

$$a_{lm,\alpha}^{PMN,L}(r_\alpha)=\left\{I_{2,\alpha}^{PMN,L}\,\dot{u}_{l\alpha}(r_\alpha,E_{l\alpha})-I_{1,\alpha}^{PMNL}\,\dot{u}'_{l\alpha}(r_\alpha,E_{l\alpha})\right\}, \tag{3.34}$$

$$b_{lm,\alpha}^{PMN,L}(r_\alpha)=\left\{I_{1,\alpha}^{PMN,L}\,u'_{l\alpha}(r_\alpha,E_{l\alpha})-I_{2,\alpha}^{PMN,L}\,u_{l\alpha}(r_\alpha,E_{l\alpha})\right\}. \tag{3.35}$$

Finally, $I_{1,\alpha}^{PMN,L}$ and $I_{2.\alpha}^{PMN,L}$ are integrals of the augmented Legendre polynomials $P_l^{|m|}$

$$I_{1,\alpha}^{PMN}(r_\alpha,L)=2\int_0^{\pi/2}\exp\left\{i\left(\mathrm{k}+\mathrm{k}_P-(L+nM)\frac{\omega}{h}\right)r_\alpha\cos\theta\right\}\times$$

$$J_m\left(\kappa_{|L+nM|,N}r_\alpha\sin\theta\right)P_l^{|m|}(\cos\theta)\sin\theta d\theta \tag{3.36}$$

and

$$I_{2,\alpha}^{PMN}(r_\alpha,L)=2\int_0^{\pi/2}\exp\left\{i\left[\left(\mathrm{k}+\mathrm{k}_P-(L+nM)\frac{\omega}{h}\right)r_\alpha\cos\theta\right]\right\}\times$$

$$\left[i\left(k+k_P-(L+nM)\frac{\omega}{h}\right)\cos\theta J_m(\kappa_{|L+nM|,N}r_\alpha\sin\theta)+(1/2)\kappa_{|L+nM|,N}\sin\theta\right]$$

$$\times[J_{m-1}(\kappa_{|L+nM|,N}r_\alpha\sin\theta)-J_{m+1}(\kappa_{|L+nM|,N}r_\alpha\sin\theta)]P_l^{|m|}(\cos\theta)\sin\theta d\theta. \quad (3.37)$$

which can be compared with analogous integrals (1.50) – (1.52) from Chapter 1.

3.1.4. Overlap and Hamiltonian Matrix Elements

The method for calculating the overlap and Hamiltonian integrals, taking into account all the symmetry properties of the tubes, is practically the same as in Chapter 1, where only the translation symmetry was taken into account. Namely, due to symmetry conditions, the overlap and Hamiltonian integrals are equal to zero,

$$\int_\Omega \Psi^*_{P_2M_2N_2}(\mathbf{r}\,|\,k_2,L_2)\Psi_{P_1M_1N_1}(\mathbf{r}\,|\,k_1,L_1)dV=0, \quad (3.38)$$

and

$$\int_\Omega \Psi^*_{P_2M_2N_2}(\mathbf{r}\,|\,k_2,L_2)\hat{H}\Psi_{P_1M_1N_1}(\mathbf{r}\,|\,k_1,L_1)dV=0, \quad (3.39)$$

if $k_1 \neq k_2$ or $L_2 \neq L_1$. Otherwise,

$$\left\langle\Psi_{P_2M_2N_2}(\mathbf{r}\,|\,k,L)\,|\,\Psi_{P_1M_1N_1}(\mathbf{r}\,|\,k,L)\right\rangle=\delta_{M_2,M_1}\delta_{N_2,N_1}\delta_{P_2,P_1}-$$
$$\sum_\alpha\int_{\Omega_\alpha}\Psi^*_{II\alpha,P_2M_2N_2}(\mathbf{r}\,|\,k,L)\Psi_{II\alpha,P_1M_1N_1}(\mathbf{r}\,|\,k,L)dV+$$
$$\sum_\alpha\int_{\Omega_\alpha}\Psi^*_{I\alpha,P_2M_2N_2}(\mathbf{r}\,|\,k,L)\Psi_{I\alpha,P_1M_1N_1}(\mathbf{r}\,|\,k,L)dV, \quad (3.40)$$

and

$$\left\langle\Psi_{P_2M_2N_2}(\mathbf{r}\,|\,k,L)\,|\,\hat{H}\,|\,\Psi_{P_1M_1N_1}(\mathbf{r}\,|\,k,L)\right\rangle=\left[E_{P_2M_2N_2}(k,L)E_{P_1M_1N_1}(k,L)\right]^{1/2}$$
$$\times\delta_{M_2,M_1}\delta_{N_2,N_1}\delta_{P_2,P_1}-\sum_\alpha\int_{\Omega_\alpha}\Psi^*_{II\alpha,P_2M_2N_2}(\mathbf{r}\,|\,k,L)(-\Delta)\Psi_{II\alpha,P_1M_1N_1}(\mathbf{r}\,|\,k,L)dV$$
$$+\sum_\alpha\int_{\Omega_\alpha}\Psi^*_{I\alpha,P_2M_2N_2}(\mathbf{r}\,|\,k,L)\hat{H}_{MT,\alpha}\Psi_{I\alpha,P_1M_1N_1}(\mathbf{r}\,|\,k,L)dV. \quad (3.41)$$

With the use of Eqs (3.29) and (3.30) for the $\Psi_{II\alpha,PMN}(\mathbf{r}\,|\,k,\ L)$ and $\Psi_{I\alpha,PMN}(\mathbf{r}\,|\,k,L)$ functions, we finally obtain the following algebraic formula for calculating the overlap matrix elements:

$$\left\langle \Psi_{P_2M_2N_2} \mid \Psi_{P_1M_1N_1} \right\rangle|_{\mathrm{k},L} = \delta_{M_2,M_1}\delta_{N_2,N_1}\delta_{P_2,P_1} - \frac{n}{h}(-1)^{n(M_2+M_1)} \times$$
$$\sum_{\alpha} \exp\left\{ i\left[\left(\mathrm{k}_{P_1} - \mathrm{k}_{P_2} - n(M_1-M_2)\frac{\omega}{h}\right)Z_\alpha + n(M_1-M_2)\Phi_\alpha\right]\right\} \times$$
$$\sum_{m=-\infty}^{\infty}\left[C^{J,L}_{M_2N_2}J_{m-(L+nM_2)}\left(\kappa_{|L+nM_2|,N_2}R_\alpha\right) + C^{Y,L}_{M_2N_2}Y_{m-(L+nM_2)}\left(\kappa_{|L+nM_2|,N_2}R_\alpha\right)\right]$$
$$\times\left[C^{J,L}_{M_1N_1}J_{m-(L+nM_1)}\left(\kappa_{|L+nM_1|,N_1}R_\alpha\right) + C^{Y,L}_{M_1N_1}Y_{m-(L+nM_1)}\left(\kappa_{|L+nM_1|,N_1}R_\alpha\right)\right] \times$$
$$\left\{ I^{P_2M_2N_2,P_1M_1N_1}_{3,m\alpha}(r_\alpha, L) - r_\alpha^4 \sum_{l=|m|}^{\infty} \frac{(2l+1)\left[(l-|m|)!\right]}{2\left[(l+|m|)!\right]} S^{P_2M_2N_2,P_1M_1N_1}_{lm,\alpha}(r_\alpha, L)\right\}, \tag{3.42}$$

and

$$\left\langle \Psi_{P_2M_2N_2} \mid \widehat{H} \mid \Psi_{P_1M_1N_1} \right\rangle|_{\mathrm{k},L} = \left[\mathrm{k} + \mathrm{k}_{P_2} - (L+nM_2)\frac{\omega}{h} + \kappa_{|L+nM_2|,N_2}\right] \times$$
$$\left[\mathrm{k} + \mathrm{k}_{P_1} - (L+nM_1)\frac{\omega}{h} + \kappa_{|L+nM_1|,N_1}\right]\delta_{P_2M_2N_2,P_1M_1N_1} - \frac{n}{h}(-1)^{n(M_2+M_1)} \times$$
$$\sum_{\alpha} \exp\left\{ i\left[\left(\mathrm{k}_{P_1} - \mathrm{k}_{P_2} - n(M_1-M_2)\frac{\omega}{h}\right)Z_\alpha + n(M_1-M_2)\Phi_\alpha\right]\right\} \times$$
$$\sum_{m=-\infty}^{\infty}\left[C^{J,L}_{M_2,N_2}J_{m-M_2}\left(\kappa_{|L+nM_2|,N_2}R_\alpha\right) + C^{Y,L}_{M_2,N_2}Y_{m-M_2}\left(\kappa_{|L+nM_2|,N_2}R_\alpha\right)\right] \times$$
$$\left[C^{J,L}_{M_1,N_1}J_{m-M_1}\left(\kappa_{|L+nM_1|,N_1}R_\alpha\right) + C^{Y,L}_{M_1,N_1}Y_{m-M_1}\left(\kappa_{|L+nM_1|,N_1}R_\alpha\right)\right] \times$$
$$\left\{\left[\mathrm{k} + \mathrm{k}_{P_2} - (L+nM_2)\frac{\omega}{h}\right]\left[\mathrm{k} + \mathrm{k}_{P_1} - (L+nM_1)\frac{\omega}{h}\right] I^{P_2M_2N_2,P_1M_1N_1}_{3,m\alpha}(r_\alpha, L) + \right.$$
$$\kappa_{|L+nM_2|,N_2}\kappa_{|L+nM_1|,N_1} I'^{P_2M_2N_2,P_1M_1N_1}_{3,m\alpha}(r_\alpha, L) + m^2 I^{P_2M_2N_2,P_1M_1N_1}_{4,m\alpha}(r_\alpha, L) -$$
$$\left. r_\alpha^4 \sum_{l=|m|}^{\infty} \frac{(2l+1)\left[(l-|m|)!\right]}{2\left[(l+|m|)!\right]}\left(E_{l,\alpha} S^{P_2M_2N_2,P_1M_1N_1}_{lm,\alpha}(r_\alpha, L) + \gamma^{P_2M_2N_2,P_1M_1N_1}_{lm,\alpha}(r_\alpha, L)\right)\right\}, \tag{3.43}$$

where

$$I^{P_2M_2N_2,P_1M_1N_1}_{3,m\alpha}(r_\alpha, L) = 2\int_0^{\pi/2}\int_0^{r_\alpha} \cos\left[r\left(\mathrm{k}_{P_1} - \mathrm{k}_{P_2}\right)\cos\theta\right] J_m\left(\kappa_{|L+nM_2|,N_2} r\sin\theta\right) \times$$
$$J_m\left(\kappa_{|L+M_1|,N_1} r\sin\theta\right) r^2 \sin\theta \, d\theta dr, \tag{3.44}$$

$$S^{P_2M_2N_2,P_1M_1N_1}_{lm,\alpha}(r_\alpha L) = \bar{a}^{P_2M_2N_2,L}_{lm,\alpha} a^{P_1M_1N_1,L}_{lm,\alpha} + N_{l,\alpha}\bar{b}^{P_2M_2N_2,L}_{lm,\alpha} b^{P_1M_1N_1,L}_{lm,\alpha}, \tag{3.45}$$

$$I'^{P_2M_2N_2,P_1M_1N_1}_{3,m\alpha}(r_\alpha L)=2\int_0^{\pi/2}\int_0^{r_\alpha}\cos\left[r\left(k_{P_1}-k_{P_2}\right)\cos\theta\right]J'_m\left(\kappa_{|L+nM_2|,N_2}r\sin\theta\right)$$
$$\times J'_m\left(\kappa_{|L+M_1|,N_1}r\sin\theta\right)r^2\sin\theta\,d\theta dr, \tag{3.46}$$

$$I^{P_2M_2N_2,P_1M_1N_1}_{4,m\alpha}(r_\alpha,L)=2\int_0^{r_\alpha}\int_0^{\pi/2}\cos\left[r(k_{P_1}-k_{P_2})\cos\theta\right]J_m(\kappa_{|L+nM_2|,N_2}r\sin\theta)\times$$
$$J_m(\kappa_{|L+nM_1|,N_1}r\sin\theta)(\sin\theta)^{-1}drd\theta, \tag{3.47}$$

$$\gamma^{P_2M_2N_2,P_1M_1N_1}_{lm,\alpha}(r_\alpha,L)=\left(I_2^*I_1+I_1^*I_2\right)\dot{u}_{l\alpha}u'_{l\alpha}-I_2^*I_2\,\dot{u}_{l\alpha}u_{l\alpha}-I_1^*I_1\,\dot{u}'_{l\alpha}u'_{l\alpha}. \tag{3.48}$$

3.1.5. Dispersion Curves and Densities of States

Expanding the electronic wave functions $\psi_i(\mathbf{r}\,|\,\mathrm{k},L)$ in terms of symmetrized basic functions $\Psi_{PMN}(\mathbf{r}\,|\,\mathrm{k},L)$

$$\psi_i(\mathbf{r}\,|\,\mathrm{k},L)=\sum_{PMN}a_{i,PMN}(\mathrm{k},L)\Psi_{PMN}(\mathbf{r}\,|\,\mathrm{k},L) \tag{3.49}$$

and applying the variational principle then yield the secular equations for the dispersion curves $E_i(\mathrm{k}, L)$ and eigenvectors $a_{i,PMN}(\mathrm{k},L)$ for any particular values of the k and L

$$\det\left\|\left\langle\Psi_{P_2M_2N_2}\middle|\hat{H}\middle|\Psi_{P_1M_1N_1}\right\rangle\Big|_{\mathrm{k},L}-E_i(\mathrm{k},L)\left\langle\Psi_{P_2M_2N_2}\middle|\Psi_{P_1M_1N_1}\right\rangle\Big|_{\mathrm{k},L}\right\|=0, \tag{3.50}$$

$$\sum_{P_1M_1N_1}\left[\left\langle\Psi_{P_2M_2N_2}\middle|\hat{H}\middle|\Psi_{P_1M_1N_1}\right\rangle\Big|_{\mathrm{k},L}-E_i(\mathrm{k},L)\left\langle\Psi_{P_2M_2N_2}\middle|\Psi_{P_1M_1N_1}\right\rangle\Big|_{\mathrm{k},L}\right]$$
$$\times a_{i,PMN}(\mathrm{k},L)=0. \tag{3.51}$$

We also give formulas for calculating the partial charges for MT

$$Q_l^\alpha(i\,|\,\mathrm{k},L)=\frac{2nr_\alpha^4}{h}\sum_{P_2M_2N_2}\sum_{P_1M_1N_1}\bar{a}_{i,P_2M_2N_2}(\mathrm{k},L)a_{i,P_1M_1N_1}(\mathrm{k},L)(-1)^{n(M_1+M_2)}\times$$
$$\exp\left\{i\left[\left(k_{P_1}-k_{P_2}-n(M_1-M_2)\frac{\omega}{h}\right)Z_\alpha+n(M_1-M_2)\Phi_\alpha\right]\right\}\times$$
$$\sum_{m=-l}^{l}\left[C^{J,L}_{M_2N_2}J_{m-M_2}\left(\kappa_{|L+nM_2|,N_2}R_\alpha\right)+C^{Y,L}_{M_2N_2}Y_{m-M_2}\left(\kappa_{|L+nM_2|,N_2}R_\alpha\right)\right]\times$$
$$\left[C^{J,L}_{M_1N_1}J_{m-M_1}\left(\kappa_{|L+nM_1|,N_1}R_\alpha\right)+C^{Y,L}_{M_1N_1}Y_{m-M_1}\left(\kappa_{|L+nM_1|}R_\alpha\right)\right]\times$$
$$\frac{(2l+1)(l-|m|!)}{2(l+|m|!)}S^{P_2M_2N_2,P_1M_1N_1}_{3,lm,\alpha}(r_\alpha,L) \tag{3.52}$$

and interspherical regions

$$Q^{IS}(i\,|\,\mathrm{k},L)=\frac{2nr_\alpha^4}{h}\sum_{P_2M_2N_2}\sum_{P_1M_1N_1}\bar{a}_{i,P_2M_2N_2}(\mathrm{k},L)a_{i,P_1M_1N_1}(\mathrm{k},L)(-1)^{n(M_1+M_2)}\times$$
$$\exp\left\{i\left[\left(\mathrm{k}_{P_1}-\mathrm{k}_{P_2}-n(M_1-M_2)\frac{\omega}{h}\right)Z_\alpha+n(M_1-M_2)\Phi_\alpha\right]\right\}\times$$
$$\sum_{m=-l}^{l}\left[C_{M_2N_2}^{J,L}J_{m-M_2}\left(\kappa_{|L+nM_2|,N_2}R_\alpha\right)+C_{M_2N_2}^{Y,L}Y_{m-M_2}\left(\kappa_{|L+nM_2|,N_2}R_\alpha\right)\right]\times$$
$$\left[C_{M_1N_1}^{J,L}J_{m-M_1}\left(\kappa_{|L+nM_1|,N_1}R_\alpha\right)+C_{M_1N_1}^{Y,L}Y_{m-M_1}\left(\kappa_{|L+nM_1|,N_1}R_\alpha\right)\right]\times$$
$$\frac{(2l+1)(l-|m|!)}{2(l+|m|!)}I_{3,m\alpha}^{P_2M_2N_2,P_1M_1N_1}(r_\alpha,L). \tag{3.53}$$

necessary for calculating the partial densities of the state (1.69) and (1.72).

It must be emphasized that the obtained equations for the dispersion curves and DOS are applicable not only to the carbon nanotubes but also to any tubule with rotational and helical symmetries. In the case of carbon tubule, there are only two terms in the sums over α in the equations for overlap and Hamiltonian integrals and partial charges. Moreover, the parameters R_α, r_α, $S_{lm'\alpha}$, $E_{l'\alpha}$, $\gamma_{lm'\alpha}$, and I_i (i = 1, ... , 4) are the same for atoms α = 1 and 2.

3.2. Applications

3.2.1. Semiconducting and Metallic Chiral Carbon Nanotubes

By way of the first example, let us consider the total band structures and DOS of the tubules (13, 0), (12, 2), (11, 3), (10, 5), (9, 6), and (8, 7) with virtually equal diameters d =10.15 ± 0.15 Å, as well as the analogous data for the (7, 7) and (12, 4) nanotubes with slightly different d = 9.48 and 10.70 Å, respectively. The nanotubes are known to be characterized by the "family index" $p = (n_1 - n_2) \bmod 3$. The tubules with p = 0 are expected to be metallic or semimetallic (quasi-metallic), and those with p = 1 and p = 2 are semiconductors. Thus, there are chiral and achiral, semiconducting, semimetallic, and metallic nanotubes in this representative series.

Figure 3.3 shows the band structure of the chiral (11, 3) p = 1 nanotube; the Γ and K points correspond to the Brillouin zone center (k = 0) and boundary (k = π/h), respectively. The convergence of the calculated band structures as a function of a number of basic functions was investigated too; Fig. 3.4 illustrates the results for the minimum optical energy gap E_g of the (11, 3) nanotube.

One can see that about 150 functions produced convergence better than 0.01 eV. The same is true for other systems studied independent of the number of atoms in the translational unit cell. There is no rotational symmetry in this system. The translational unit cell contains as many as 652 atoms; however, the total band structure is seen to be very simple due to the account of the screw symmetry of the tubule. Particularly, there are only four dispersion curves in a valence band corresponding to the doubly occupied predominantly s, $p_{1\sigma}$, $p_{2\sigma}$, and p_π electronic states and only one low-energy unoccupied p_π*-type dispersion curve in the conduction band. The same is true for the (8, 7) $p = 1$ tubule with 676 atoms per translational unit cell (Fig. 3.5).

According to the LACW data, the nanotubes (11, 3) and (8, 7) are the semiconductors with the direct energy gaps E_{11} equal to 0.656 at k ≈ 0.24 π/h and 0.711 eV at k ≈ 0.089 π/h, respectively.

In the case of the (12, 2) $p = 2$ nanotube, the largest common divisor of n_1 and n_2 indices is $n = 2$; therefore, the structure of this tubule is characterized by the second order rotational axis C_2, and the eigenstates depend on the two quantum numbers, namely, the wave vector k and the rotational quantum number $L = 0$ and $L = 1$ (Fig. 3.6).

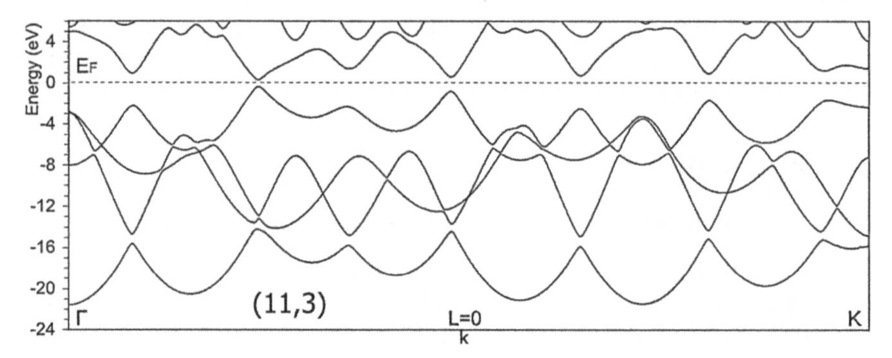

Fig. 3.3. Band structure of the (11, 3) nanotube.

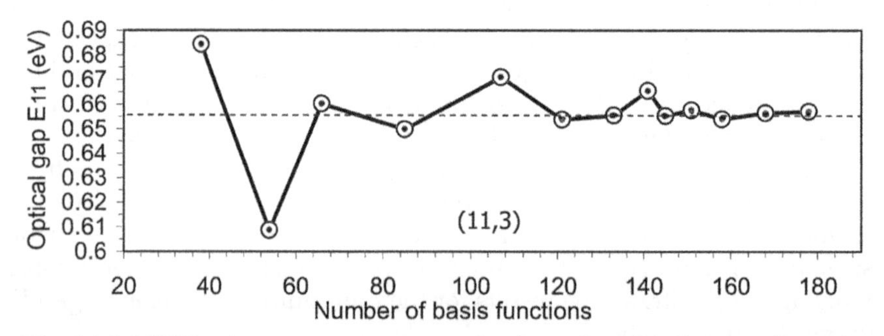

Fig. 3.4. LACW basic set convergence test for the carbon (11, 3) nanotube with 652 atoms in the translational unit cell.

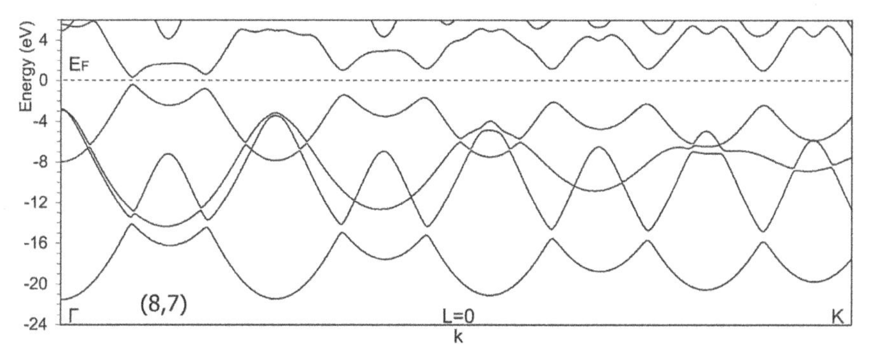

Fig. 3.5. Band structure of the (8, 7) nanotube.

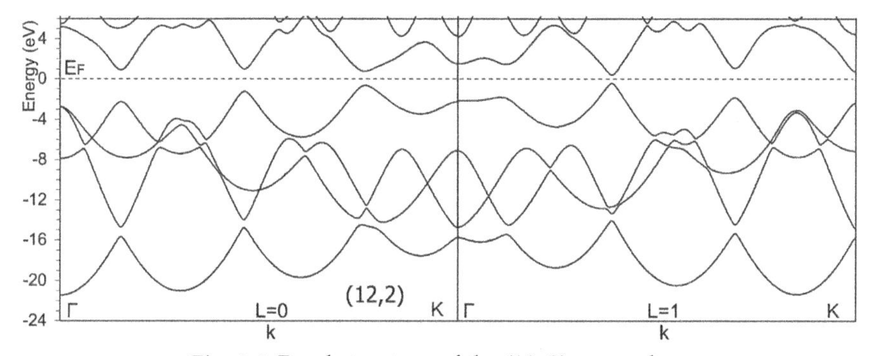

Fig. 3.6. Band structure of the (12, 2) nanotube.

As one goes from the (11, 3) and (8, 7) nanotubes to the (12, 2) tubule, the Brillouin zone is reduced approximately by one-half, Eq. (3.4). Now, there are four dispersion curves in the valence band and one curve in the low-energy region of the conduction band for every L value, and the band structure can be best demonstrated using the doubly repeated zone scheme. In this case, the dispersion curves for $L = 1$ appear as the extensions of the curves for $L = 0$, and the band diagrams of the (12, 2), (11, 3), and (8, 7) nanotubes become rather similar. In the (12, 2) tube, the $E_{11} = 0.855$ eV corresponds to the direct transition between the states with $L = 1$ about k point equal to $0.38(\pi/h)$.

In the case of the (12, 4) single-wall nanotube of the semiconducting $p = 2$ family (Fig. 3.7), the integer L takes the values of 0, ..., 3.

In the band structure, there are two pairs of continuous dispersion curves corresponding to $L = 0$ and 2 and $L = 1$ and 3, respectively, and the minimum gap $E_{11} = 0.46$ eV corresponds to the direct transition between the states with $L = 3$ at k ≈ 0.79 (π/h).

Figures 3.8 and 3.9 show the band structures of the chiral (10, 5) $p = 2$ and achiral zigzag (13, 0) $p = 1$ nanotubes, which are characterized by the higher (5th and 13th) order rotational axes.

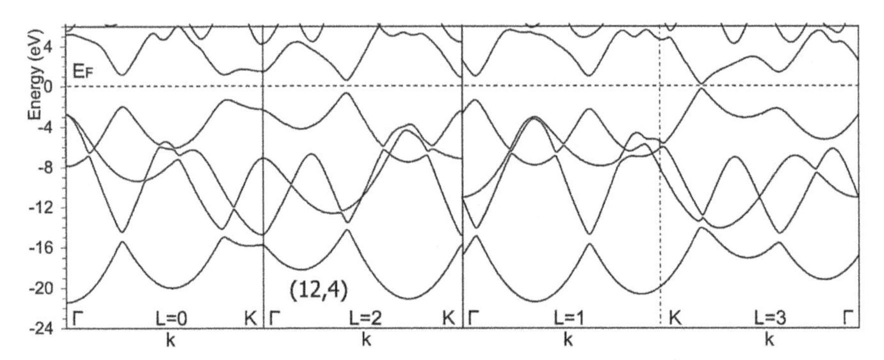

Fig. 3.7. Band structure of the (12, 4) nanotube.

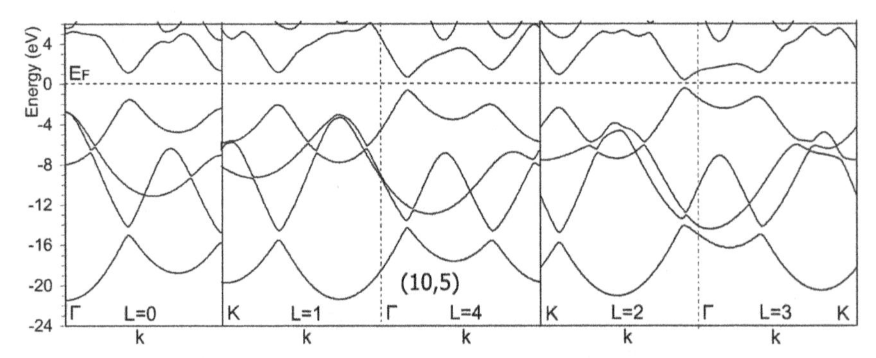

Fig. 3.8. Band structure of the (10, 5) nanotube.

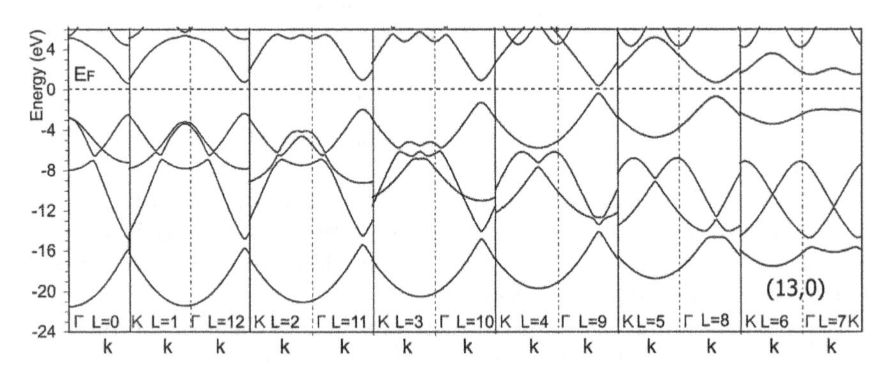

Fig. 3.9. Band structure of the (13, 0) nanotube.

In the repeated zone scheme, there are three and seven sets of dispersion curves, respectively. The bands for $n - L$ are the continuous extensions of the bands for L. The $E_g = 0.816$ eV for the $L = 2$ in the case of the (10, 5) tubule and $E_g = 0.799$ eV for the $L = 9$ in the case of the (13, 0) nanotube. Note that, in the zigzag tubule, if one takes into account the translation symmetry only, the minimum gap E_g corresponds to the direct transition at the Brillouin zone k = 0. With due account of the screw symmetry of the (13, 0) nanotube, the E_g corresponds to the point k ≈ 0.7, $L = 9$.

Figure 3.10 shows the total DOS of the five semiconducting nanotubes with virtually the same diameters $d = 10.15 \pm 0.15$ Å. Although the band structures of these tubules look different, the total DOS are very similar. An example of the partial DOSs of semiconducting tubule is presented in Fig. 3.11.

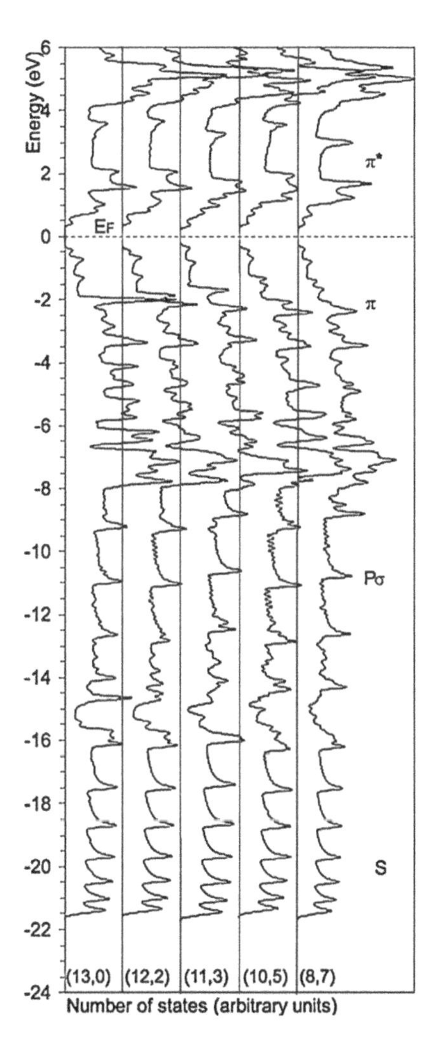

Fig. 3.10. Total DOS of the nanotubes with $d = 10.15 \pm 0.15$ Å.

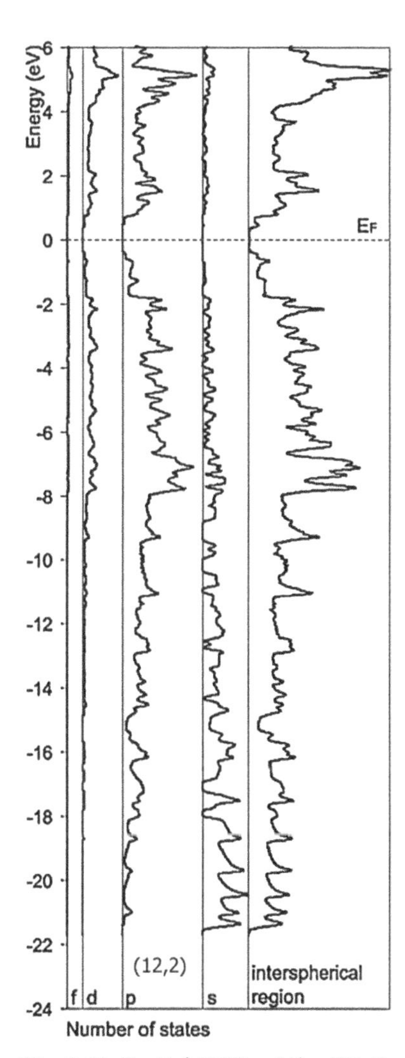

Fig. 3.11. Partial DOSs of the (12, 2) tubule.

Obviously, the armchair (7, 7) single-walled nanotube is characterized by the metallic band structure with Fermi level at $L = 0$ and $k \approx (2/3)(\pi/h)$ (Fig. 3.12).

The (9, 6) tubule with the rotational C_3 axis belongs to the $p = 0$ family. In the repeated zone scheme, there are two sets of dispersion curves corresponding to $L = 0$, 1, and 2, the curves for the $L = 2$ being the extensions of the curves for the $L = 1$. The simple zone-folding model predicts the metallic type of band structure in this case. However, the LACW method predicts a formation of minigap between the bonding and antibonding π states equal to 0.035 eV in this case (Fig. 3.13).

This tubule is an example of the quasi-metallic system, which is of great importance for terahertz applications; the specific features of their band structure are discussed in the next section.

As a final demonstration of the possibilities of this version of the LACW theory, let us look at the results of the band structure calculations of the (100, 99) nanotube containing a total of 118,804 atoms per translational unit cell (Figs. 3.14 – 3.16).

Even for this system with huge unit cell, the band structure is quickly converged and tubule is predicted to be the semiconductor with the gap $E_{11} = 0.04$ eV near Brillouin zone center.

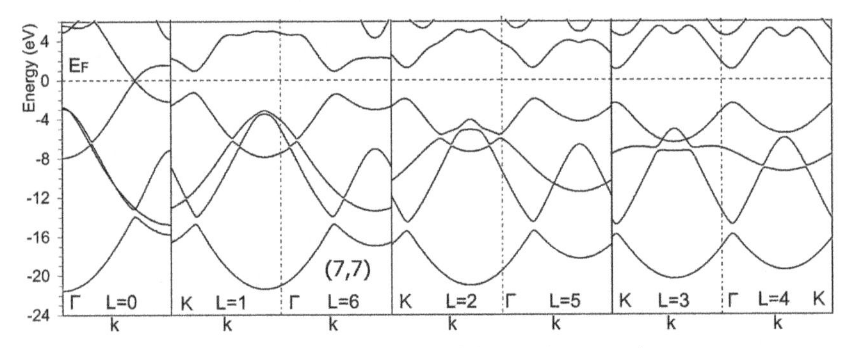

Fig. 3.12. Band structure of the (7, 7) nanotube.

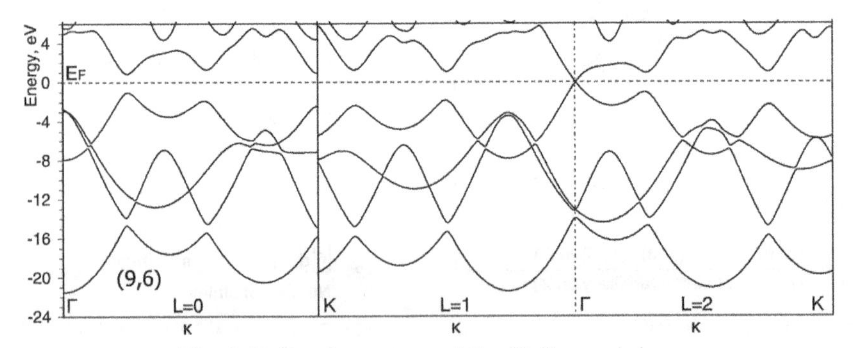

Fig. 3.13. Band structure of the (9, 6) nanotube.

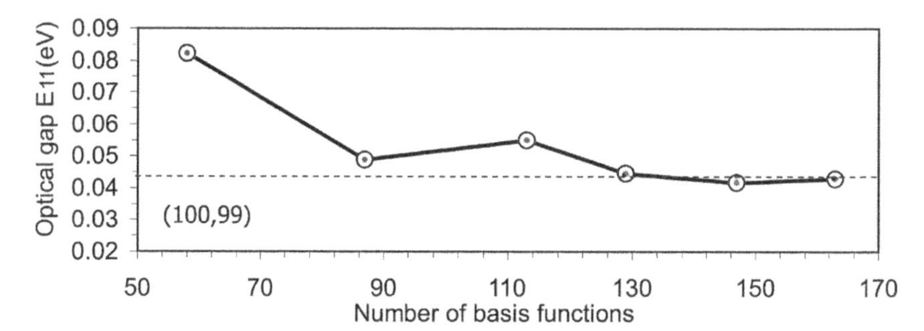

Fig. 3.14. LACW basic set convergence test for the (100, 99) nanotube with 118, 804 atoms in the translational unit cell.

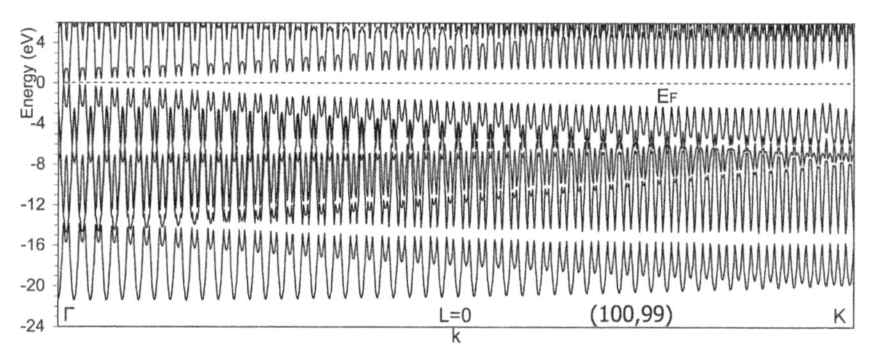

Fig. 3.15. Total band structure of the (100, 99) nanotube.

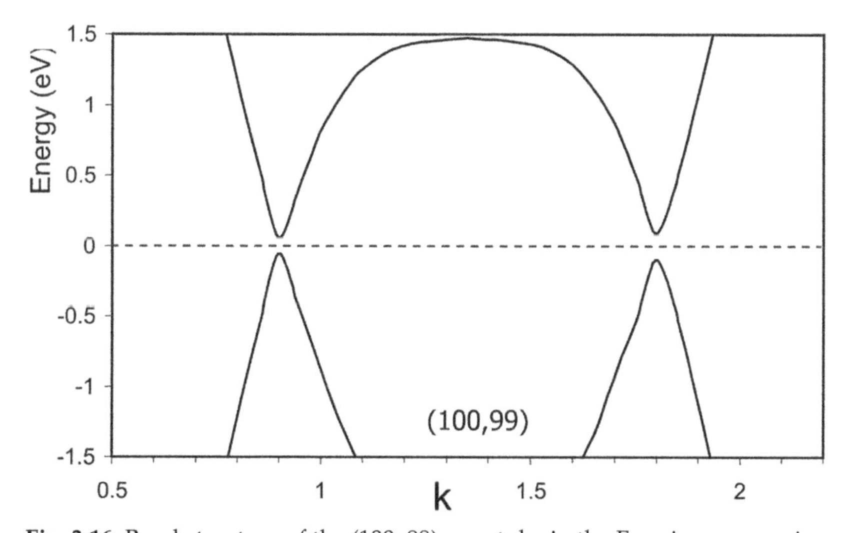

Fig. 3.16. Band structure of the (100, 99) nanotube in the Fermi energy region.

Finally, the highest occupied and lowest unoccupied LACW bands are in reasonable agreement with the standard π tight-binding data (White et al. 1993) calculated using the following equation

$$\varepsilon(\mathrm{k},L)=\pm V_0\left[3+2\cos\left(\frac{n_1\mathrm{k}-2\pi Lp_1}{n}\right)+2\cos\left(\frac{n_2\mathrm{k}-2\pi Lp_2}{n}\right)+\right.$$
$$\left.2\cos\left(\frac{(n_1+n_2)\mathrm{k}-2\pi L(p_1+p_1)}{n}\right)\right]^{1/2} \tag{3.54}$$

and plotted using the same quantum numbers L and k (Fig. 3.17); here, V_0 is the matrix element between the nearest neighboring π-orbitals.

3.2.2. Quasi-Metallic Carbon Nanotubes

Neglecting the surface curvature of a nanotube makes it possible to determine the π-electron band structure of nanotubes as a result of

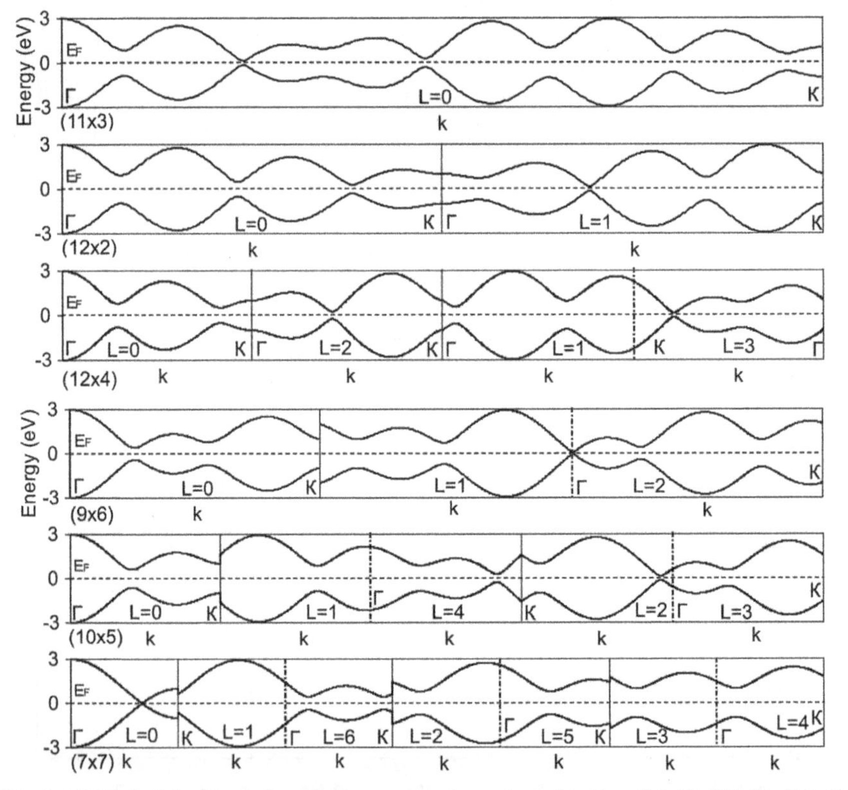

Fig. 3.17. Tight-binding π-band structures of semiconducting (11, 3), (12, 2), (12, 4), (10, 5), metallic (7, 7), and quasi-metallic (9, 6) nanotubes plotted as the functions of two quantum numbers k and L.

projecting the π-bands of graphene onto the cylinder. This approximation demonstrates that if the $n_1 - n_2$ difference is a multiple of three and $n_1 \neq n_2$, the top of the valence band and the bottom of the conduction band must touch each other and, hence, there should not be a band gap in the nanotube. Sometimes such nanotubes are known to belong to the quasi-metallic family. However, in this case, taking into account the nanotube surface curvature sharply changes the prediction and demonstrates that nanotubes with $n_1 - n_2 = 3q$ can be actually the narrow-gap semiconductors with optical band gaps that can fall into the terahertz (THz) range 0.1–40 meV (Blasé et al. 1994, Kane and Mele 1997, Chibotaru et al. 2002, Hartmann and Portnoi 2015, Ouyang et al. 2001, Itkis et al. 2002, Borondocs et al. 2006, Pekker and Kamara 2008, Kampfrath et al. 2008, Bushmaker et al. 2009, Mantsch and Naumann 2010), which enables their use as terahertz emitters, detectors, polarizers, and multipliers, as well as *pn*-junctions and transistors with properties that can be tailored by applying weak external electromagnetic field (Nemilentsau et al. 2010, Hartmann and Portnoi 2016, Portnoi et al. 2008, Hartmann et al. 2014, Mann et al. 2007, Mueller et al. 2010, Wang et al. 2007, Chang et al. 2015, Zhong et al. 2008, Cui et al. 2003, Choi 2014, Santavicca and Prober 2008, Fu et al. 2008).

It has been noted that the minigap energies in such quasi-metallic nanotubes are determined not only by purely band factors (Hartmann and Portnoi 2016, Portnoi et al. 2008, Deshpande et al. 2009). For example, the band gap energies measured for the nanotubes located on a substrate and in a free state can differ up to 10 times due to the charge carrier trapping in the substrate (Lin et al. 2010). It is not quite clear whether one-electron or collective electronic excitations determine the experimentally observed band gaps (Hartmann and Portnoi 2016, Portnoi et al. 2008, Amer et al. 2012). Finally, synthesis of nanotubes can generate mechanical strains, which can also lead to the formation of band gaps near the Fermi level.

In the work of D'yachkov and Bochkov (2018), the band structure of 50 chiral and achiral (n_1, n_2) nanotubes with $4 \leq n_1 \leq 18$ and $n_2 = n_1 - 3q$ were calculated by the LACW method. In terms of this data, the nanotubes for which the optical band gaps fall in the terahertz range have been identified. We here discuss the purely band structure effects of the energy gaps formation in the quasi-metallic nanotubes. Note that the standard quantum-chemical calculations using basic sets of plane waves or atomic orbitals with inclusion of translational symmetry of nanotubes turned out to be inefficient for determining band gap energies in chiral narrow-gap nanotubes with giant unit cells; only a small number of such calculations dealing only with zigzag (n, 0) nanotubes are available in literature.

The computation results are summarized in Table 3.1, which gives the band gap width in the nanotubes and their structural characteristics—screw axis parameters h and ω, radii R_{NT}, and chirality angles of nanotubes.

Table 3.1. Energies of gaps (eV) in carbon nanotubes (n_1, n_2) with $n_1 - n_2$ divisible by three.

Nano-tube	Structure parameters				Energy gap, eV		
	h, a.u.	ω, rad	*R*, Å	θ, grad	LACW	Expe-riment	*Ab initio* LCAO
(4, 1)	0.878	1.35	1.79	10.5	0		0 [a]
(5, 2)	0.643	2.65	2.44	16.1	0		0 [a]
(6, 0)	4.02	0	3.35	0	0		0 [a]
(6, 3)	1.52	1.34	3.11	19.1	0		0.056 [a]
(7, 1)	0.532	0.826	2.96	6.59	0		0.138 [a]
(7, 4)	0.417	1.72	3.77	21.1	0.125		0.067 [a]
(8, 2)	0.878	2.47	3.59	10.9	0.32		0.069 [a]
(8, 5)	0.354	2.41	4.45	22.4	0.058		
(9, 0)	4.02	0	3.52	0	0.19	0.08 [c]	0.17 [b] 0.17 [d] 0.08 [d] 0.093 [e] 0.096 [a] 0.20 [f]
(9, 3)	1.12	0.564	4.32	13.9	0.11		0.02 [g]
(9, 6)	0.922	0.827	5.12	23.4	0.035		
(10, 1)	0.382	0.594	4.12	4.72	0.20		
(10, 4)	0.644	1.33	4.89	16.1	0.13		
(10, 7)	0.272	1.85	5.79	24	0.038		
(11, 2)	0.332	2.89	4.75	8.21	0.13		
(11, 5)	0.284	1.17	5.55	17.8	0.074		
(11, 8)	0.243	2.31	6.47	24.8	0.026		
(12, 0)	4.02	0	4.70	0	0.042	0.042 [c]	0.040 [e] 0.078 [a] 0.08 [a]
(12, 3)	0.877	0.449	5.38	10.9	0.076		
(12, 6)	1.52	0.374	6.21	19.1	0.078		
(12, 9)	0.661	0.594	7.14	25.3	0.013		
(13, 1)	0.297	0.464	5.30	3.67	0.14	0.18 [h]	
(13, 4)	0.261	1.47	6.03	13.0	0.084		
(13, 7)	0.229	0.946	6.89	20.1	0.049		
(13, 10)	0.201	1.91	7.82	25.7	0.015		

(Contd.)

(14, 2)	0.532	0.413	5.91	6.59	0.13		
(14, 5)	0.236	1.33	6.68	14.7	0.062		
(14, 8)	0.417	0.861	7.55	21.1	0.045		
(14, 11)	0.185	2.26	8.50	26.0	0.011		
(15, 0)	4.02	0	5.87	0	0.022	0.029 [c]	0.023 [a] 0.028 [e] 0.030 [f]
(15, 3)	0.722	0.372	6.54	8.95	0.050		
(15, 6)	0.644	0.886	7.33	16.1	0.029		
(15, 9)	0.574	0.791	8.22	21.8	0.015		
(15, 12)	0.515	0.464	9.17	26.3	0.006		
(16, 1)	0.243	0.380	6.47	3.00	0.080		
(16, 4)	0.877	0.337	7.18	10.9	0.089		
(16, 7)	0.197	2.73	7.99	17.3	0.058		
(16, 10)	0.354	1.21	8.98	22.4	0.025		
(16, 13)	0.160	1.95	9.85	26.9	0.011		
(17, 2)	0.222	2.97	7.08	5.50	0.065		
(17, 5)	0.201	2.57	7.81	12.5	0.057		
(17, 8)	0.182	0.751	8.66	18.3	0.022		
(17, 11)	0.165	1.12	9.56	22.9	0.015		
(17, 14)	0.150	2.23	10.5	26.8	0.009		
(18, 0)	4.02	0	7.045	0	0.009		0.02 [e]
(18, 3)	0.612	0.317	7.70	7.59	0.038		
(18, 6)	1.12	0.282	8.47	13.90	0.030		
(18, 9)	1.52	0.249	9.32	19.1	0.034		
(18, 12)	0.922	0.413	10.24	23.41	0.018		
(18, 15)	3.41	0.380	11.20	27.0	0.003		

[a]Zólyomi and Kürti 2004; [b]Blase et al. 1994; [c]Ouyang et al. 2001; [d]Miyake and Saito 2005; [e]Gülseren and Yildirim et al. 2002; [f]Sun et al. 2003; [g]Reich et al. 2002; [h]Amer et al. 2013.

Let us begin with discussing eight nanotubes with $4 \leq n_1 \leq 8$ and small radii R_{NT} (Fig. 3.18).

Because of the large surface curvature of these tubes, the Hückel π-electron model is not applicable here. In particular, nanotubes with $n_1 =$ 4, 5, and 6 as well as the (7, 1) nanotubes turn out to be metallic, rather than semiconducting, because of the overlap of different dispersion curves at the Fermi level. The (7, 4), (8, 2), and (8, 5) nanotubes of larger radius and smaller surface curvature have the expected band gaps, but they are too large (0.125, 0.32, and 0.058 eV) for using these tubes in THz technologies.

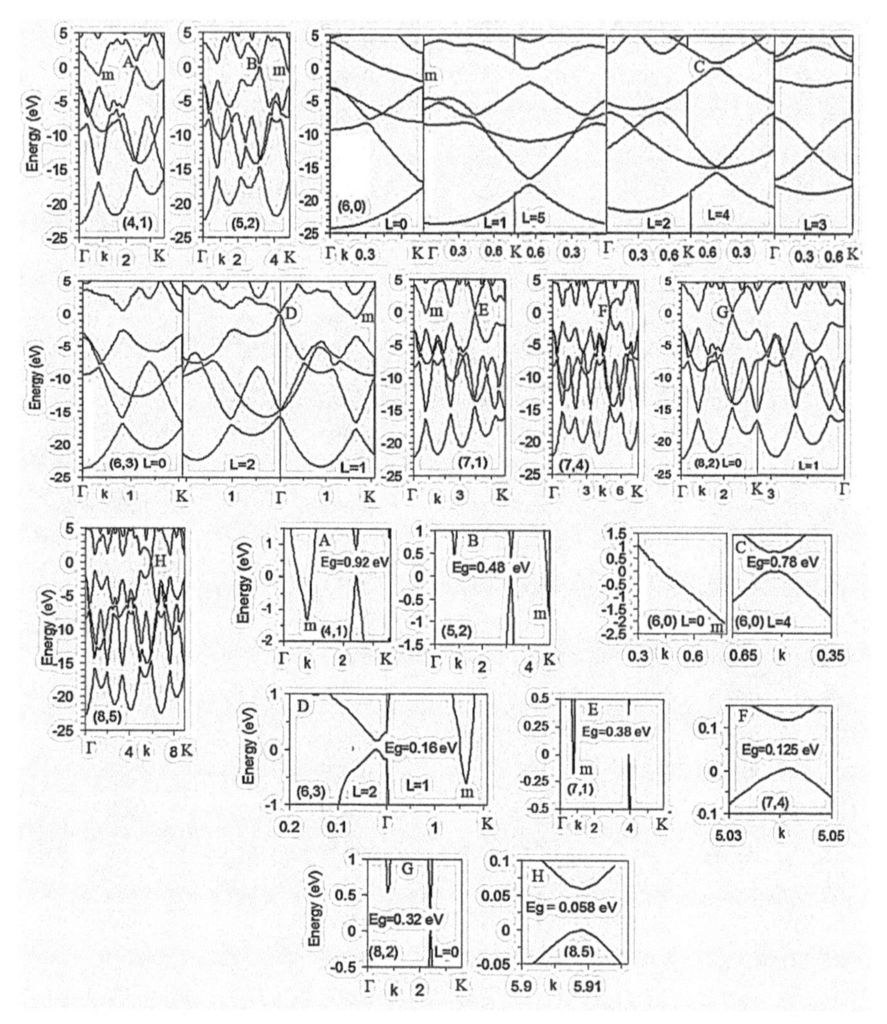

Fig. 3.18. Band structures of the (n_1, n_2) nanotubes with $4 \leq n_1 \leq 8$. Here – dispersion curves in the minimum gap regions (A, B, ...) are presented in enlarged energy scale.

Figure 3.19 shows the band structures of the (9, n_2) tubules series (n_2 = 0, 3, and 9) written in the terms of repeated zone scheme.

For the (9, 0) and (9, 3) tubes, the minimum energy gaps between the conduction and valence bands correspond to the points L = 6, k = 0.52 a.u.$^{-1}$ $\approx 2\pi/3h$ and L = 2, k = 1.88 a.u^{-1} $\approx 2\pi/3h$, respectively. In the case of (9, 6) tubule, the minigap is located at k = 0.007 a.u.$^{-1}$ that is virtually at the Brillouin zone Γ point on the boundary between L = 1 and L = 2 regions with slight shift in the L = 2 direction. The increase in the index n_2 is accompanied by the decrease of the band gap energy E_g from 0.19 to

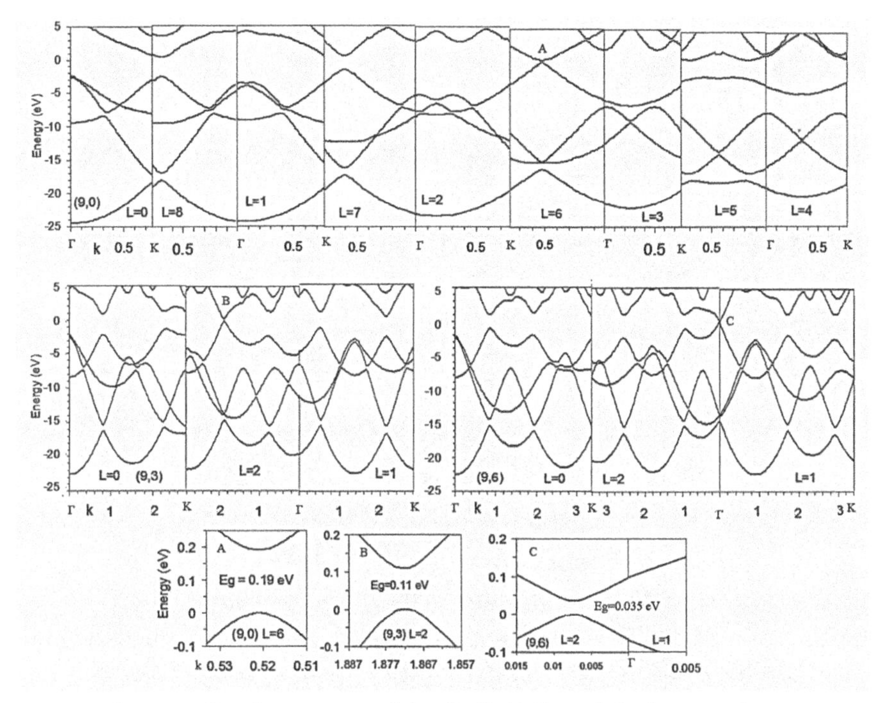

Fig. 3.19. Band structures of the (9, 0), (9,3), and (9, 6) nanotubes.

0.11 and 0.035 eV. The band gap narrowing agrees qualitatively with the simple π-electron LCAO ideas predicting that increasing the nanotubes radius R_{NT} and chirality angle θ should result in a reduction of the minigap (Kane and Mele 1997, Hartmann et al. 2014)

$$E_g = \frac{\hbar v_F d_{C-C}}{8R_{NT}^2}\cos 3\theta\,, \tag{3.55}$$

where $v_F = 7.25{\cdot}10^5$ m/s is the Fermi velocity of graphene. The increase of the n_2 index leads to increase of both R_{NT} and θ. For the (9, n_2) nanotubes, the π electron E_g values are about three times underestimated compared with the LACW data; the similar significant underestimation of minigap energies in the π electron model is also true for majority of other nanotubes studied. Note that the experimental data for the suspended tubules (Chang et al. 2015, Lin et al. 2010) also show that empirical tight-binding models underestimate the gap about 3-4 times. For the (9, 0) tubule, the LACW E_g = 0.19 eV value can be compared with the experimental gap E_g = 0.08 eV for nanotube deposited on a gold substrate and with the following gaps obtained earlier by using the LCAO Local-Density Approximation (LDA): 0.0843, 0.09344, 0.09645, 0.177, 0.1743 (LDA with many-body corrections), and 0.20 eV (Sun et al. 2003). For the (9, 3) tubule, the LACW E_g = 0.11 eV;

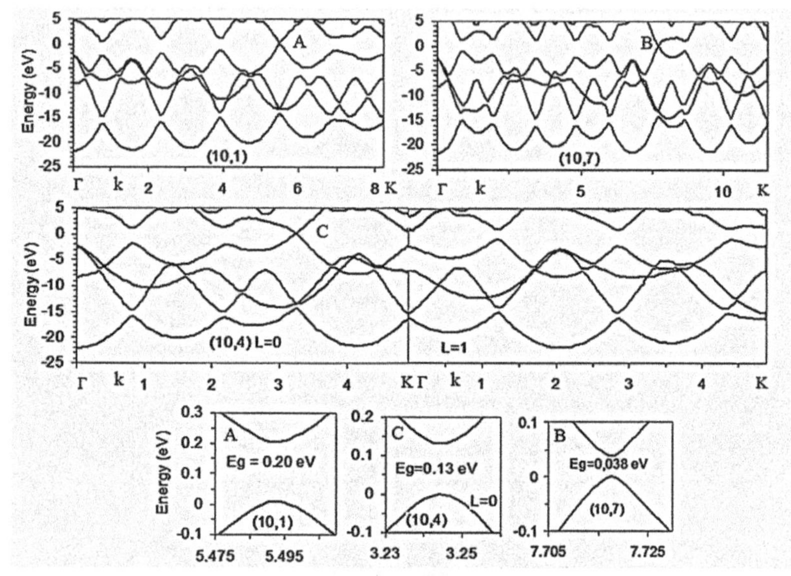

Fig. 3.20. Band structures of the (10, n_2) nanotubes.

the LDA pseudopotential gap equal to 0.02 eV appears to be much too low even from the point of view of the π electron theory (Reich et al. 2002).

The transition from zigzag (9, 0) tubule with 9th order rotational axis to the low-symmetry chiral (10, 1) tubule is accompanied by dramatic simplification of band structure, but it almost does not affect the band gap. The same is true for the pairs (9, 3) and (10, 4), (9, 6) and (10, 7) (Fig. 3.20).

In the (10, 1) and (10, 7) tubes with C_1 axis, the gaps are located at k ≈ $2\pi/3h$; in the case of (10, 4) tubule with C_2 rotational symmetry, the gap is near the k ≈ $2\pi/3h$ point for the $L = 0$.

There is no rotational symmetry in the tubules (11, n_2) with $n_2 = 2, 5$, and 8. The gaps are located near the k = $2\pi/3h$ point for $L = 0$ in all cases (Fig. 3.21).

The nanotubes radii and chirality angles of these tubules are noticeably larger than those of the corresponding (9, n_2) tubules, and the gaps of the (11, n_2) tubes are about 30% smaller. Among the nine nanotubes of (9, n_2), (10, n_2), and (11, n_2) series, the E_g values are less than 0.040 eV, i.e., the optical gap is in the THz region only in the case of the three tubes (9, 6), (10, 7), and (11, 8) having the largest R_{NT} and θ values. In these three series, the gaps decrease approximately linearly as functions of the n_2 index. There are no experimental or *ab initio* data on the band structure and gap energies for these chiral tubules. In contrast to the naive π electron model, according to which the growth of the index n_2 must always lead to a narrowing of the gap, the LACW method predicts that the dependence of E_g on the n_2 is non-monotonic among the nanotubes (12, n_2) with $n_2 = 0$, 3, 6, and 9 (Fig. 3.22).

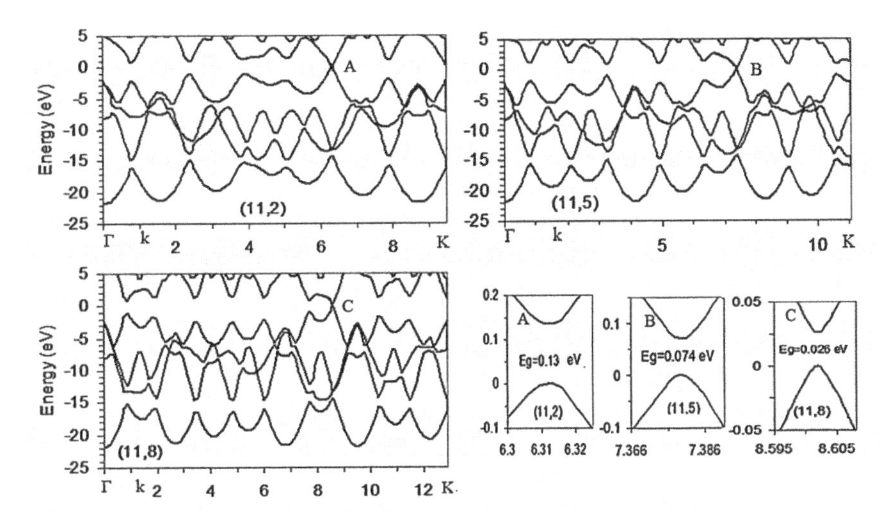

Fig. 3.21. Band structures of the (11, n_2) nanotubes.

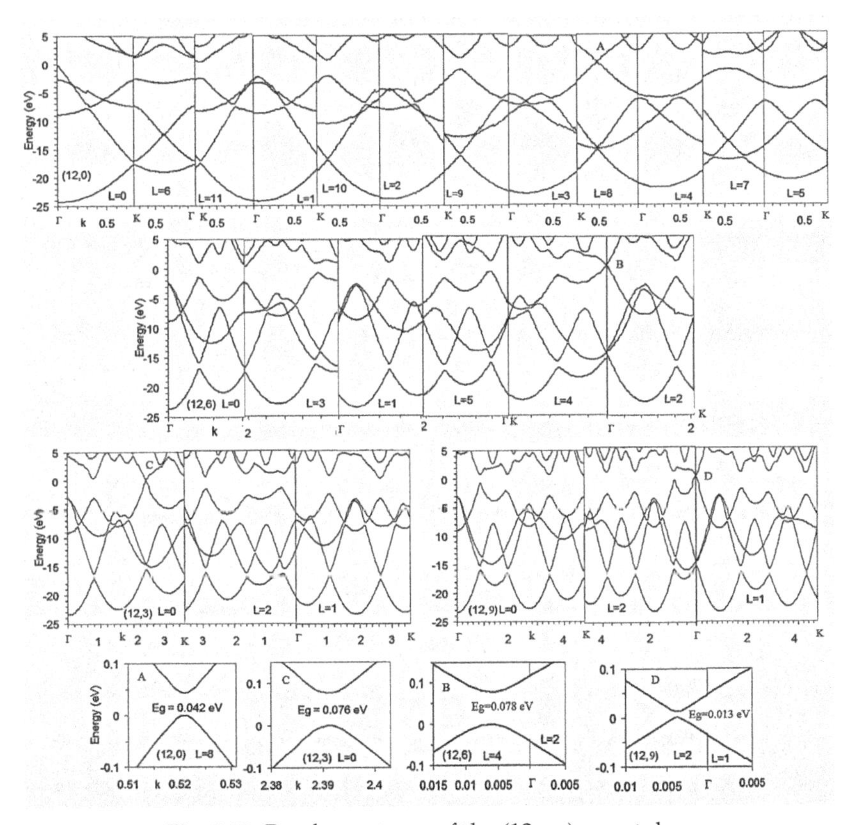

Fig. 3.22. Band structures of the (12, n_2) nanotubes.

The maximum forbidden gaps E_g = 0.076 and 0.077 eV are observed for n_2 = 3 and 6; they are almost two times larger than the gap equal to 0.042 eV for the tube (12, 0). The gaps are near the point k = $2\pi/3h$ for L = 8 and L = 0 in the case of the (12, 0) and (12, 3) tubes, respectively, and they are near Brillouin zone center k = 0 in the boundary between L = 2, L = 4 and L = 2, L = 1 regions in the case of the (12, 6) and (12, 9) tubes, respectively. Again, only one tubule (12, 9) with E_g = 0.013 eV falls in the THz range. The LACW energy gap coincides with experimental value E_g = 0.042 eV for the (12, 0) nanotube deposited on a gold substrate (Ouyang et al. 2001), and can be compared with the values of 0.040, 0.078, and 0.08 eV obtained using LDA calculations (Gülseren et al. 2002, Zólyomi and Kürti 2004, Sun et al. 2003).

According to the equation (3.55), the transition from the (12, 0) tube to the (13, 1) and (14, 2) tubes should be accompanied by a decrease of E_g. However, instead, the LACW method shows the threefold gap increase up to 0.14 and 0.13 eV, respectively. To the best of our knowledge, the (13, 1) tubule is the only chiral system, for which the experimental gap was identified and shown to be equal to the 0.18 eV using the device with suspended nanotubes (Amer et al. 2013). Figures 3.23 and 3.24 show that the LACW band gaps decrease monotonically with increasing n_2 index in the series (13, n_2) with n_2 = 1, 4, 7, 10 and (14, n_2) with n_2 = 2, 5, 8, 11 in accordance with simple equation (3.55).

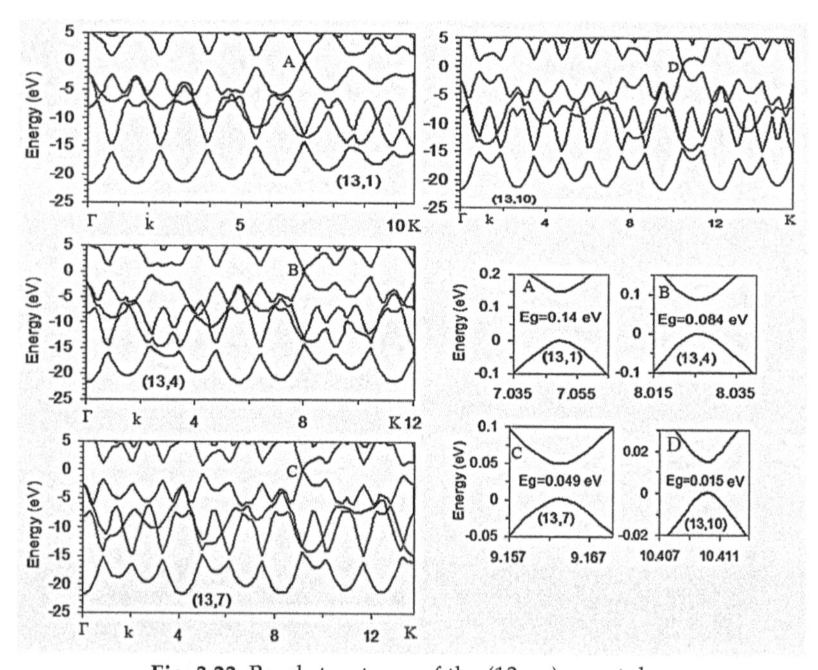

Fig. 3.23. Band structures of the (13, n_2) nanotubes.

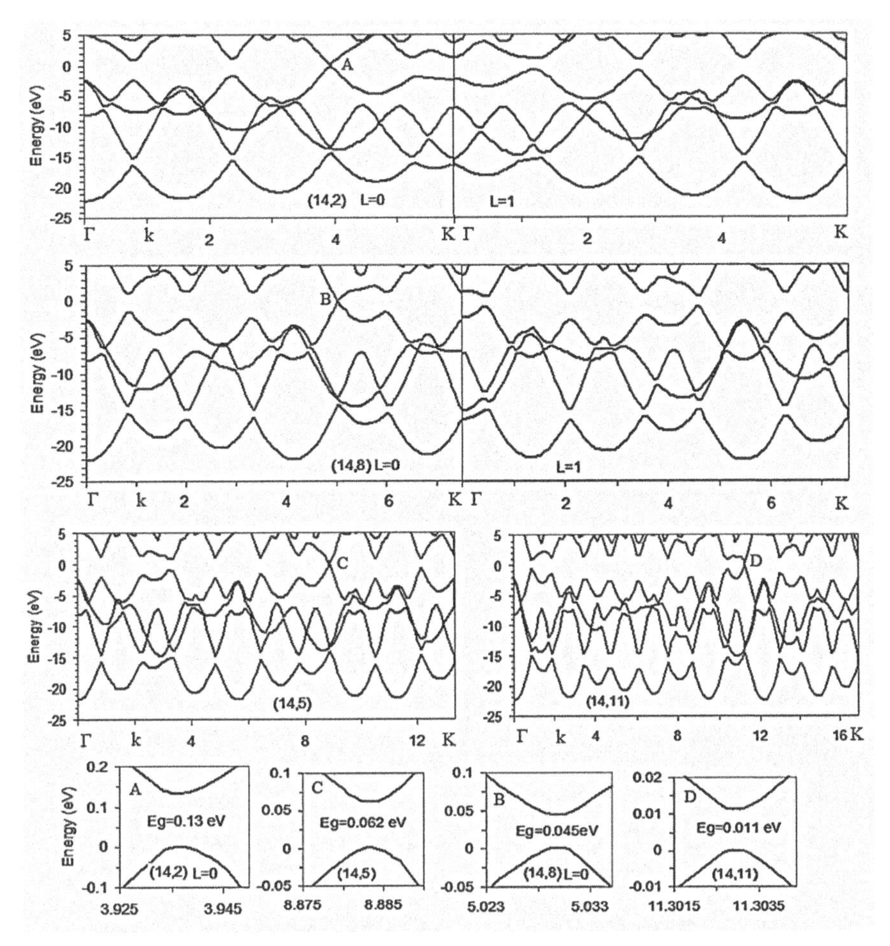

Fig. 3.24. Band structures of the (14, n_2) nanotubes.

Only the (13, 10) and (14, 11) tubules with the maximum n_2 values and, therefore, minimum gaps equal to 0.015 and 0.011 eV are in the TIIz frequency range. Similar to the (12, n_2) case and in contrast to the predictions of π electron model, the gap width increases from 0.022 to 0.050 eV at the transition from the zigzag tube (15, 0) to the chiral tube (15, 3) (Fig. 3.25).

Further increase of the n_2 index in the (15, n_2) series results in the gap decrease down to a value of 0.029, 0.015, and 0.006 eV for the (15, 6), (15, 9), and (15, 12) tubules, respectively. The gaps correspond to the Brillouin zone points k $\approx 2\pi/3h$ for the tubules (15, 0) ($L = 10$), (15, 6) ($L = 0$), (15, 9) ($L = 2$), or to the Γ point separating $L = 2$ and $L = 1$ regions in the case of (15, 3) and (15, 12) tubes. Only for the (15, 0) tubule, the LACW $E_g = 0.022$ eV can be compared the experimental gap equal to 0.029 eV (Ouyang et

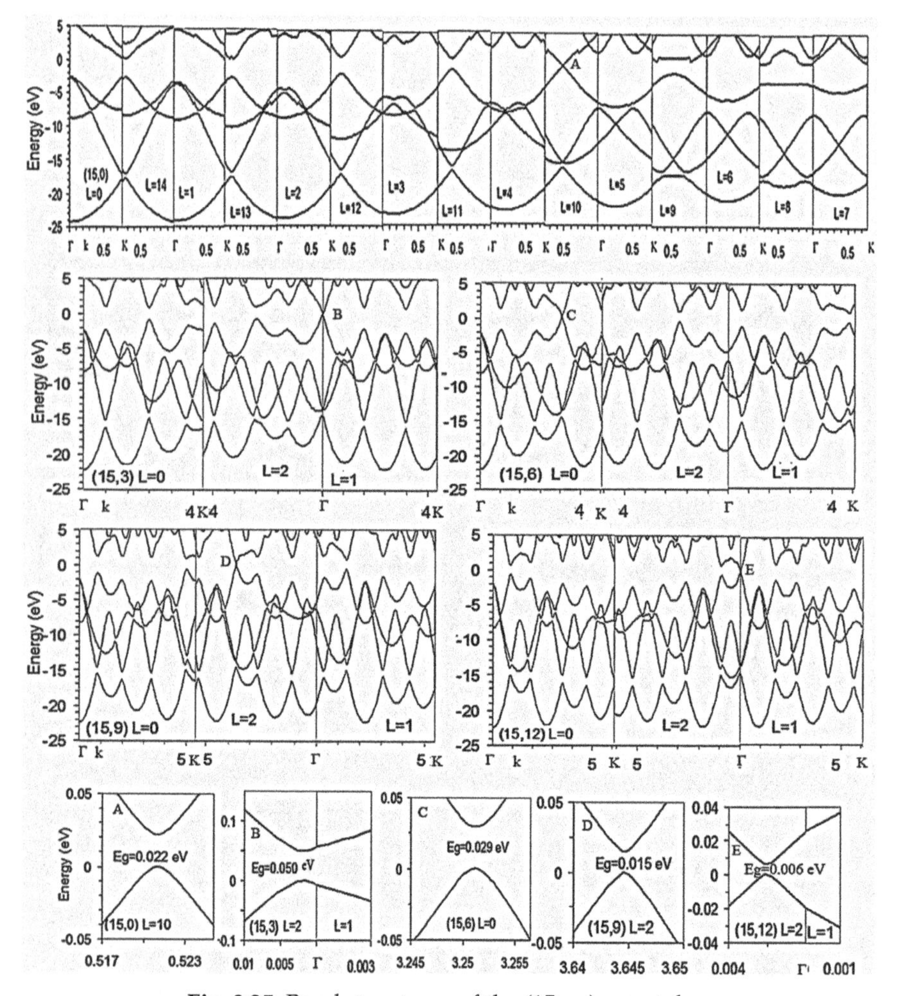

Fig. 3.25. Band structures of the (15, n_2) nanotube.

al. 2001) for the tube on Au substrate and with following LDA data: 0.023 (Zólyomi and Kürti 2004), 0.028 (Gülseren et al. 2002), and 0.030 eV (Sun et al. 2003).

Figures 3.26 and 3.27 show that the gaps of (16, n_2) and (17, n_2) nanotubes are larger than those of the (15, n_2) counterparts in the case of small n_2 indices; as expected, increasing n_2 is accompanied by the reduction of gaps. About half of these tubes have the gap in the THz region.

In the (18, n_2) tubes, the gaps range from the 0.003 to 0.038 eV, that is, they all are in the THz region. With the growth of n_2 index, initially the gap increases from the E_g = 0.009 eV to the E_g = 0.030-0.038 eV and then drops to the value equal to the 0.003 eV for n_2 = 15 (Fig. 3.28).

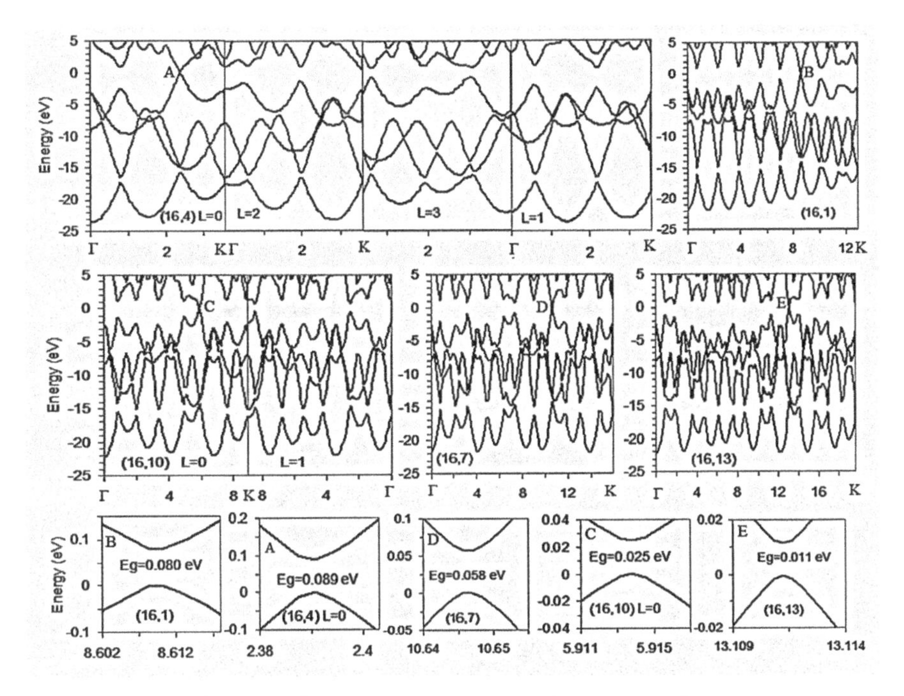

Fig. 3.26. Band structures of the (16, n_2) nanotubes.

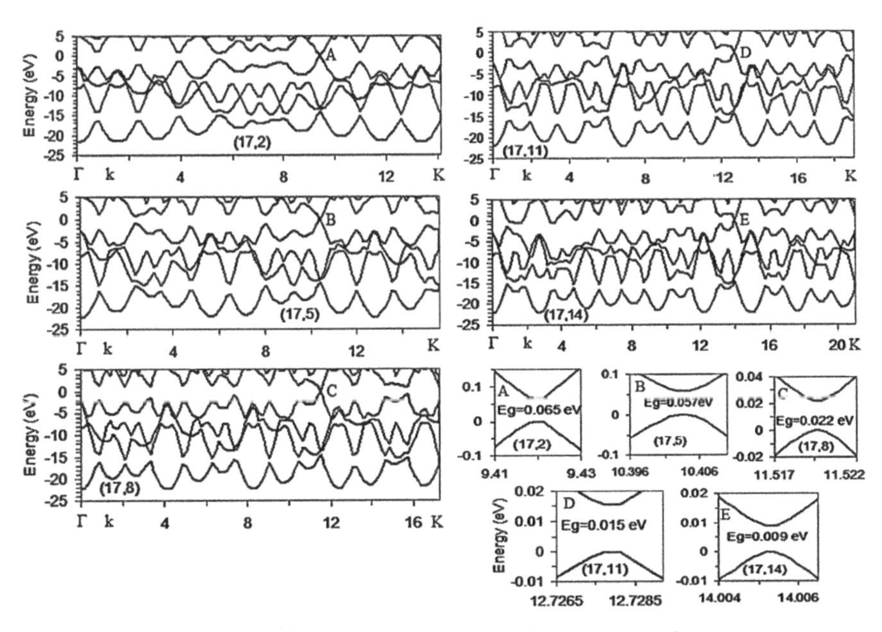

Fig. 3.27. Band structures of the (17, n_2) nanotubes.

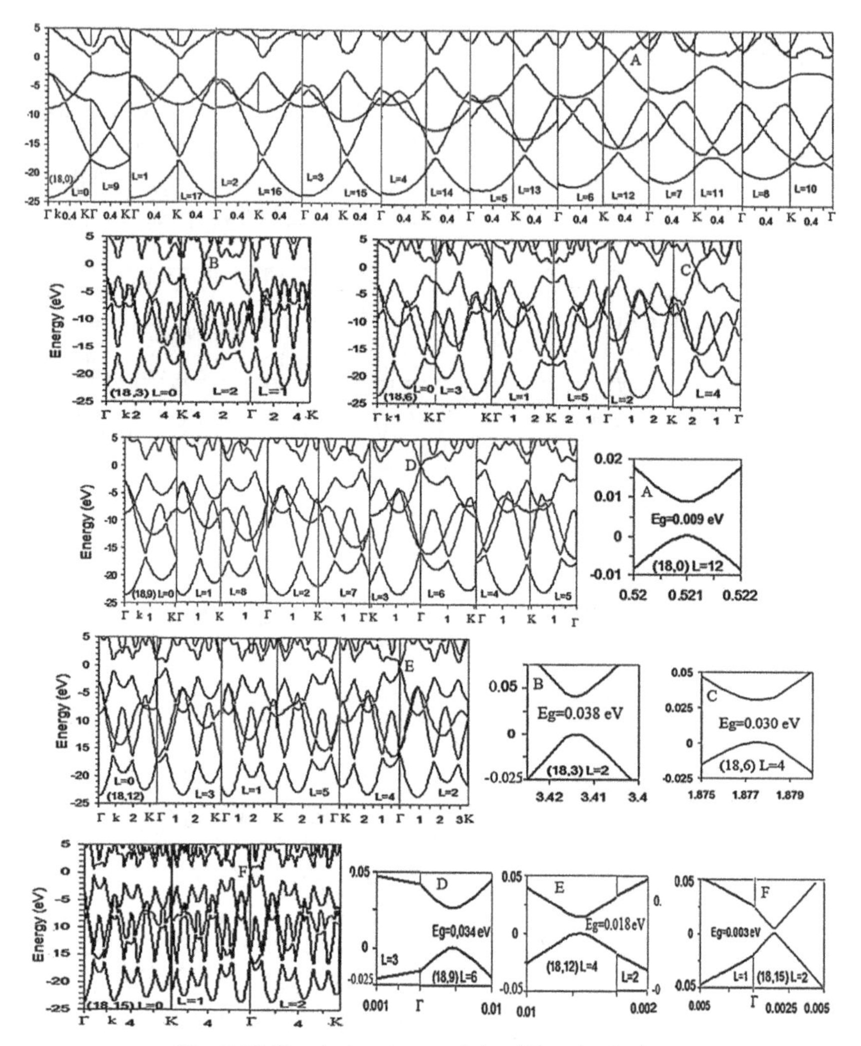

Fig. 3.28. Band structures of the (18, n_2) tubules.

Figure 3.29 shows the plots, where the trends in E_g versus the tubules radii and chirality can be easily followed.

In conclusion, the band structures of many chiral and achiral nanotubes are determined. In the Fermi energy region, the formation of energy gaps depending on the tubule's geometry is demonstrated. The nanotubes with optical gaps falling within the THz range are identified that can be of importance for application of nanotubes in the THz nanodevices. The further developments of this theory to study the point defects as well as an account of spin-orbit coupling effect and applications to the noble metal's nanotubes are presented in the next two chapters.

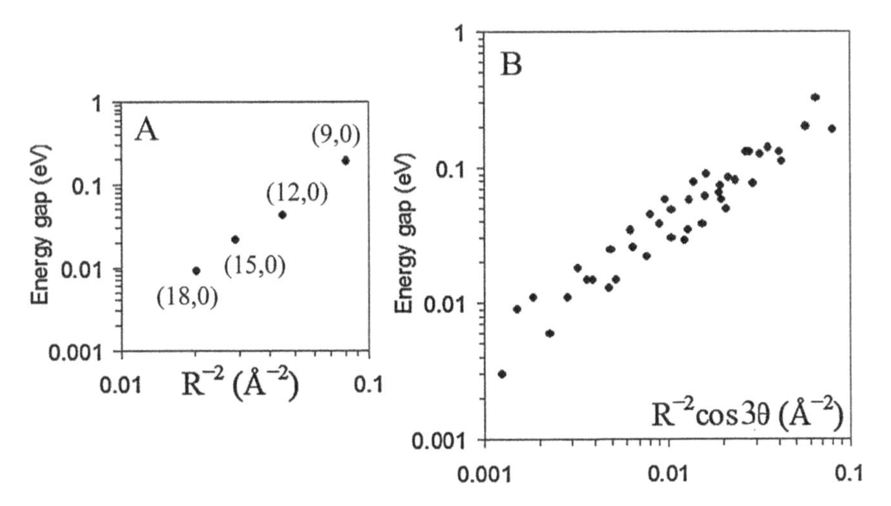

Fig. 3.29. The trends in E_g versus radii and chirality of tubules. (a) zigzag tubules; (b) chiral tubules.

CHAPTER

4

Augmented Cylindrical Wave Green's Function Technique for Impurities

4.1. Defects in Nanotubes

In their simplest form, the defect-free single-walled carbon nanotubes are the hexagonal networks of covalently bound carbon atoms in various cylindrical structures with helical and rotational symmetries. The nanotubes with perfect honeycomb carbon arrangements have emerged as the attractive materials for molecular electronic applications because the nanotubes can be either one-dimensional metals, semiconductors, or even quantum wires depending on their diameter and chirality (Ebbesen 1996, Dekker 1999, Tans et al. 1997). Based on the nanotubes, the single-electron and field-effect transistors, chemical sensors, emission, electromechanical, electromagnetic devices, nanotube radio, and computer have been realized experimentally (Tans et al. 1998, Postma et al. 2001, Kong et al. 2000, Modi et al. 2003, Jensen et al. 2007, 2008, Xiao et al. 2008, Rinkiö et al. 2009, Kuemmeth et al. 2008, Shulaker et al. 2013). Carbon nanotubes are famous for their almost perfect structure; the high-quality nanotubes are confirmed to contain only one defect per 4 mm on average (Fan et al. 2005). However, some defects are still present in the nanotubes. The nanotubes may have various atomic-scale point defects such as the vacancies and impurities in the carbon network, adatoms on the tubule's walls, kinks, junctions, as well as the different topological defects such as pentagon-heptagon pairs, all of which can appear during the nanotube growth or can be created by external action. The electron and light irradiations are successful in creating individual atomic-scale nanotube vacancies (Krasheninnikov 2007, Hashimoto et al. 2004, Osváth et al. 2006, 2007); it has been suggested to dope nanotubes with the H, B, and N atoms using the ion irradiation too (Kotakoski et al. 2005). The experimental reversible creation and annihilation of defects on the nanotubes with the tip of a Scanning Tunnelling Microscope (STM) have been reported (Berthe et al. 2007). The presence of structural defects results in a change in the nanotubes electronic structure, transport properties, optical absorption, specific heat, magnetic susceptibility, and even a single defect can have tremendous electronic effects in one-dimensional conductors (Bockrath et al. 2001, Odom et al. 2000). A local modification of the electronic structure of carbon nanomaterials is important for the development of the carbon-based nanoelectronics. Particularly, the point defects in the nanotubes can act as the gate-tunable electron scatters, and nanotubes with defects can be the basis for new types of electronic devices (Tans et al. 1998, Odom et al. 2000). Inducing defect sites in the nanotubes structure may give rise to complex functional devices such as single-electron transistors operable at room temperature (Postma et al. 2001). The defects can control the operation of nanotube-based chemical sensors (Robinson et al. 2006). Moreover, the defects can impede the adsorption of gases inside a bundle of carbon nanotubes (Gordillo 2006) and give rise to irradiation-mediated pressure build up inside nanotubes (Sun et al. 2006). The defective nanotubes could be used as catalysts and could facilitate thermal dissociation of water (Kostov et al. 2005). The dangling bonds of vacancy can provide active sites for atomic absorption or serve as a bridge of chemical connection between two tubes. Finally, the structural defects play a major role in toxicity of the nanotubes (Fenoglio et al. 2008).

Understanding how imperfections influence the electronic behavior of materials is of fundamental importance. The non-ideal nanotubes are therefore intensively studied. In individual metallic nanotubes, defects can be successfully characterized by transport measurements and scanned gate microscopy because a resonant electron scattering by defects in the nanotube is varied by the gate voltage (Bockrath et al. 2001). STM and Scanning Tunneling Spectroscopy (STS) are the methods which provide direct information both on the atomic and local electronic structure of the carbon nanotubes (Tapasztó et al. 2008, Tolvanen et al. 2009). Defects are easily detected in Raman spectroscopy, because they break the symmetry of the nanotube, relax the momentum conservation rules that govern Raman scattering processes, and locally stiffen the lattice. These spectra can be used to differentiate between positively and negatively charged defects. It also reveals that phonons and electrons associated with doped tubes behave differently from their brethren in unperturbed carbon nanotubes (Maciel et al. 2008, Freitag 2008). In the individual nanotube, using a selective electrochemical method, one can label the point defects and make them easily visible for quantitative analysis; a sequence of electrochemical potentials applied to the nanotube can selectively nucleate a metal deposition at the sites of highest chemical reactivity. The signatures of defects can be detected with X-ray photoelectron spectroscopy by monitoring changes in the 1*s* C peak shape, which is sensitive to the type of carbon bonding, and with the electron spin resonance method (Krasheninnikov 2007).

The foundation stones of the nanotubes defects electron-structure theory were laid down in terms of the π-electronic tight-binding studies (Charlier et al. 1996a, b, Crespi et al. 1997, Chico et al. 1998, Kostyrko et al. 1999a, b). The calculations based on the first models have shown that structural defects in the underlying carbon lattice can substantially modify the electronic properties of nanotubes due to the formation of the defect states and resonant electron scattering at corresponding energies. For example, it was predicted using the tight-binding π-band approximation with surface Green's function matching method that introduction of isolated pentagon, heptagon, or the single pentagon-heptagon pair defects into the hexagonal network of the nanotube can change the helicity of the tube and result in a formation of nanoscale metal/semiconductor or semiconductor/semiconductor junctions (Lambin et al. 1995). The tight-binding recursion model has shown that the one, two, and three pentagon-heptagon pair topological defects in the hexagonal network of the nanotubes form resonant states sharp peaks in the DOS and govern the electronic behavior around the Fermi level E_F (Charlier et al. 1996 a, b). A similar approach shows that a pure carbon quantum dot can be designed by introducing several pentagon-heptagon defects in the nanotubes and predicts the energies of discrete levels in these systems. According to the tight-binding supercell data, the topological bond rotation Stone-Wales defects close the gap in large-gap nanotubes, open the gap in small-gap nanotubes, and increase the DOS in metallic tubules (Chico et al. 1996, Crespi 1997 et al.). For the simplest possible defect, a single vacancy, the quantum conductance was calculated as a function of tube radius within the Landauer formalism in the π-tight-binding scheme (Igami et al. 1999, Choi and Ihm 1999, Neophytou et al. 2007). Investigated in terms of the π-electronic recursion and Green's function approach, the calculated DOS and STS images of both metallic and semiconducting nanotubes predict the vacancy-induced states at the Fermi energy and hillock-like features in the atomically resolved STM images (Krasheninnikov 2001, Krasheninnikov et al. 2001). The π-tight-binding transfer-matrix method was used to calculate the reflection coefficient for a barrier created by point defects in the armchair and zigzag nanotubes (Kostyrko et al. 1999a, b).

The π-electronic and transport properties of the nanotubes with defects were also studied within a *kp* scheme and Green's function scattering formalism for a model strong short-range potential (Ando et al. 1999). The result obtained in this scheme is shown to agree essentially with the tight-binding models (Ando 2005).

Going beyond the π-electron theory, the effects of vacancies, substitutional boron, or nitrogen impurities, and local topological defects on the DOS and quantum conductance of the metallic (10, 10) nanotube were first calculated within the framework of DFT using an *ab initio* pseudopotential method and a plane-wave basis set (Choi et al. 2000). The DOS and conductance have shown quite different behavior than the prediction from the π-electron model. In the case of vacancies, the tight-binding model predicts the single DOS peak exactly at the Fermi level (Anantram and Govindan 1998); however, the electron-hole symmetry is no longer valid in the realistic calculation, and the position of the peak moves. Moreover, the two narrow peaks originating from the broken σ bonds around the vacancy not found in the π-tight-binding model were obtained in the pseudopotential calculations. For the substitutional impurities and Stone-Wales defect, the DOS peaks are located away from the E_F too, and the conductance close to E_F is not significantly affected by the defects, thus showing that the conductance at the E_F is quite robust with respect to the intratube local defects. According to similar *ab initio* pseudopotential calculations combined with the STM data, there is no significant modification of the π-band DOS due to B doping; the B-related acceptor states could only be detected in the σ band (Carroll et al. 1998). In this case, the observed prominent acceptor-like feature near the Fermi level and closing of band gap of the semiconducting tubes are explained in terms of nanodomains of dopant islands, but not by the isolated B substitutional atoms. Presumably, the N-donor structure is located around 0.35 eV above the Fermi level (Carroll et al. 1998). The DFT pseudopotential all-electron code with atom-centered numerical and plane-wave basic functions applied to vacancies in the armchair (n, n) $n = 4, 6, 8$ and zig-zag (10, 0) nanotubes shows that the positions of vacancies states, their population, and electrical activity depend on the nanotube's diameter (Carlsson 2006). In the work (Tolvanen et al. 2009), the local DOS of the single, double, and triple vacancies, of the one and two adatoms on wall of tubule, of the vacancy/adatom complexes, as well as of the Stone-Wales defects in the semiconducting (10, 0) nanotube were studied using virtually the same first-principles DFT technique. The results are compared with the low-temperature scanning tunnelling microscopy and spectroscopy of the Ar^+-irradiated nanotubes. According to these data, in some cases, not the new states in gap can be observed due to the impacts of energetic ions, but changes in the local DOS in the valence and conduction bands only. On the contrary, other defects give rise to the single and double peaks in the band-gap region or complex multipeak configurations with nonzero intensity almost in the entire gap region. Finally, the pseudopotential supercell calculations for electric field dependence of the electronic and structural properties of the defective and radially deformed nanotubes are reported. The calculations show that band structure of the defective nanotubes varies quite differently on the applied electric field and strain from that of the perfect nanotubes (Tien et al. 2005, Shtogun and Woods 2009a,b, 2010). The nonlinear elastic properties of the radially deformed and defective (8, 0) tubule were also investigated.

All the *ab initio* calculations have serious disadvantages because they were performed using a supercell model exhibiting a periodic arrangement of defects in the nanotube, the tubule with defects being arranged in a bulk periodic lattice. As a typical example, the nanotubes composed of 2–6 unit cells and having 80–240 atoms were used for simulations of the (10, 0) nanotube with defects in the work (Tolvanen et al. 2009). The supercell calculations necessarily include interaction effects between the periodically arranged defects and tubules. Moreover, the chiral nanotubes have prohibitively large unit cells which made impossible simulations of defects in these tubules using the standard *ab initio* supercell models. Finally, the ideas of the Green's function theory for electron structure of defects in the tubules were lost in the *ab initio* supercell calculations. The Green's function approach used already in the original π-electron models is more appropriate to treat the problem since it takes full advantage of the periodicity of the host nanotube structure and short range of the defect potential. Note that the Green's function method is widely used in the *ab initio* calculations of defects electronic structure in the bulk materials (Clogston 1962, Beeby 1967, Zeller and Dederichs 1979, Braspenning et al. 1984, Stepanyuk et al. 1990, Dederichs et al. 2006, Farberovich et al. 2008).

In this chapter, we present a method for treating the electronic structure of point defects in the nanotubes which is a generalization of the LACW theory of the band structure of perfect nanotubes above presented. The method avoids using the supercell and superlattice geometries and combines the advantages of density-functional *ab initio* theory with the Green's function approach to the point defects electronic structure theory of bulk materials (D'yachkov et al. 2010).

The proposed method of calculating nanotubes with point impurities is that the band structure of the ideal nanotube is first calculated using the LACW technique. The results are used to calculate the Green's function of the perfect nanotube using a spectral representation of this function. The Green's function of point impurity is then evaluated in terms of the matrix Dyson equation. Knowledge of the Green's function of an impurity system allows one to directly determine its electronic properties. The electron characteristics of the boron and nitrogen impurities in the carbynes, chiral and achiral metallic, semimetallic, and semiconducting carbon nanotubes show the applications of this new technique.

4.2. Theory

4.2.1. Geometry

The perfect defect-free nanotubes can be constructed by rolling up a single graphite sheet (Fig. 1.1), and they can be labeled by the pair of integers (n_1, n_2). At the same time, their geometry can be obtained by the repeated screw operations $S(h, \omega)$ representing the translations h (Eq. 3.4) together with the rotations ω (Eq. 3.5) plus $\nu - 1$ successive $2\pi/\nu$ rotations about the C_ν rotational axis of tubule, where ν is the largest common divisor of n_1 and n_2. There are only two atoms in the minimum cell of the perfect nanotubes, and in this chapter we use the two indices (n, α) to specify a particular carbon atom of the tubule, where n is the number of the two-atomic cell and $\alpha = 1, 2$ corresponds to the number of the atom in the cell. As previously, we apply the one-electron Hamiltonian and cylindrical MT approximation for electron potential. In calculations of the nanotubes with substitutional boron and nitrogen impurities, we neglect a possible lattice relaxation in the defect regions because the covalent radii of the boron (0.82 Å) and nitrogen (0.75 Å) atoms differ not too much from that of the carbon (0.77 Å) and the carbon nanotubes lattice is known to be very rigid. In this approximation, the atomic coordinates of the ideal tubule can be also used for the nanotubes with point impurities. Note that according to the pseudopotential data (Nevidomskyy et al. 2003), the equilibrium position of nitrogen atom is almost unchanged with respect to the corresponding C atoms in the undoped nanotubes, being moved by at most 0.01 Å. Here we also do not change the radii a and b of the potential barriers Ω_a and Ω_b in comparison to the radii used for the pristine carbon systems.

4.2.2. Cylindrical Waves in MT Regions

We assume that the band structure of ideal tubule has already been calculated by the MT LACW method, as described in the previous chapter; namely, solutions of the Schrödinger equation in the form (3.50-3.52)

$$\hat{H}\psi_\lambda(\mathbf{r}|\mathrm{k},\Lambda) = E_\lambda(\mathrm{k},\Lambda)\psi_\lambda(\mathbf{r}|\mathrm{k},\Lambda), \tag{4.1}$$

$$\psi_\lambda(\mathbf{r}|\mathrm{k},\Lambda) = \sum_{PMN} a^\lambda_{PMN}(\mathrm{k},\Lambda)\Psi_{PMN}(\mathbf{r}|\mathrm{k},\Lambda) \tag{4.2}$$

are obtained. The one-electron eigenfunctions $\psi_\lambda(\mathbf{r}|\mathrm{k},\Lambda)$ of the ideal nanotube corresponding to the energy eigenvalues $E_\lambda(\mathrm{k},\Lambda)$ are written as the linear combinations of the basic functions $\Psi_{PMN}(\mathbf{r}|\mathrm{k},\Lambda)$, the coefficients a^λ_{PMN} being determined from the secular equations. Here, λ is a band index; $0 \leq \mathrm{k} \leq \pi/h$ and $\Lambda = 0, 1, ..., \nu - 1$ are the helical quantum numbers; the indices $P = 0, \pm 1, \ldots, M = 0, \pm 1, \ldots$ and $N = 1, 2, \ldots$ number the basic cylindrical waves. (To avoid misunderstandings, we note that only in this chapter will we use the letter ν to denote the order of rotation axis of an ideal nanotube, the letter Λ for the rotational quantum number, and the letter λ for the zone number; in other sections, the letters n, L, and i or

j are used). The Hamiltonian $\hat{H} = -\Delta + V(\mathbf{r})$ contains the kinetic-energy operator $-\Delta$ and the operator $V(\mathbf{r})$ describing the summed action on the electron in consideration of all the other electrons in the system and all its nuclei. Similar to the band structure calculations of ideal nanotubes presented in previous chapters, we also apply here the cylindrical MT and local density approximations with Slater exchange for $V(\mathbf{r})$ (Slater 1974, Hohenberg and Kohn 1964, Kohn and Sham 1965). The MT spheres touch but do not overlap; the sphere radius is kept the same for the carbon, and impurity boron, and nitrogen atoms.

The formulas (3.29-3.37) for calculating basic functions are given in previous chapter. Here, we need the form $\Psi_{PMN}(\mathbf{r}\,|\,\mathrm{k},\Lambda)\big|_{\mathbf{r}\in\Omega_{n\alpha}}$ of these functions only for the regions of MT spheres of ideal tubes. In terms of spherical coordinates r, θ, ϕ and notations used in this chapter, these functions are written explicitly as follows:

$$\Psi_{PMN}(\mathbf{r}\,|\,\mathrm{k},\Lambda)\big|_{\mathbf{r}\in\Omega_{n\alpha}} = \frac{\left(r_{n\alpha}^{MT}\right)^2}{\sqrt{2\pi h / v}}(-1)^{\Lambda+vM} \exp i\left\{\left[\mathrm{k}+\mathrm{k}_P-(\Lambda+vM)\frac{\omega}{h}\right]Z_{n\alpha}\right.$$
$$\left.+\left(\Lambda+vM\right)\Phi_{n\alpha}\right\}\times\sum_{m=-\infty}^{+\infty}\left[C_{M,N}^{J,\Lambda}J_{m-(\Lambda+vM)}\left(\kappa_{|\Lambda+vM|,N}R_{n\alpha}\right)\right.$$
$$\left.+C_{M,N}^{Y,\Lambda}Y_{m-(\Lambda+vM)}\left(\kappa_{|\Lambda+vM|,N}R_{n\alpha}\right)\right]$$
$$\sum_{l=-|m|}^{\infty}(-1)^{0.5(m+|m|)+l}\,i^{l}\left[\frac{(2l+1)(l-|m|)!}{(l+|m|)!}\right]^{1/2}$$
$$\times\left[a_{lm,n\alpha}^{PMN}\left(r_{n\alpha}^{MT}\,\middle|\,\mathrm{k},\Lambda\right)u_l^{n\alpha}\left(r,E_l^{n\alpha}\right)+\right.$$
$$\left.b_{lm,n\alpha}^{PMN}\left(r_{n\alpha}^{MT}\,\middle|\,\mathrm{k},\Lambda\right)\dot{u}_l^{n\alpha}\left(r,E_l^{n\alpha}\right)\right]Y_{lm}(\hat{r}). \quad (4.3)$$

4.2.3. Scattering Matrix

In this chapter, a scattering t matrix

$$t_l^{n\alpha}(E) = \int_0^{r_{n\alpha}} j_l\left(\sqrt{E}r\right)V_{n\alpha}(r)u_l^{n\alpha}(r,E)r^2dr, \quad (4.4)$$

which characterizes the potential $V_{n\alpha}(r)$ within the MT spheres in terms of angular momentum representation, will also be used. Here, $u_l^{n\alpha}(r,E)$ is the energy dependent solution of the radial Schrödinger equation (1.28) in the MT region of (n, α) atom. As usually in the linear methods (Andersen 1975, Koelling and Arbman 1975), we calculate the energy dependent functions $u_l^{n\alpha}(r,E)$ in (Eq. 4.4) using the exact solutions $u_l^{n\alpha}\left(r,E_l^{n\alpha}\right)$ at the fixed energy $E_l^{n\alpha}$ (Eq. 1.27) and their energy derivative $\dot{u}_l^{n\alpha}\left(r,E_l^{n\alpha}\right)$ at $E_l^{n\alpha}$ as follows.

$$u_l^{n\alpha}(r,E)=u_l^{n\alpha}\left(r,E_l^{n\alpha}\right)+\left(E-E_l^{n\alpha}\right)\dot{u}_l^{n\alpha}\left(r,E_l^{n\alpha}\right). \tag{4.5}$$

As an example, Figs. 4.1 and 4.2 show the numerically calculated radial dependences of the potentials inside the MT sphere and the energy and angular momentum dependences of the scattering t matrices of the carbon, boron, and nitrogen atoms for the particular case of (5, 5) nanotube.

The MT potential $V(r)$ decreases as one goes from the boron to carbon and from carbon to nitrogen, and the product $V(r)r$ is equal to the nuclear charge at $r = 0$. The MT spheres of the carbon, boron, and nitrogen atoms contain 3.62, 2.74, and 4.91 electrons, respectively. Only about 10% of the electron density falls on the interspherical region. Figure 4.2 shows that the absolute values of the scattering matrix elements differences for the C, B, and N atoms decrease strongly with increase in the l. In the interspherical region Ω_{IS}, the potential is constant (zero-point energy); it does not change as one goes from the pristine and doped tubules.

4.2.4. One-Electron Green's Function

In the band-structure calculations, the solution of the one-electron Schrödinger equation $\hat{H}\Psi_j = E_j\Psi_j$ for the single-particle wave functions and corresponding energies represents the central problem. However, the calculation of the wave functions and energies can be avoided, if instead the single-particle Green's function $G(\mathbf{r},\mathbf{r}';E)$, which is the solution of the Schrödinger equation with a source at position $\mathbf{r}'$

$$\{-\Delta + V(\mathbf{r}) - E\}G(\mathbf{r},\mathbf{r}'; E) = -\delta(\mathbf{r} - \mathbf{r}') \tag{4.6}$$

is determined (Dederichs et al. 2006, Mavropoulos and Papanikolaou 2006).

In terms of a complete set of the eigenfunctions $\Psi_j(\mathbf{r})$ corresponding to the eigenvalues E_j, the following spectral representation for the retarded Green's function can be obtained:

$$G(\mathbf{r},\mathbf{r}',E)=\sum_j \frac{\Psi_j(\mathbf{r})\Psi_j^*(\mathbf{r}')}{E-E_j+i\varepsilon} \tag{4.7}$$

representing, in the limit of $\operatorname{Im}E=\varepsilon\to 0^+$, an outgoing wave at $\mathbf{r}$ with a source term at $\mathbf{r}'$. From the above equation and identity

$$\frac{1}{E-E_j+i\varepsilon}=-i\pi\delta\left(E-E_j\right)+\tilde{P}\frac{1}{E-E_j} \tag{4.8}$$

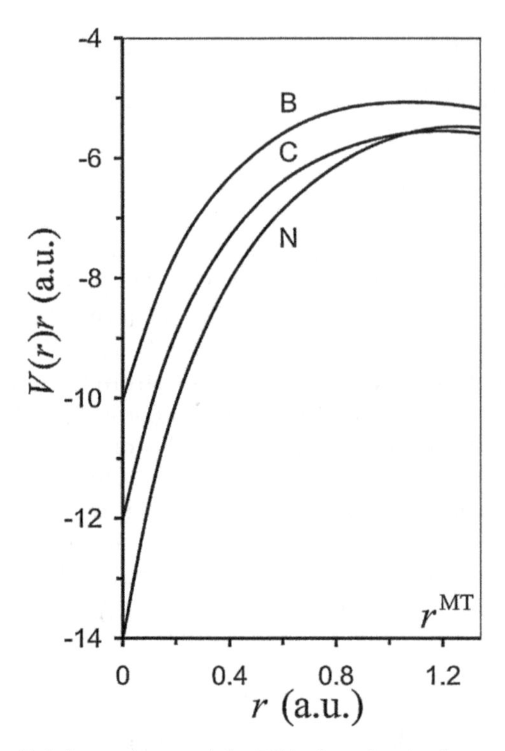

Figure 4.1. Radial dependence of the *V*(*r*)*r* function in the region of MT spheres of the carbon, boron, and nitrogen atoms of the (5, 5) nanotube. Here and in Fig. 4.2, Rydberg atomic units (a.u.) are used.

where $\tilde{P}\left(E-E_j\right)^{-1}$ is called the Cauchy principal value defined as follows:

$$\int_{-\infty}^{+\infty} f(E')\tilde{P}(E-E')^{-1}\,dE' = \tilde{P}\int_{-\infty}^{+\infty} \frac{f(E')}{(E-E')}dE'$$
$$= \lim_{\varepsilon\to 0}\left(\int_{-\infty}^{E-\varepsilon} \frac{f(E')}{(E-E')}dE' + \int_{E+\varepsilon}^{+\infty} \frac{f(E')}{(E-E')}dE'\right) \quad (4.9)$$

(Simplifying somewhat, we can assume that the function $\tilde{P}(x)^{-1}$ has the following property: $\tilde{P}(x)^{-1} = x^{-1}$ for $x \neq 0$ and $\tilde{P}(x)^{-1} = 0$ for $x = 0$). The expectation value of any physical quantity represented by an operator $\hat{A}$ can be harvested via the relation (Dederichs et al. 2006, Mavropoulos and Papanikolaou 2006)

$$\langle A\rangle = -\frac{1}{\pi}\mathrm{Im}\int_{-\infty}^{E_F} \mathrm{Tr}\left[AG(E)\right]dE\,, \quad (4.10)$$

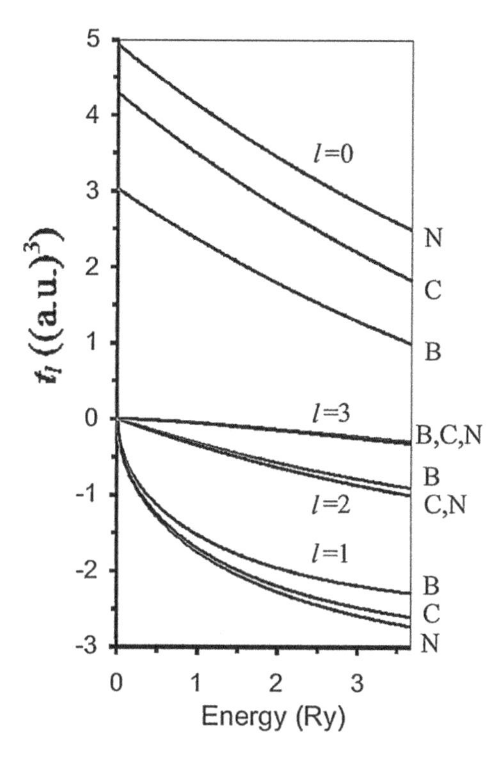

Figure 4.2. Energy dependence of the scattering t matrix of the carbon, boron, and nitrogen atoms of the (5, 5) nanotube.

if the matrix elements and trace in the right-hand side of this equation are calculated using the full basic set. Therefore, the Green's function contains all information which is given by the eigenfunctions, and if the Green's function can be computed, then all physical properties of the system can be found. Particularly, the imaginary part of $G(\mathbf{r}, \mathbf{r}; E)$ is directly related to the spectrally and space-resolved density of states

$$\rho(\mathbf{r},E) = -\frac{1}{\pi}\operatorname{Im} G(\mathbf{r},\mathbf{r};E) \tag{4.11}$$

and the local density of states

$$\rho_V(E) = -\frac{1}{\pi}\int_V \operatorname{Im} G(\mathbf{r},\mathbf{r};E)d\mathbf{r} \tag{4.12}$$

of a volume V is obtained by integrating over this volume. The real problem is the evaluation of the Green's function for the system of interest; in our case, it is the nanotube with point defect.

The equation for the Green's function $\tilde{G}(\mathbf{r},\mathbf{r}';E)$ of the nanotube with impurity can be written as

$$(\hat{H}-E)\tilde{G}(\mathbf{r},\mathbf{r}';E) = -\delta(\mathbf{r}-\mathbf{r}')-\Delta V(\mathbf{r})\tilde{G}(\mathbf{r},\mathbf{r}';E) \quad (4.13)$$

Here, $\hat{H}$ is the Hamiltonian of the perfect nanotube and $\Delta V(\mathbf{r}) = \tilde{V}(\mathbf{r})-V(\mathbf{r})$ is the difference between the potentials of the impurity $\tilde{V}(\mathbf{r})$ and perfect tubule $V(\mathbf{r})$. It follows from this equation that the Green's function $\tilde{G}(\mathbf{r},\mathbf{r}';E)$ corresponding to the new Hamiltonian $\hat{H}+\Delta V(\mathbf{r})$ is related to the Green's function $G(\mathbf{r},\mathbf{r}';E)$ corresponding to $\hat{H}$ via the Dyson integral equation

$$\tilde{G}(\mathbf{r},\mathbf{r}';E)=G(\mathbf{r},\mathbf{r}';E)+\int G(\mathbf{r},\mathbf{r}'';E)\Delta V(\mathbf{r}'')\tilde{G}(\mathbf{r}'',\mathbf{r}';E)d\mathbf{r}''. \quad (4.14)$$

Most important is that the perturbed potential $\Delta V(\mathbf{r})$ is well localized near the impurity, while the perturbed wave functions $\tilde{\Psi}(\mathbf{r};E)$ accurately described by the Lippmann-Schwinger equation

$$\tilde{\Psi}(\mathbf{r};E)=\Psi(\mathbf{r};E)+\int G(\mathbf{r},\mathbf{r}';E)\Delta V(\mathbf{r}')\tilde{\Psi}(\mathbf{r}';E)d\mathbf{r}' \quad (4.16)$$

are not localized.

4.2.5. Green's Function for Array of MT Spheres

Consider the Green's function of an array of spherically symmetric non overlapping potentials. The potential is given by

$$V(\mathbf{r}-\mathbf{R}_{n\alpha}) = V(Z+t_s h, \Phi + t_s\omega + t_\nu\omega_\nu) = V_{n\alpha}(\mathbf{r}), \quad (4.17)$$

$$t_s = 0, \pm 1, \pm 2, \ldots; t_\nu = 0, 1, \ldots, \nu - 1,$$

and the Green's function is defined via

$$(-\Delta + V_{n\alpha}(\mathbf{r}) - E)G(\mathbf{r}+\mathbf{R}_{n\alpha}, \mathbf{r}'+\mathbf{R}_{n'\alpha'};E) = -\delta_{n\alpha,n'\alpha'}\delta(\mathbf{r}-\mathbf{r}'). \quad (4.18)$$

In the mixed site angular momentum representation, by introducing the atom-centered coordinates, we can present the Green's function in the form of a series expansion based on the solutions of the Schrödinger equation in the MT spheres (Korringa 1947, Kohn and Rostoker 1954, Zeller and Dederichs 1979, Braspenning et al. 1984, Stepanyuk et al. 1990. Dederichs et al. 2006, Mavropoulos and Papanikolaou 2006)

$$G(\mathbf{r}+\mathbf{R}_{n\alpha},\mathbf{r}'+\mathbf{R}_{n'\alpha'};E) = -i\delta_{n,n'}\delta_{\alpha,\alpha'}\sqrt{E}\sum_L u_l^{n\alpha}(r_<,E)Y_L(\check{\mathbf{r}})H_l^{n\alpha}(r_>,E)Y_L^*(\hat{\mathbf{r}}')+$$

$$\sum_{L,L'} u_l^{n\alpha}(r,E)Y_L(\hat{\mathbf{r}})G_{L,L'}^{n\alpha,n'\alpha'}(E)u_{l'}^{n'\alpha'}(r',E)Y_{L'}^*(\hat{\mathbf{r}}'). \quad (4.19)$$

Here, $\mathbf{R}_{n\alpha}$ and $\mathbf{R}_{n'\alpha'}$ are the positions of atoms α and α' in the cells n and n', respectively; $\mathbf{r}$ and $\mathbf{r}'$ restricted to the MT spheres are the local coordinate vectors of atoms (n,α) and (n',α'); $r_< = \min(r,r')$ and $r_> = \max(r,r')$; $L = (l,m)$

are the orbital quantum numbers; $G_{L,L'}^{n\alpha,n'\alpha'}(E)$ are the energy-dependent coefficients of the Green's function; $H_l^{n\alpha}(r,E)=u_l^{n\alpha}(r,E)+iN_l^{n\alpha}(r,E)$. Finally, the $u_l^{n\alpha}$ is regular (converging at $r\to 0$) solution of the radial Schrödinger Eq. (1.28) and the $N_l^{n\alpha}$ is irregular (diverging at $r\to 0$) solution of this equation.

In Eq. (4.19), the first term represents the Green's function for the scattering problem by the central potential in vacuum and the second one characterizes the effects of the nanotube structure. By construction, the expression in Eq. (4.19) for the Green's function satisfies in each MT sphere (n, α) the general solution of the Schrödinger Eq. (4.18) for the Green's function while the matrix $G_{L,L'}^{n\alpha,n'\alpha'}(E)$, the so-called structural Green's function, describes the connection of the solutions in the different spheres and thus contains all the information about the multiple-scattering problem, which is in this way reduced to the solution of an algebraic problem.

4.2.6. Structural Green's Function

4.2.6.1. Perfect Nanotube with Screw and Rotational Symmetries

Now, let us determine the coefficients of the Green's function $G_{L,L'}^{n\alpha,n'\alpha'}(E)$ for the perfect nanotube having the screw and rotational symmetries, the band structure and wave functions of the tubule being determined in terms of the symmetrized LACW method described in previous chapter. Let us multiply Eq. (4.19) by $u_l^{n\alpha}(r,E)r^2drY_{L_1}^*(\hat{H})d\hat{r}$ and $u_{l'}^{n'\alpha'}(r',E)r'^2dr'Y_{L_2'}^*(\hat{r}')d\hat{r}'$ and integrate over MT regions of the atoms (n, α) and (n', α'). Then, we have

$$G_{L,L'}^{n\alpha,n'\alpha'}(E)\alpha_l^{n\alpha}(E)\alpha_{l'}^{n'\alpha'}(E)=i\delta_{n,n'}\delta_{\alpha,\alpha'}\delta_{L,L'}\sqrt{E}\left[\alpha_l^{n\alpha}(E)\right]^2+$$
$$\int_{\Omega_{n\alpha}}\int_{\Omega_{n'\alpha'}} G(\mathbf{r}+\mathbf{R}_{n\alpha},\mathbf{r}'+\mathbf{R}_{n'\alpha'};E)Y_L^*(\hat{r}')Y_{L'}(\hat{r}')$$
$$\times u_l^{n\alpha}(r,E)u_{l'}^{n'\alpha'}(r',E)r^2r'^2drdr'd\hat{r}d\hat{r}', \qquad (4.20)$$

where

$$\alpha_l^{n\alpha}(E)=\int_0^{r_{n\alpha}}[u_l^{n\alpha}(r,E)]^2r^2dr=1+(E-E_l^{n\alpha})^2N_{l,\alpha.}, \qquad (4.21)$$

$$N_{l,\alpha.}=\int_0^{r_{n\alpha}}[\dot{u}_l^{n\alpha}(r,E_l^{n\alpha})]^2r^2dr\,. \qquad (4.22)$$

In order to calculate the integral in Eq. (4.20), let us apply the spectral representation of Green's function

$$G(\mathbf{r},\mathbf{r}';E)=\frac{h}{2\pi}\sum_{\lambda}\sum_{\Lambda=0}^{\nu-1}\int_{-\pi/h}^{\pi/h}\frac{\psi_\lambda(\mathbf{r}\,|\,\mathrm{k},\Lambda)\psi_\lambda^*(\mathbf{r}'\,|\,\mathrm{k},\Lambda)}{E-E_\lambda(\mathrm{k},\Lambda)+i\varepsilon}d\mathrm{k}. \tag{4.23}$$

Substituting in Eq. (4.21) the wave functions (4.3) for $\mathbf{r}\in\Omega_{n\alpha}$ and $\mathbf{r}'\in\Omega_{n'\alpha'}$, we finally obtain

$$G_{L,L'}^{n\alpha,n'\alpha'}(E)=i\delta_{n,n'}\delta_{\alpha,\alpha'}\delta_{L,L'}\sqrt{E}+\frac{\nu\left(r_{n\alpha}^{MT}r_{n'\alpha'}^{MT}\right)^2}{4\pi\alpha_l^{n\alpha}(E)\alpha_{l'}^{n'\alpha'}(E)}i^{l-l'}(-1)^{0.5(m+|m|+m'+|m'|)+l+l'}$$
$$\times\left[\frac{(2l+1)(l-|m|)!(2l'+1)(l'-|m'|)!}{(l+|m|)!(l'+|m'|)!}\right]^{1/2}\sum_{PMN}\sum_{P'M'N'}(-1)^{\nu(M+M')}\times$$
$$\sum_{\Lambda=0}^{\nu-1}\exp\left\{i\left[\left(\mathrm{k}_P-(\Lambda+\nu M)\omega/h\right)Z_{n\alpha}+(\Lambda+\nu M)\Phi_{n\alpha}\right]\right\}\times$$
$$\exp\left\{-i\left[\left(\mathrm{k}_{P'}-(\Lambda+\nu M')\omega/h\right)Z_{n'\alpha'}+(\Lambda+\nu M')\Phi_{n'\alpha'}\right]\right\}\times$$
$$\left[C_{M,N}^{J,\Lambda}J_{m-(\Lambda+\nu M)}\left(\kappa_{|\Lambda+\nu M|,N}R_{n\alpha}\right)+\right.$$
$$\left.C_{M,N}^{Y,\Lambda}Y_{m-(\Lambda+\nu M)}\left(\kappa_{|\Lambda+\nu M|,N}R_{n\alpha}\right)\right]\times$$
$$\left[C_{M',N'}^{J,\Lambda}J_{m'-(\Lambda+\nu M')}\left(\kappa_{|\Lambda+\nu M'|,N}R_{n'\alpha'}\right)+\right.$$
$$\left.C_{M',N'}^{Y,\Lambda}Y_{m'-(\Lambda+\nu M')}\left(\kappa_{|\Lambda+\nu M'|,N}R_{n'\alpha'}\right)\right]\times$$
$$\int_{-\pi/h}^{\pi/h}\exp\left[i\mathrm{k}\left(Z_{n\alpha}-Z_{n'\alpha'}\right)\right]\left[a_{lm,n\alpha}^{PMN}(r_{n\alpha}\,|\mathrm{k},\Lambda)+\right.$$
$$\left.\left(E-E_l^{n\alpha}\right)N_l^{n\alpha}b_{lm,n\alpha}^{PMN}(r_{n\alpha}\,|\mathrm{k},\Lambda)\right]\times\left[\bar{a}_{l'm',n'\alpha'}^{P'M'N'}(r_{n'\alpha'}\,|\mathrm{k},\Lambda)+\right.$$
$$\left.\left(E-E_{l'}^{n'\alpha'}\right)N_{l'}^{n'\alpha'}\bar{b}_{l'm',n'\alpha'}^{P'M'N'}(r_{n'\alpha'}\,|\mathrm{k},\Lambda)\right]\times$$
$$\sum_{\lambda}\left[-i\pi\delta\left[E-E_\lambda(\mathrm{k},\Lambda)\right]+\tilde{P}\left(\frac{1}{E-E_\lambda(\mathrm{k},\Lambda)}\right)\right]$$
$$a_{PMN}^{\lambda}(\mathrm{k},\Lambda)\bar{a}_{P'M'N'}^{\lambda}(\mathrm{k},\Lambda)d\mathrm{k}. \tag{4.24}$$

We also give without deriving the equation for the calculation of the structural Green's function for nanotubes without a screw, but with translational symmetry, as well as for monatomic wires.

4.2.6.2. Perfect Nanotubes with Translational Symmetry

When taking into account only the translational symmetry of the nanotube, the structural Green's function of perfect nanotube takes the following form

$$G_{L,L'}^{n\alpha,n'\alpha'}(E)=i\delta_{n,n'}\delta_{\alpha,n'}\delta_{L,L'}\sqrt{E}+\frac{\left(r_{n\alpha}r_{n'\alpha'}\right)^2}{2\pi\alpha_{l,n\alpha}(E)\alpha_{l',n'\alpha'}(E)}i^{l-l'}$$
$$\times(-1)^{0.5(m+|m|+m'+|m'|)+l+l'}\times\left[\frac{(2l+1)(l-|m|)!(2l'+1)(l'-|m'|)!}{(l+|m|)!(l'+|m'|)!}\right]^{1/2}$$
$$\times\sum_{PMN}\sum_{P'M'N'}(-1)^{M+M'}\times\exp\left[i\left(\mathrm{k}_pZ_{n\alpha}-\mathrm{k}_{p'}Z_{n'\alpha'}+M\Phi_{n\alpha}-M'\Phi_{n'\alpha'}\right)\right]$$
$$\times\left[C_{MN}^{J}J_{m-M}\left(\kappa_{|M|,N}R_{n\alpha}\right)+C_{MN}^{Y}Y_{m-M}\left(\kappa_{|M|,N}R_{n\alpha}\right)\right]$$
$$\times\left[C_{M'N'}^{J}J_{m'-M'}\left(\kappa_{|M'|N'}R_{n'\alpha'}\right)+C_{M'N'}^{Y}Y_{m'-M'}\left(\kappa_{|M'|N'}R_{n'\alpha'}\right)\right]$$
$$\times\int_0^{\pi/c}dk\exp\left[ik(Z_{n\alpha}-Z_{n'\alpha'})\right]\left[a_{lm,n\alpha}^{MNP,\mathrm{k}}(r_{n\alpha})+\left(E-E_{l,n\alpha}\right)\right.$$
$$\left.\times N_{l,n\alpha}b_{lm,n\alpha}^{MNP,\mathrm{k}}(r_{n\alpha})\right]\times\left[\bar{a}_{l'm',n'\alpha'}^{M'N'P',\mathrm{k}}(r_{n'\alpha'})+\left(E-E_{l',n'\alpha'}\right)\right.$$
$$\left.\times N_{l',n'\alpha'}\bar{b}_{l'm',n'\alpha'}^{M'N'P',\mathrm{k}}(r_{n'\alpha'})\right]\times\sum_{\lambda}\left[-i\pi\delta(E-E_{k\lambda})+\right.$$
$$\left.\tilde{P}\left(E-E_{k\lambda}\right)^{-1}\right]\bar{a}_{M'N'P'}^{k\lambda}a_{MNP}^{k\lambda}. \tag{4.25}$$

4.2.6.3. Perfect Linear Atomic Chains

There is no inner cylindrical cavity in the linear atomic chains; however, the structural Green's function for the achiral case (4.25) can be easily rewritten for this simpler case

$$G_{L,L'}^{n\alpha,n'\alpha'}(E)=i\delta_{n,n'}\delta_{\alpha,n'}\delta_{L,L'}\sqrt{E}+\frac{c\left(r_{n\alpha}r_{n'\alpha'}\right)^2}{\Omega\alpha_{l,n\alpha}(E)\alpha_{l',n'\alpha'}(E)}i^{l-l'}\times$$
$$(-1)^{0.5(m+|m|+m'+|m'|)+l+l'}\left[\frac{(2l+1)(l-|m|)!(2l'+1)(l'-|m'|)!}{(l+|m|)!(l'+|m'|)!}\right]^{1/2}\times$$
$$\sum_{PMN}\sum_{P'M'N'}(-1)^{M+M'}\left[J'_M\left(\kappa_{|M|,N}a\right)J'_{M'}\left(\kappa_{|M'|,N'}a\right)\right]^{-1}$$
$$\times\exp\left[i\left(k_pZ_{n\alpha}-k_{p'}Z_{n'\alpha'}\right)\right]\times\int_0^{\pi/c}dk\exp\left[ik(Z_{n\alpha}-Z_{n'\alpha'})\right]\left[a_{lm,n\alpha}^{MNP,k}(r_{n\alpha})\right.$$
$$\left.+\left(E-E_{l,n\alpha}\right)N_{l,n\alpha}b_{lm,n\alpha}^{MNP,k}(r_{n\alpha})\right]\times\left[\bar{a}_{l'm',n'\alpha'}^{M'N'P',k}(r_{n'\alpha'})+\left(E-E_{l',n'\alpha'}\right)\right.$$
$$\left.\times N_{l',n'\alpha'}\bar{b}_{l'm',n'\alpha'}^{M'N'P',k}(r_{n'\alpha'})\right]\times\sum_{\lambda}\left[-i\pi\delta(E-E_{k\lambda})+\tilde{P}\left(E-E_{k\lambda}\right)^{-1}\right]\bar{a}_{M'N'P'}^{k\lambda}a_{MNP}^{k\lambda}. \tag{4.26}$$

4.2.6.4. Nanotubes with Impurity

Once the structural Green's function $G_{L,L'}^{n\alpha,n'\alpha'}(E)$ of the perfect nanotube is known, the Green's function for the nanotube with impurity can be evaluated in terms of the matrix Dyson equation

$$\tilde{G}_{L,L'}^{n\alpha,n'\alpha'}(E) = G_{L,L'}^{n\alpha,n'\alpha'}(E) + \sum_{n'',\alpha''}\sum_{L''} G_{L,L''}^{n\alpha,n''\alpha''}(E)\Delta t_{l''}^{n''\alpha''}(E)\tilde{G}_{L'',L'}^{n''\alpha'',n'\alpha'}(E). \quad (4.27)$$

The $\Delta t_{l''}^{n''\alpha''}(E) = \tilde{t}_{l''}^{n''\alpha''} - t_{l''}^{n''\alpha''}$ are the differences between the t matrices in the perturbed and perfect nanotubes determined by the perturbation of the potential well localized in the MT spheres of the impurity atoms. Since this difference is restricted to the vicinity of the impurity, the Green's function in this subspace can be easily determined in real space by matrix inversion. The rank of the matrices to be inverted is given by $n_d(l_{\max}+1)^2$; here, n_d is the number of perturbed MT potentials in a region of point impurity, and $l_{\max}$ is the maximum angular momentum used in the calculations. In our work, for the single impurities, we neglect the perturbation of the neighboring host atoms and take into account in Eq. (4.27) only the perturbation due to the impurity potential; this so-called single-site approximation is known to give a quite reasonable description of the electronic structure of the impurities in crystals (Zeller and Dederichs 1979, Braspenning et al. 1984, Stepanyuk et al. 1990, Dederichs et al. 2006). Figure 4.2 shows that the absolute values of the scattering matrix elements perturbations $\Delta t_l^B = \tilde{t}_l^B - t_l^C$ and $\Delta t_l^N = \tilde{t}_l^N - t_l^C$ decrease strongly with increase in the angular momentum l. This results in rapid convergence of the Green's function and electronic properties of the perturbed nanotubes. Particularly, $l_{\max} = 2$ or even $l_{\max} = 1$ give good convergence of the calculated electronic DOS in the regions of impurities.

Substituting the structural function $G_{L,L'}^{n\alpha,n'\alpha'}(E)$ to Eq. (4.19), we can obtain the Green's function $\tilde{G}_{L,L'}^{n\alpha,n'\alpha'}(E)$ for calculating the physical properties of the nanotube with impurity. In this work, we apply this approach to the particular case of the local electronic DOS corresponding to the MT region of the impurity atoms,

$$\rho_{MT,n\alpha}(E) = \frac{1}{\pi}\sum_L \alpha_l^{n\alpha}(E)\left[\sqrt{E} - \operatorname{Im}\tilde{G}_{L,L'}^{n\alpha,n'\alpha'}(E)\right]. \quad (4.28)$$

4.3. Applications

4.3.1. Boron and Nitrogen Impurities in Carbon Chains

Before turning to the calculations of the single-walled nanotubes, let us look at the impurity-related levels in a carbyne, which is the simplest carbon nanowire with cylindrical symmetry. The polyynic carbyne is a linear chain of carbon atoms with alternating single and triple bonds. The

LACW band structure shows that the polyynic carbyne is the direct gap semiconductor. Similar to the semiconducting nanotubes, the π bonding and antibonding states of the polyynic carbyne form the top of valence and the bottom of conduction bands.

The local DOS in the carbon MT region of the perfect polyynic carbyne is visualized in Fig. 4.3 as the imaginary part of the Green's function calculated by the LACW method.

The local DOS in the MT regions of the boron and nitrogen impurities of the doped polyynic carbyne calculated in terms of the Dyson equation are plotted in this figure too. Both the boron and nitrogen impurities close the gap between the valence and conduction bands, the local DOS in this region being larger for the boron atom in comparison with the nitrogen one. The nitrogen impurity virtually does not influence the Van Hove

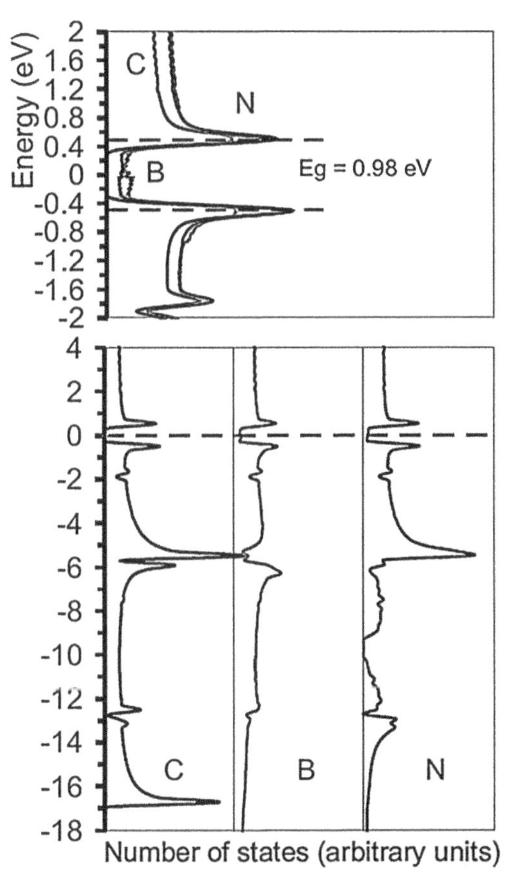

Figure 4.3. Local DOS of polyynic carbyne in the band-gap region (upper panel) and from the bottom of s band up to conduction band (lower panel). Here and below: perfect system (C), the boron (B), and nitrogen (N) impurities.

singularities located at +0.5 and –0.5 eV relative to the Fermi energy and corresponding to the gap edges of the ideal system, but the introduction of the boron atom results in a decrease in these peaks. In addition to the states near the gap, the DOS of the perfect polyynic carbyne forms a double peak centered at –5.5 and –6 eV. In the case of the local DOS of boron impurity, there is not the peak, but a well-defined dip in this region; for the nitrogen, a noticeable smoothing of the resonance is observed. The peak of the local DOS at –17 eV corresponding to the bottom of the s band is absent in the case of both impurities.

The cumulenic carbyne is linear chain of carbon atoms with the double bonds having the metallic band structure and DOS (Fig. 4.4).

Similar to the case of metallic single-walled nanotube, the Fermi level crosses the band separating the low-energy bonding and high-energy antibonding π states. If the boron or nitrogen atom takes the place of carbon, the local DOS at the Fermi level increases to 27 or 16%, respectively. Only the nitrogen defect gives rise to the new extremely narrow and high peak

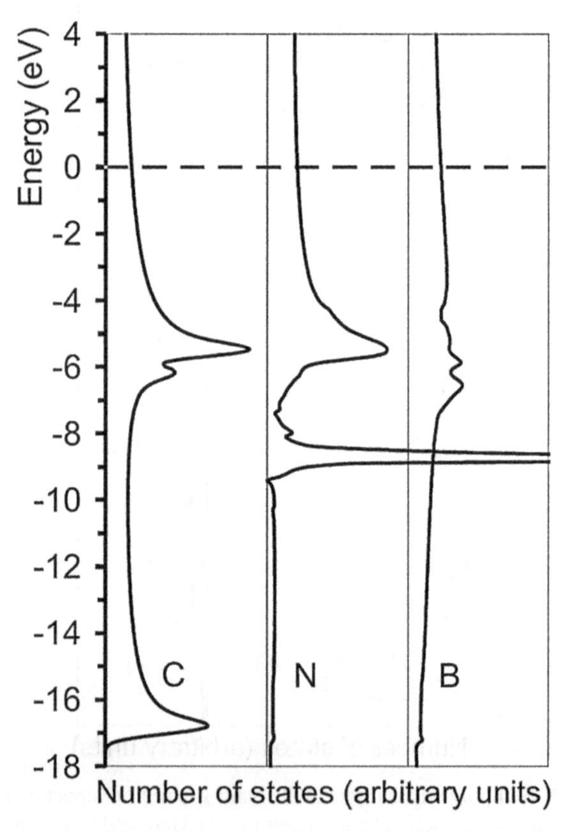

Figure 4.4. Local DOS of pure and B- and N-doped cumulenic carbyne.

about −8.5 eV. A smearing of band between 4 and 7 eV specifies the effect of boron.

4.3.2. Boron and Nitrogen Impurities in Carbon Tubules

Now let us discuss the effects of the boron and nitrogen impurities in a series of the chiral and achiral, wide-gap and low-gap semiconducting, semimetallic, and metallic nanotubes; the band structures of these tubules were already plotted in our previous chapter. The purely carbon (7, 7) nanotube with armchair geometry has a metallic electronic structure with constant DOS in the energy region between −0.7 and +0.7 eV relative to the Fermi level (Fig. 4.5).

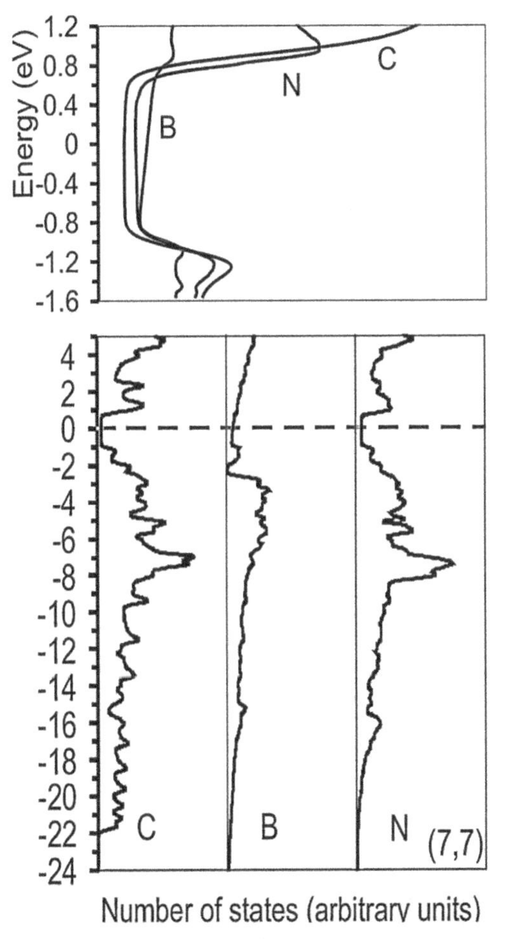

Figure 4.5. Local DOS of the perfect and B- and N-doped (7, 7) armchair nanotubes. Local DOS in the Fermi energy region (upper panel) and from the bottom of *s* band up to conduction band (lower panel).

Close to the E_F, the electronic structure is significantly affected by the impurities. However, the boron and nitrogen defects do not destroy the metallic character of DOS. In this region, the main effect of the nitrogen impurity is about 50% virtually constant increase in the DOS; in the case of boron, there is further growth of the DOS. Comparison of these results with data for (5, 5) and (10, 10) armchair tubule shows that the impurity-induced perturbations of the DOSs are very similar independent of the diameter of armchair nanotubes (Figs. 4.6. and 4.15).

The defect-free chiral (9, 6) nanotube belongs to the quasi-metallic family. According to the LACW band-structure data, there is a gap of 0.035 eV between the occupied and unoccupied states due to the nanotube curvature effects (Fig. 4.7).

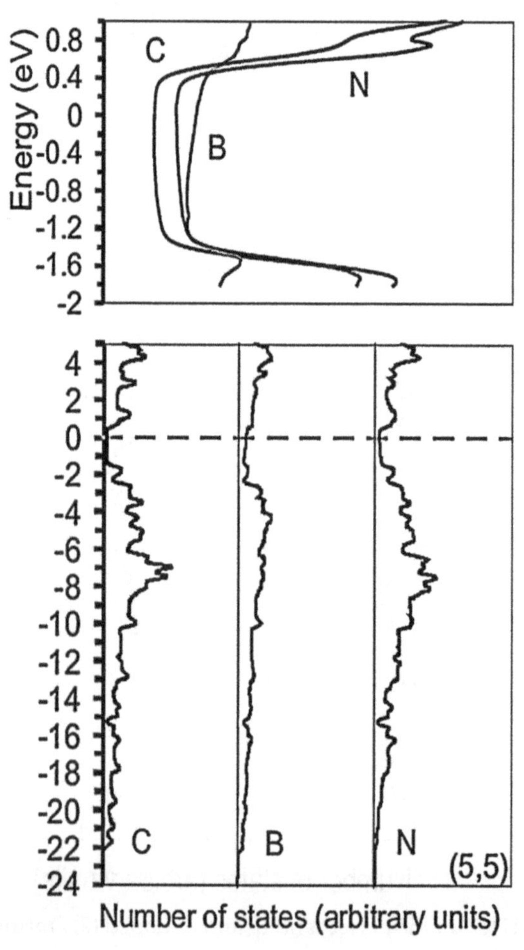

Figure 4.6. Local DOS of the perfect and B- and N-doped (5, 5) armchair nanotubes.

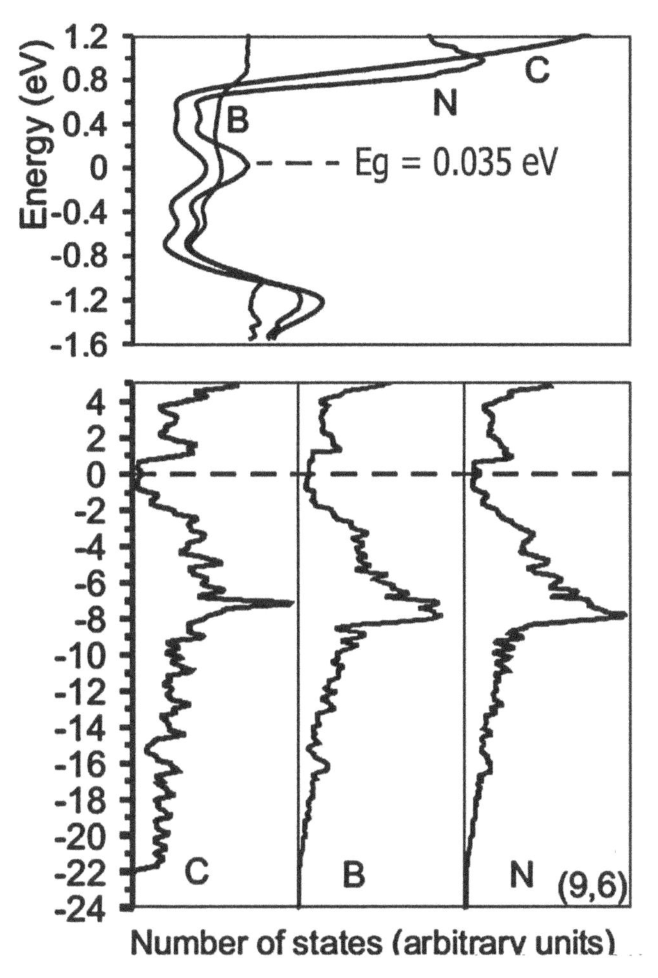

Figure 4.7. Local DOS of the perfect chiral (9, 6) quasi-metallic nanotubes and its boron- and nitrogen-doped derivatives.

On the boron and nitrogen substitution, the local DOS in a region of E_F increases. The nitrogen impurity gives the largest DOS at E_F. The boron defect smoothens the three-peak structure between −0.5 and +0.5 eV. The perfect chiral (8, 2) nanotube also belongs to the quasi-metallic family; however, in this case, a gap with $E_g = 0.32$ eV is formed because of the very large curvature of small-diameter tubule (Ouyang et al. 2001). As a result, there is no peak, but a dip in the DOS of the ideal system in Fig. 4.8. The dip at the E_F retains in the DOS of both the boron and nitrogen dopants in spite of total increase in the local DOS in the region between −1.0 and +0.5 eV relative to the Fermi energy.

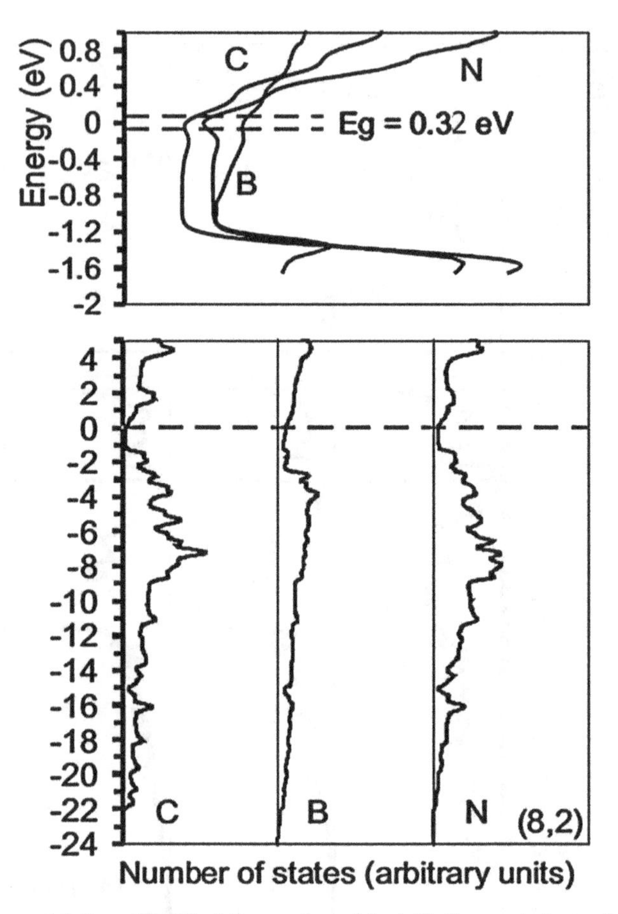

Figure 4.8. Local DOS of the perfect chiral (8, 2) nanotube and its boron- and nitrogen-doped derivatives.

Figures 4.9-4.12 exhibit the influences of the boron and nitrogen impurities on the electronic DOSs of the four chiral semiconducting nanotubes (10, 5), (12, 4), (11, 3), and (8, 7) with the diameters of 10 Å and different chirality angles, the perfect chiral (11, 3) and (8, 7) nanotubes having as many as 652 and 676 atoms in the translational unit cells.

In all these semiconducting nanotubes, in the vicinity of optical gap, a drastic difference between the effects of the two types of impurities should be emphasized. The boron-related states clearly close but those of the nitrogen atoms do not close the gap of the perfect tubules. In the gap region, the effects of nitrogen atom are restricted with a minor growth of the local DOSs just below and above the E_F.

Figures 4.13 and 4.14 show that this is also true for the semiconducting zigzag nanotubes (10, 0) and (13, 0).

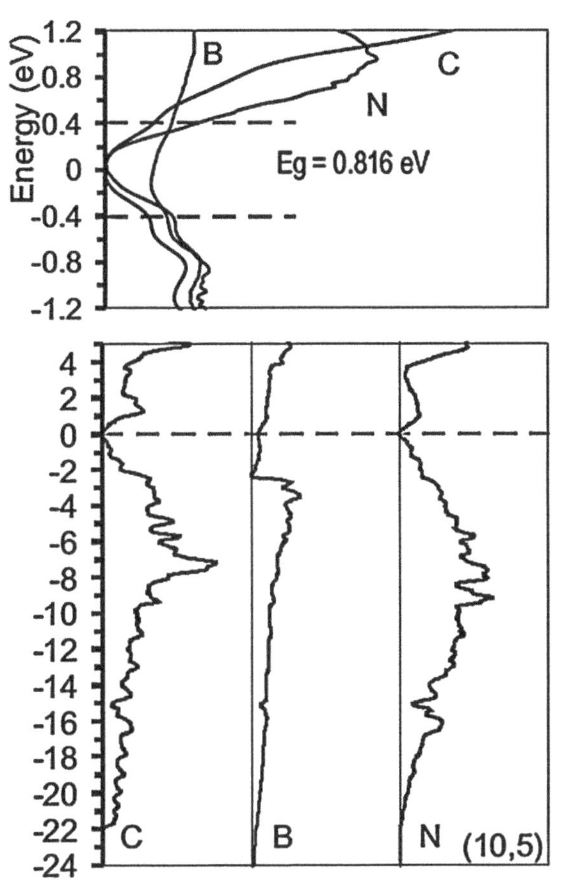

Figure 4.9. Local DOS of the perfect chiral semiconducting (10, 5) nanotube and its boron- and nitrogen-doped derivatives.

Beyond the Fermi-energy region up to the s bottom of the valence bands, the effects of impurities are more or less similar in all the tubules. As one goes from carbon to the boron, the local DOS within the MT sphere decreases, and the peaks almost disappear. Generally, the effects of nitrogen defect are opposite; the nitrogen local DOS is somewhat greater than that of the carbon, and there is no significant smoothing of the DOS picture.

It would be very interesting to compare the results of the Green's function LACW technique designed for a single defect to those obtained by the usual supercell plane-wave pseudopotential DFT method for array of impurities. Unfortunately, the possibilities of this comparison are greatly limited because there are the pseudopotential calculations of the achiral nanotubes only. The unit cells of the chiral nanotubes contain too

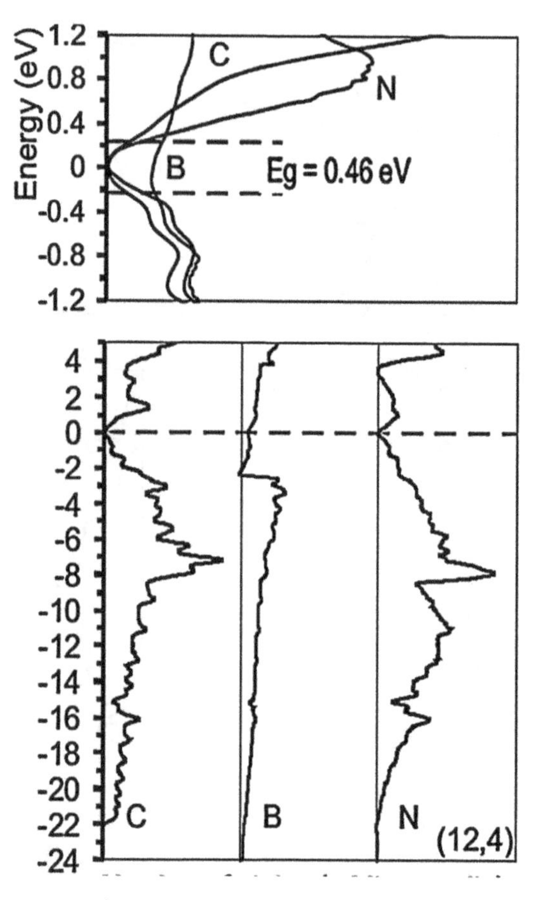

Figure. 4.10. Local DOS of the perfect chiral semiconducting (12, 4) nanotubes and their boron- and nitrogen-doped derivatives.

many atoms for the plane-wave pseudopotential method, which suffers from a slow convergence and an unfavorable scaling (Skylaris et al. 2005). For example, in the paper by Choi et al. (2000), the electronic structure of the (10, 10) nanotube with boron and nitrogen impurities was calculated in terms of the *ab initio* plane-wave nonlocal pseudopotential method, the supercell of 10–20 Å in each direction being used.

Figure 4.15 shows the local DOS for the Fermi-energy region of the boron- and nitrogen-doped (10, 10) nanotubes calculated using this approach together with the Green's function LACW data.

The overall structures of the local DOSs are similar; particularly, both methods predict that introduction of impurities does not result in a formation of the forbidden gaps. However, there are important differences. The pseudopotential calculation of the local DOS around the

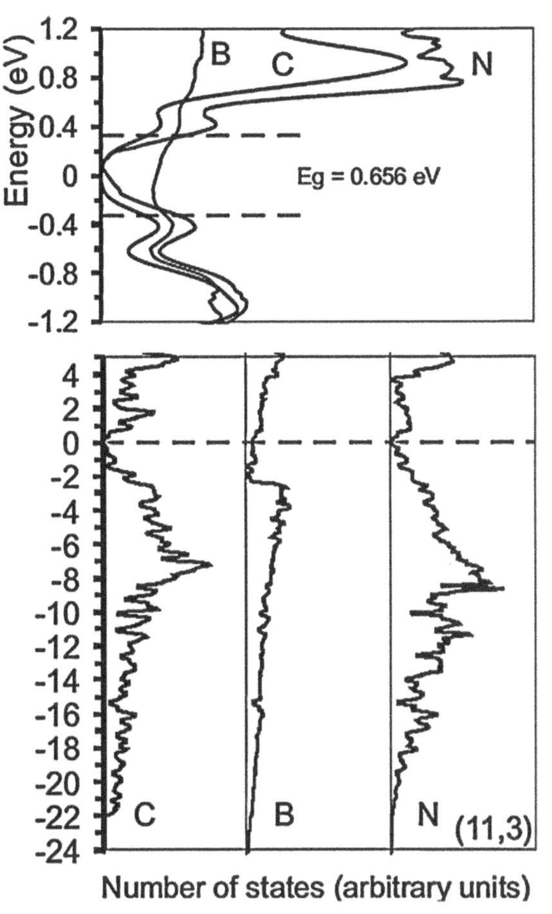

Figure 4.11. Local DOS of the perfect chiral semiconducting (11,) nanotubes and their boron- and nitrogen-doped derivatives.

boron impurity predicts a double peak with maximums at –0.7 and –0.8 eV below the Fermi level as well as a sharp peak above this level at 0.75 eV. In the pseudopotential local DOS of the nitrogen impurity, a position of these bands is opposite to the boron case, the double-peak band is located above and a single peak below the Fermi level. Note that a form of all these bands having very sharp edges is typical for the Van Hove singularities in the one-dimensional periodic systems. This seems to be an artifact of theory determined by the interaction of defects in supercell geometry that is used. It is pointed in the paper by Choi et al. (2000) that the spatial extent of the wave functions corresponding to the impurity-related bound states with energies of –0.7 and 0.5 eV for boron and nitrogen impurities equals to 10 Å; the spatial extent of the wave functions corresponding to the states with binding energies equal to –0.8 eV and +0.8 eV for the

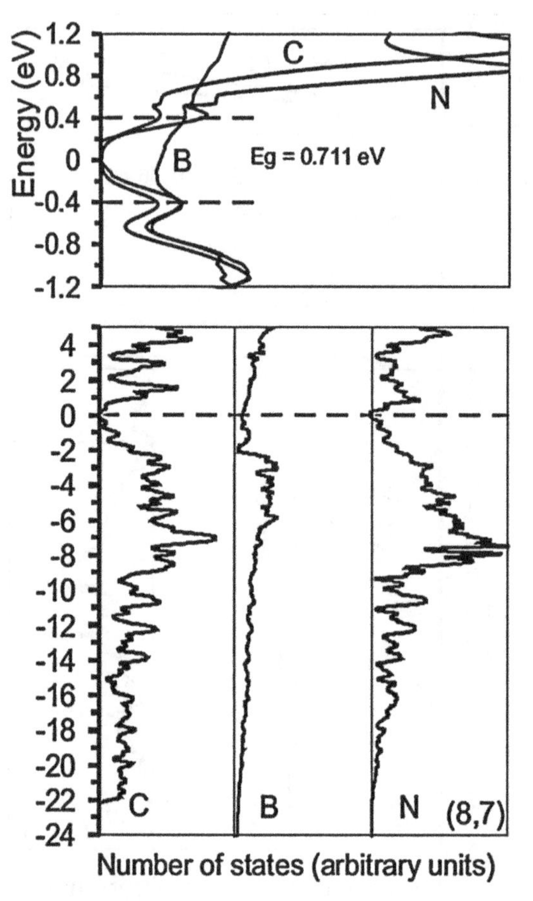

Figure 4.12. Local DOS of the perfect chiral semiconducting (8, 7) nanotube and its boron- and nitrogen-doped derivatives.

boron and nitrogen is roughly 200 Å and 50 Å, respectively. Thus, the delocalization of functions is strong enough to guarantee a coupling between the neighboring impurities. The double-peak structures in the local DOS of impurities can be understood as the bonding and antibonding combinations of the impurity orbitals. In the Green's function LACW DOS, one observes the analogous bands located in the same energy regions as in the case of the plane-wave pseudopotential theory, but the bands are obviously broadened and their splitting disappears as it is expected for the single impurity. Finally, it is to be noted that the plane-wave pseudopotential calculations are restricted with a vicinity of the Fermi level but the LACW approach permits calculating the electronic structure in a wide energy range from the conduction band up to the bottom of the valence *s* band (Fig. 4.15).

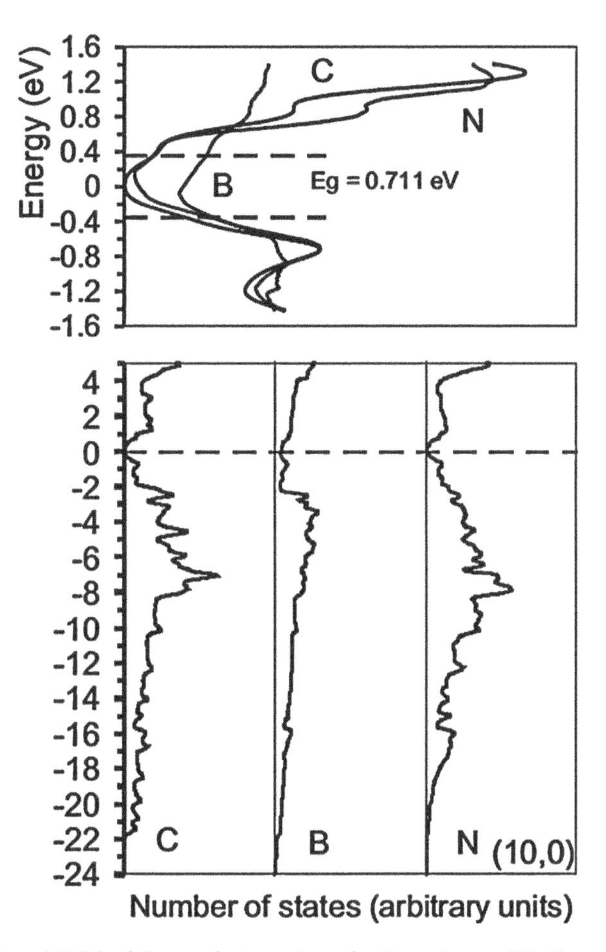

Figure 4.13. Local DOS of the perfect semiconducting zigzag (10, 0) nanotube and its boron- and nitrogen-doped derivatives.

Thus, in this chapter, we present the distinct method for the realistic theoretical studies of electronic structure of the nanotubes with point defects and performed calculations of local DOS for the substitutional boron and nitrogen impurities in the variety of the nanotubes. The calculations are based on the LACW Green's function method and the local-density functional and muffin-tin approximations for the electronic potential. It is of great importance that the *ab initio* method is developed which is applicable to any nanotube including the chiral ones with very large translational unit cells. The method realized in terms of the single-site approximation can be extended to the cases where the potentials of the neighboring atoms are also disturbed and the atomic structure relaxations are allowed.

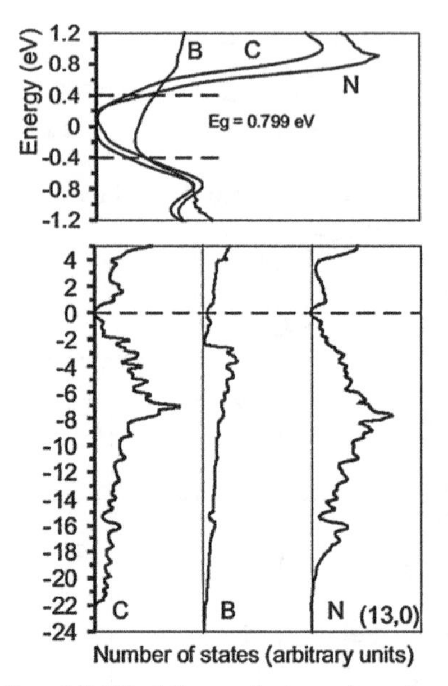

Figure 4.14. Local DOS of the perfect semiconducting zigzag (13, 0) nanotube and its boron- and nitrogen-doped derivatives.

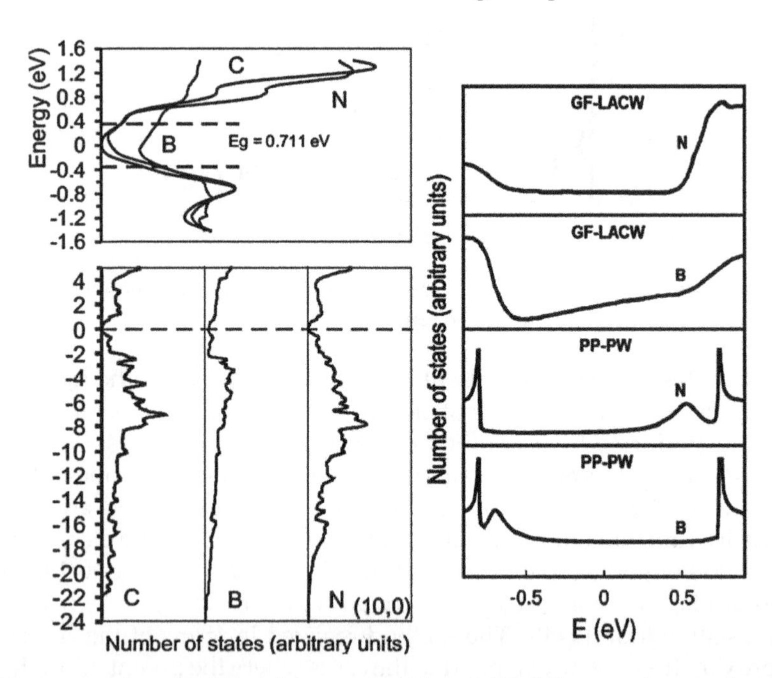

Figure 4.15. Local DOS of the pure and doped (10, 10) nanotube calculated using the plane-wave pseudopotential and Green's function LACW methods.

CHAPTER

5

Relativistic Augmented Cylindrical Wave Method

5.1. Spin-Orbit Interaction in Nanotubes

The single-walled carbon nanotubes are the cylinders composed of the rolled graphene ribbons on which the carbon hexagons are arranged in a helical fashion around the axis. The large amount of theoretical and experimental investigations of nanotubes has brought to light many interesting properties of these materials. Particularly, the interaction of the spin of electrons with their orbital motion has become a focus of great attention in nanotubes research. Due to the cylindrical structure of the carbon nanotubes, the electron states near the gap correspond to the semiclassical clockwise and anticlockwise electron orbits encircling the tubule. In addition to the unique two-fold degree of freedom leading to the doubly degenerate orbits, the electrons can have different spins. In the absence of a spin-orbit coupling, the two-fold orbital degeneracy and two-fold spin degeneracy should yield a four-fold degenerate electronic spectrum in the carbon nanotubes.

Based on low atomic number $Z = 6$ of carbon, the spin-orbit effects were commonly assumed to be negligible in carbon materials; in carbon atom, the spin-orbit splitting $\Delta_{S\text{-}O}$ equals to 6 meV and for flat graphene $\Delta_{S\text{-}O} = 10^{-2}\text{-}10^{-3}$ meV. However, already in 2000, the first theoretical study of spin-orbit interaction in a carbon nanotube was performed using the *kp*-method and simple tight-binding model with a single π orbital at each carbon atom, the spin-orbit effect being considered to the lowest order (Ando 2000). It was shown that the spin-orbit term gives rise to a splitting of energy levels and formation of gaps about 0.1-1 meV. In more recent works (Huertas-Hernando et al. 2006, Chiko et al. 2004, 2009, Izumida et al. 2009, Wunsch 2009, Zhou et al. 2009, Liu 2011), the spin-orbit effects were studied in the curved graphene, fullerene, and nanotubes also in the terms of simple π-electron model and perturbation theory, the importance of curvature of a carbon material surface for larger spin-orbit coupling being indicated again. In these works, the calculations of the curvature induced spin-orbit splitting in the Fermi level region were performed using an empirical tight-binding Hamiltonian.

Remarkably, the electrical measurements of the individual tubules in the presence of axial magnetic field show that the values of orbital magnetic moment μ_{orb} are 10-20 times larger than the Bohr magneton and scale with the tubule diameter D (Minot et al. 2004). The reason of this large moment is the large size of electron orbits encircling the carbon nanotubes compared to the radii of atomic orbitals. Later, the ultra-low-temperature (30 mK) measurements of the intrinsic electronic spectrum of the ultraclean nanotube quantum dot with D = 5 nm have shown that the expected four-fold symmetry is broken by spin-orbit splitting $\Delta_{S\text{-}O}$ equal to 0.37 and 0.21 meV for the electrons and holes, respectively (Kuemmeth et al. 2008, 2010, Ilani and McEuen 2010). More recently, the spin-orbit coupling was revealed by direct current magnetoconductance measurements and spectroscopy studies for the complementary situations of the open carbon nanotubes-quantum wire and of the nanotubes quantum dot in the general multielectron regime and in the presence of finite disorder (Jhang et al. 2010, Jaspersen et al. 2011).

The situation is promising for applications of the spin-orbit interaction effects in carbon nanotubes devices. For example, it is demonstrated that carbon nanotubes are promising candidates for memory devices with fast magnetization switching; they are excellent spin-current waveguides (Guimaraẽs et al. 2010, Wang et al. 2013) and allow spin control (Flensberg and Marcus 2010) and spin filtering (Gunlycke et al. 2006). The spin-orbital coupling enables fast electrical spin manipulation in carbon nanotube spin qubits (Flensberg and Marcus 2010, Bulaev et al. 2008); it can influence the Josephson supercurrents in nanotube superconducting junctions (Lim et al. 2011), allows the spin to couple with the high-quality vibrational modes of nanotubes (Paĺyi et al. 2012, Ohm et al. 2012) and can affect electron spin decoherence in nanotube quantum dots (Kuemmeth et al. 2010).

In the cited theoretical works, the atomic spin-orbit interaction term included in Hamiltonian was given by $H_{S\text{-}O} = \lambda\mathbf{LS}$, where λ is the atomic spin-orbit coupling constant and **L** and **S** stands for the orbital and spin orbital angular momentum of the electron, respectively. The parameters describing the strength of the spin-orbit interaction cannot be determined in these calculations, and the simple tight-binding models used can describe the band structure only qualitatively. The exact value of the spin-orbit coupling parameter is under discussion (Chiko et al. 2009, Izumida et al. 2009). In works by Chico et al. (2004, 2009) some results are presented in the units of λ, and an artificially large value of λ equal to 0.2 eV was chosen in the band structure diagrams; in the Refs (Izumida et al. 2009, Zhou et al. 2009), the λ constants were varied between 0 and 0.2 meV.

For this reason, it is felt that a development of the first-principles technique with due account of the spin-orbit interactions and applicable to any single-walled nanotube independent on diameter and chirality can be of great interest. It is desirable that the new method be closely related to the solid-state band theory; in the case of crystals, the *ab initio* band structure calculations with the relativistic effects have become routine work already some decades ago. Such an approach based on the LACW technique is presented in this chapter. We study the spin-orbit coupling effects in the band structures of a variety of nanotubes including the carbon and noble metal tubules and atomic chains. We start from the first-principle calculations of the Fermi energy gaps in the carbon armchair nanotubes, which are very interesting because formation of the gap between the valence and conduction bands is totally relativistic effect here.

5.2. Theory

A relativistic version of LACW method can be readily obtained on the grounds of the relativistic Augmented Plane-Wave (APW) techniques (Conklin et al. 1965, Loucks 1965, MacDonald et al. 1980, Koelling and Harmon 1977). In Conklin et al. (1965), where the relativistic APW method was developed, let us use the one-electron approximation and start from the two-component relativistic Hamiltonian written using Rydberg units

$$H = -\Delta + V + \frac{1}{c^2}\sigma \cdot \left[(\nabla V) \times \mathbf{p} \right]. \tag{5.1}$$

It is obtained from the Dirac Hamiltonian by application of the Foldy-Wouthuysen transformation and neglecting the Darwin and mass-velocity corrections (Schiff 1949, Davydov 1965). Here, $c = 274$ is the velocity of light in the Rydberg units, V is the electron potential, σ are the Pauli matrices:

$$\sigma_1 = \begin{pmatrix} 0 & 1 \\ 1 & 0 \end{pmatrix}, \ \sigma_2 = \begin{pmatrix} 0 & -i \\ i & 0 \end{pmatrix}, \ \sigma_3 = \begin{pmatrix} 1 & 0 \\ 0 & -1 \end{pmatrix}. \tag{5.2}$$

The sum $H_0 = -\Delta + V(\mathbf{r})$ is the nonrelativistic Hamiltonian used above for the LACW band structure studies. The third term of the Hamiltonian (5.1) is the familiar spin-orbit coupling term $H_{S\text{-}O}$, which may split the degenerate levels and cause mixing of levels thus altering the band picture.

For electron potential $V(\mathbf{r})$, we apply again the Muffin-Tin (MT) and Slater $\rho^{1/3}$ exchange approximations, which dramatically simplify both the nonrelativistic calculations and account of the spin-orbit corrections. Particularly, the $H_{S\text{-}O} = 0$ in the interspherical space, where the potential $V(\mathbf{r})$ is constant. Inside the MT spheres, where the potential

$V(\mathbf{r}) = V_{\alpha_{MT}}(r) \equiv V_{\alpha_{MT}}(\rho)$ is spherically symmetric, it can be written using the angular momentum operators L_+, L_- and L_z

$$H_{S-O} = \frac{1}{c^2}\frac{1}{\rho}\frac{dV_{\alpha_{MT}}(\rho)}{d\rho}\sigma\mathbf{L} = \frac{1}{c^2}\frac{1}{\rho}\frac{dV_{\alpha_{MT}}(\rho)}{d\rho}\left(\frac{1}{2}\sigma_+ L_- + \frac{1}{2}\sigma_- L_+ + \sigma_Z L_Z\right). \quad (5.3)$$

The effect of spin operator on the functions α and β is the following

$$\sigma_+\alpha = 0,\ \sigma_-\alpha = 2\beta,\ \sigma_z\alpha = \alpha,\ \sigma_+\beta = 2\alpha,\ \sigma\beta = 0,\ \sigma_z\beta = -\beta. \quad (5.4)$$

The effect of the angular momentum operators on the spherical harmonics is

$$L_z Y_{lm}(\theta\varphi) = m Y_{lm}(\theta\varphi), \quad (5.5)$$

$$L_\pm Y_{lm}(\theta\varphi) = \left[l(l+1) - m(m\pm1)\right]^{1/2} Y_{lm}(\theta\varphi), \quad (5.6)$$

they do not affect the radial part of the wave function.

5.3. Applications

5.3.1. Achiral Nanotubes

5.3.1.1. Secular Equation

In order to take into account the spin-orbit term in the energy-band calculation for the nanotube, it is necessary to extend the formalism of the LACW method to include the matrix elements of the spin-orbit term between the spinor wave functions. Because the nonrelativistic terms make the largest contribution to the energy, the following procedure is used: First, the eigenfunctions $\Psi^0_{n,k}(\mathbf{r})$ and the eigenvalues $E^0_n(k)$ of the nonrelativistic Hamiltonian are found by the LACW method described in Chapter 1 for the achiral nanotubes. The results of such nonrelativistic calculation are the energies for various bands and the eigenfunctions of the nanotube expressed as a linear combination of the $\Psi^k_{PMN}(\mathbf{r})$ basic functions

$$\Psi^0_{n,k}(\mathbf{r}) = \sum_{PMN} a^{nk}_{PMN}\Psi^k_{PMN}(\mathbf{r}). \quad (5.7)$$

Next, the spinor basic partners $\Psi^0_{n,k}(\mathbf{r})\alpha$ and $\Psi^0_{n,k}(\mathbf{r})\beta$ are to be formed, where α and β are the pure spin functions. While the matrix elements remain to be calculated the spin-orbit operator between the spinor basic functions $\Psi^0_{n,k}(\mathbf{r})\chi$

$$\left\langle \Psi^0_{n_2,k}(\mathbf{r})\chi_2 \middle| H \middle| \Psi^0_{n_1,k}(\mathbf{r})\chi_1 \right\rangle = \varepsilon_{n_1}\delta_{n_1,n_2} + \left\langle \Psi^0_{n_2,k}(\mathbf{r})\chi_2 \middle| H_{S-O} \middle| \Psi^0_{n_1,k}(\mathbf{r})\chi_1 \right\rangle$$
$$= \varepsilon_{n_1}\delta_{n_1,n_2} + \sum_{P_2M_2N_2}\sum_{P_1M_1N_1} \bar{a}^{n_2k}_{P_2M_2N_2} a^{n_1k}_{P_1M_1N_1}$$
$$\left\langle \Psi^k_{P_2M_2N_2}(\mathbf{r})\chi_2 \middle| H_{S-O} \middle| \Psi^k_{P_1M_1N_1}(\mathbf{r})\chi_1 \right\rangle. \tag{5.8}$$

Application of the equations for $\Psi^k_{PMN}(\mathbf{r})$ obtained in Chapter 1 for the case of achiral nanotubes together with Eqs. (5.3-5.6) and orthogonality and normalization of the spin functions α and β and of the spherical harmonics, allows the angular integration to be performed analytically to give finally (D'yachkov and Kutlubaev 2012)

$$\left\langle \Psi^k_{P_2M_2N_2}(\mathbf{r})\alpha \middle| H_{S-O} \middle| \Psi^k_{P_1M_1N_1}(\mathbf{r})\alpha \right\rangle = -\left\langle \Psi^k_{P_2M_2N_2}(\mathbf{r})\beta \middle| H_{S-O} \middle| \Psi^k_{P_1M_1N_1}(\mathbf{r})\beta \right\rangle, \tag{5.9}$$

$$\left\langle \Psi^k_{P_2M_2N_2}(\mathbf{r})\alpha \middle| H_{S-O} \middle| \Psi^k_{P_1M_1N_1}(\mathbf{r})\beta \right\rangle = \left\langle \Psi^k_{P_2M_2N_2}(\mathbf{r})\beta \middle| H_{S-O} \middle| \Psi^k_{P_1M_1N_1}(\mathbf{r})\alpha \right\rangle^*, \tag{5.10}$$

$$\left\langle \Psi^k_{P_2M_2N_2}(\mathbf{r})\alpha \middle| H_{S-O} \middle| \Psi^k_{P_1M_1N_1}(\mathbf{r})\alpha \right\rangle = \frac{1}{2c^2 d}(-1)^{M_1+M_2} \times$$
$$\sum_{\alpha_{MT}} r^4_{\alpha_{MT}} \sum_{l=0}^{\infty}(2l+1)\sum_{m=-l}^{l} m\frac{(l-|m|)!}{(l+|m|)!}\times\Big\{\varsigma_{l,\alpha_{MT}}\bar{a}^{P_2M_2N_2,k}_{lm,\alpha_{MT}} a^{P_1M_1N_1,k}_{lm,\alpha_{MT}} +$$
$$\ddot{\varsigma}_{l,\alpha_{MT}}\bar{b}^{P_2M_2N_2,k}_{lm,\alpha_{MT}} b^{P_1M_1N_1,k}_{lm,\alpha_{MT}} + \dot{\varsigma}_{l,\alpha_{MT}}\left(\bar{a}^{P_2M_2N_2,k}_{lm,\alpha_{MT}} b^{P_1M_1N_1,k}_{lm,\alpha_{MT}} + b^{P_2M_2N_2,k}_{lm,\alpha_{MT}} a^{P_1M_1N_1,k}_{lm,\alpha_{MT}}\Big\}\times$$
$$\exp\left\{i\left[\left(K_{p_2}-K_{p_1}\right)Z_{\alpha_{MT}} + (M_1-M_2)\Phi_{\alpha_{MT}}\right]\right\}\times$$
$$\left[C^J_{M_2N_2} J_{m-M_2}\left(\kappa_{|M_2|,N_2} R_{\alpha_{MT}}\right) + C^Y_{M_2N_2} Y_{m-M_2}\left(\kappa_{|M_2|,N_2} R_{\alpha_{MT}}\right)\right]\times$$
$$\left[C^J_{M_1N_1} J_{m-M_1}\left(\kappa_{|M_1|,N_1} R_{\alpha_{MT}}\right) + C^Y_{M_1N_1} Y_{m-M_1}\left(\kappa_{|M_1|,N_1} R_{\alpha_{MT}}\right)\right], \tag{5.11}$$

and

$$\left\langle \Psi^k_{P_2M_2N_2}(\mathbf{r})\beta \middle| H_{S-O} \middle| \Psi^k_{P_1M_1N_1}(\mathbf{r})\alpha \right\rangle = \frac{1}{2c^2 d}(-1)^{M_1+M_2} \times$$
$$\sum_{\alpha_{MT}} r^4_{\alpha_{MT}} \sum_{l=0}^{\infty}(2l+1)\sum_{m=-l}^{l}\left[l(l+1)-m(m+1)\right]^{1/2}\left[\frac{(l-|m|)!(l-|m+1|)!}{(l+|m|)!(l+|m+1|)!}\right]^{1/2}$$
$$\times(-1)^{(m+|m|)/2+[m+1+|m+1|]/2}\times\Big\{\varsigma_{l,\alpha_{MT}}\left(\bar{a}^{P_2M_2N_2,k}_{lm+1,\alpha_{MT}}\right)^*\left(a^{P_1M_1N_1,k}_{lm,\alpha_{MT}}\right)+$$
$$\ddot{\varsigma}_{l,\alpha_{MT}}\left(\bar{b}^{P_2M_2N_2,k}_{lm+1,\alpha_{MT}}\right)^*\left(b^{P_1M_1N_1,k}_{lm,\alpha_{MT}}\right) + \dot{\varsigma}_{l,\alpha_{MT}}\Big[\left(\bar{a}^{P_2M_2N_2,k}_{lm+1,\alpha_{MT}}\right)^*\left(b^{P_1M_1N_1,k}_{lm,\alpha_{MT}}\right)+$$
$$\left(\bar{b}^{P_2M_2N_2,k}_{lm+1,\alpha_{MT}}\right)^*\left(a^{P_1M_1N_1,k}_{lm,\alpha_{MT}}\right)\Big]\Big\}\times\exp\left\{i\left[\left(K_{p_2}-K_{p_1}\right)Z_{\alpha_{MT}} + (M_1-M_2)\Phi_{\alpha_{MT}}\right]\right\}\times$$
$$\left[C^J_{M_2N_2} J_{m+1-M_2}\left(\kappa_{|M_2|,N_2} R_{\alpha_{MT}}\right) + C^Y_{M_2N_2} Y_{m+1-M_2}\left(\kappa_{|M_2|,N_2} R_{\alpha_{MT}}\right)\right]\times$$
$$\left[C^J_{M_1N_1} J_{m-M_1}\left(\kappa_{|M_1|,N_1} R_{\alpha_{MT}}\right) + C^Y_{M_1N_1} Y_{m-M_1}\left(\kappa_{|M_1|,N_1} R_{\alpha_{MT}}\right)\right], \tag{5.12}$$

where

$$\varsigma_{l,\alpha_{MT}} = \int_0^{r_{\alpha_{MT}}} [u_{l,\alpha_{MT}}(\rho)]^2 \frac{dV_{\alpha_{MT}}(\rho)}{d\rho} \rho d\rho \tag{5.13}$$

$$\dot{\varsigma}_{l,\alpha_{MT}} = \int_0^{r_{\alpha_{MT}}} u_{l,\alpha_{MT}}(\rho)\dot{u}_{l,\alpha_{MT}}(\rho) \frac{dV_{\alpha_{MT}}(\rho)}{d\rho} \rho d\rho \tag{5.14}$$

$$\ddot{\varsigma}_{l,\alpha_{MT}} = \int_0^{r_{\alpha_{MT}}} [\dot{u}_{l,\alpha_{MT}}(\rho)]^2 \frac{dV_{\alpha_{MT}}(\rho)}{d\rho} \rho d\rho. \tag{5.15}$$

The meaning of all the notations in these equations is the same as in Chapter 1, except that the translation unit-cell constant here is denoted by the letter d, and not by the letter c. The letter c is used for the speed of light. Moreover, the atoms here are numbered by the index α_{MT} to avoid coincidence with the spin index α.

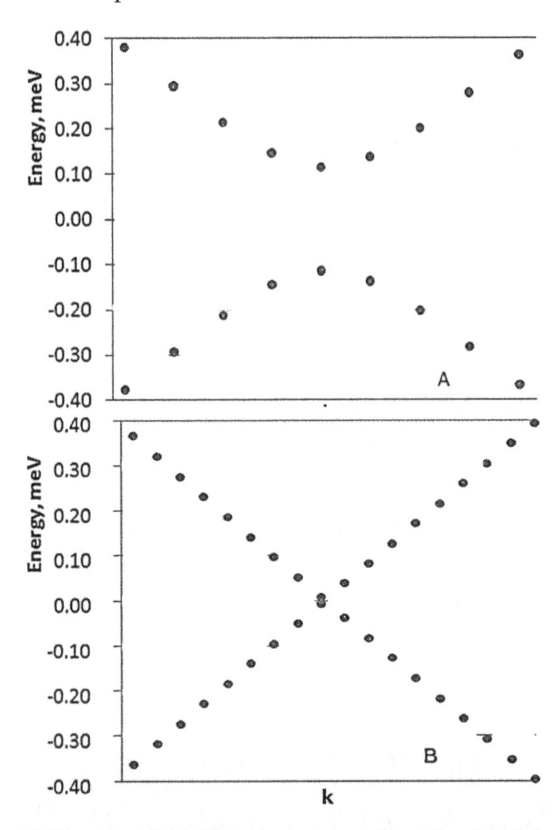

Figure 5.1. LACW bands of (5, 5) carbon nanotube calculated for Fermi energy region with (A) and without (B) spin-orbit coupling.

5.3.1.2. *Armchair Carbon Tubules*

Energy bands for the single-walled carbon armchair nanotubes (n, n) (where $4 \leq n \leq 12$) with spin-orbit interaction were calculated by solving the secular equation. As a typical example, Fig. 5.1 shows the energy bands of the (5, 5) tubule for the Fermi level region.

For comparison, the energy bands calculated without spin-orbit interaction are shown in this figure too. The bands are plotted over only half of the zone, because the bands at $-k$ are the same as those at k. For the absence of spin–orbit interaction, the two linear bonding (π) and antibonding (π^*) bands cross at the Fermi energy near Brillouin zone point $k = 2\pi/(3d)$. Due to the spin–orbit interaction, the energy gap opens between these bands, and the metallic nature for armchair nanotubes is broken. However, each energy sub band still has the two-fold degeneracy, because the spin–orbit interaction does not break the time-reversal symmetry and the achiral armchair nanotubes have the inversion symmetry.

The numerical LACW calculation shows a gap equal to 0.22 meV at the Fermi energy for the (5, 5) carbon nanotube (Table 5.1).

Table 5.1. Spin-orbit energy gap for armchair nanotubes versus the roll-up index n

(n, n)	E_g, meV	(n, n)	E_g, meV	(n, n)	E_g, meV
(4, 4)	0.537	(7, 7)	0.104	(10, 10)	0.083
(5, 5)	0.223	(8, 8)	0.086	(11, 11)	0.076
(6, 6)	0.153	(9, 9)	0.078	(12, 12)	0.086

As expected, the spin-orbit splitting decreases with growth of n or, equivalently, of the tubule diameter. The splitting varies between 0.537 and 0.086 meV for the (n, n) tubules with $4 \leq n \leq 12$.

5.3.1.3. *Armchair Silicon Tubules*

Geometry of the hexagonal silicon nanotubes can be constructed by folding a monolayer of silicon hexagons (silicene) in the form of cylinder. (Ezawa 2012a,b). Since the hexagonal silicon and carbon nanotubes have the same cylindrical structure and topology of chemical bonds, and the Si and C atoms belong to the same group of elements in the periodic table, one can expect a certain similarity between the spin properties of carbon and silicon nanotubes. The silicon nanotubes are obtained experimentally in various ways (Grünzel et al. 2013). It is possible to use them for creating field-effect transistors, waveguides, optoelectronic elements, chemical and biological sensors, for creating heterojunctions by combining carbon and silicon nanotubes (Wu et al. 2012, Park et al. 2009, Song et al. 2010, Espinosa-Soria and Martínez 2016, Ma et al. 2018). In literature, there are several examples of nonempirical calculations of the band structure of silicon nanotubes, but all neglecting the spin-orbit interaction effects and screw symmetry (Hever et al. 2012, Baňacký et al. 2013, Ezawa 2012a, b, Wang et al. 2017).

In order to detect the spin-orbit gaps in the Fermi level region, we calculated the electronic states of the armchair Si nanotubes. The Si atoms were located inside the cylindrical potential well of width 3.2 Å equal to the half-sum of the covalent and van der Waals atomic diameters of silicon, and the Si-Si bond lengths of 2.22 Å were taken as in silicene.

As an example, Fig. 5.2 shows the energy bands of the Si (5.5) nanotube.

Again, the result of the spin-orbit interaction is the formation of a gap at the Fermi level. In the absence of spin-orbit coupling, the π and π^* states overlap. Because of the spin-orbit interaction, the intersecting lines transform into two parabolas, one of which corresponds to the valence band and opens downward, and the other, to the conduction

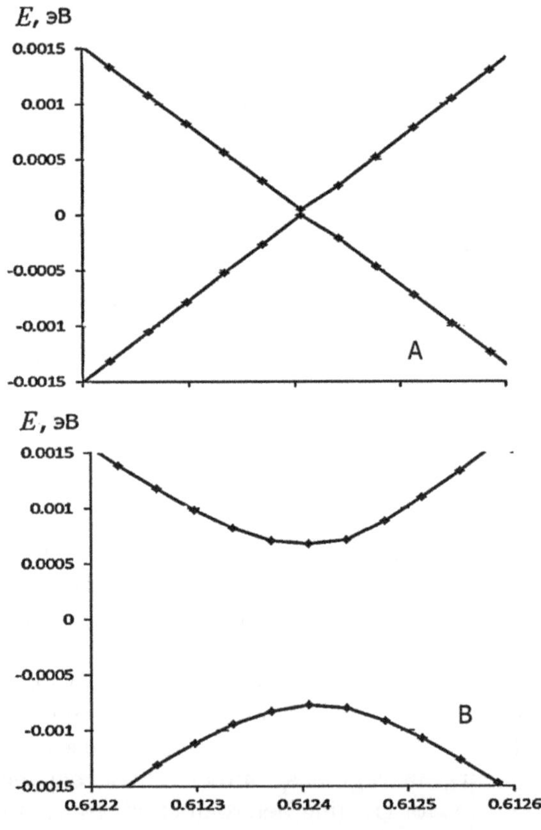

Figure 5.2. Dispersion curves of Si nanotube (5, 5) calculated neglecting (*a*) and considering the spin-orbit interaction (*b*). The Fermi level at $E = 0$.

band and opens upward. Orbital degeneracy at the point of intersection of the non-relativistic valence and conduction bands is removed, a gap is formed between these states, and the metallic nature of the nanotube band structure, predicted by non-relativistic calculations, is destroyed; the compound turns out to be a narrow-gap semiconductor. For the Si (5, 5) tube, the energy of the spin-orbit gap is equal to 1.89 meV.

Table 5.2 shows the energies of spin-orbit gaps at the Fermi level for Si (n, n) nanotubes with n = 5–20.

Table 5.2. Energies of spin-orbit gaps in armchair silicon nanotubes.

Nanotube	E_g, meV	*Nanotube*	E_g, meV
(5, 5)	1.89	(10, 10)	0.87
(6, 6)	1.44	(11, 11)	0.67
(7, 7)	1,33	(12, 12)	0.56
(8, 8)	1.21	(15, 15)	0.43
(9, 9)	1.0	(20, 20)	0.34

It can be seen that the energies of the gaps lie in the interval 1.89-0.34 meV, and an increase in the diameter of the nanotube is accompanied by a decrease in the spin-orbit gap, the energy of the gap being inversely proportional to the index n. The gaps in silicon tubes are about 10 times larger than in carbon counterparts in agreement with the ratio between the atomic spin-orbit constants: 211 cm^{-1} for silicon and 29 cm^{-1} for carbon.

5.3.2. Account of Screw and Rotational Symmetries

Now let us consider the spin-dependent LACW technique with account of the screw and rotational symmetries of tubules applicable to any nanotube independent on its composition, chirality or number of atoms in translational cells (D'yachkov and Makaev 2016). It is suggested that the nanotube has the helical symmetry $S(h, \omega)$ in the form of translation (h) coupled with rotation (ω) and the n-fold rotational symmetry, where n is the highest common factor of n_1 and n_2.

5.3.2.1. Secular Equation

Here, we start from the symmetrized non-relativistic eigenfunctions $\psi_\lambda^0(\mathbf{r}\,|\,\mathrm{k}L)$ written as the linear combinations of the basic functions $\Psi_{PMN}(\mathbf{r}\,|\,\mathrm{k}L)$

$$\psi_\lambda^0(\mathbf{r}\,|\,\mathrm{k}L) = \sum_{PMN} a_{PMN}^\lambda(\mathrm{k}L)\Psi_{PMN}(\mathbf{r}\,|\,\mathrm{k}L) \tag{5.16}$$

and from corresponding eigenvalues dependent on the wave vector k and rotational quantum number L and apply the relativistic Hamiltonian (5.1)-

(5.6). Then, the spinor basic partners are formed. The matrix elements of the $H_{S\text{-}O}$ between the spinor functions remain to be determined

$$\left\langle \psi^{0}_{\lambda_2}(\mathbf{r}\,|\,\mathrm{k}L)\chi_2 \middle| H_{S-O} \middle| \psi^{0}_{\lambda_1}(\mathbf{r}\,|\,\mathrm{k}L)\chi_1 \right\rangle = \sum_{P_2M_2N_2} \sum_{P_1M_1N_1} \bar{a}^{\lambda_2}_{P_2M_2N_2}(\mathrm{k}L) a^{\lambda_1}_{P_1M_1N_1}(\mathrm{k}L) \times \left\langle \Psi_{P_2M_2N_2}(\mathbf{r}\,|\,\mathrm{k}L)\chi_2 \middle| H_{S-O} \middle| \Psi_{P_1M_1N_1}(\mathbf{r}\,|\,\mathrm{k}L)\chi_1 \right\rangle. \quad (5.17)$$

Here

$$\left\langle \Psi_{P_2M_2N_2}(\mathbf{r}\,|\,\mathrm{k}L)\beta \middle| H_{S-O} \middle| \Psi_{P_1M_1N_1}(\mathbf{r}\,|\,\mathrm{k}L)\beta \right\rangle = -\left\langle \Psi_{P_2M_2N_2}(\mathbf{r}\,|\,\mathrm{k}L)\alpha \middle| H_{S-O} \middle| \Psi_{P_1M_1N_1}(\mathbf{r}\,|\,\mathrm{k}L)\alpha \right\rangle, \quad (5.18)$$

$$\left\langle \Psi_{P_2M_2N_2}(\mathbf{r}\,|\,\mathrm{k}L)\alpha \middle| H_{S-O} \middle| \Psi_{P_1M_1N_1}(\mathbf{r}\,|\,\mathrm{k}L)\beta \right\rangle = \left\langle \Psi_{P_2M_2N_2}(\mathbf{r}\,|\,\mathrm{k}L)\beta \middle| H_{S-O} \middle| \Psi_{P_1M_1N_1}(\mathbf{r}\,|\,\mathrm{k}L)\alpha \right\rangle^{*}, \quad (5.19)$$

$$\begin{aligned}
&\left\langle \Psi_{P_2M_2N_2}(\mathbf{r}\,|\,\mathrm{k}L)\alpha \middle| H_{S-O} \middle| \Psi_{P_1M_1N_1}(\mathbf{r}\,|\,\mathrm{k}L)\alpha \right\rangle = \frac{n}{2c^2h}(-1)^{n(M_1+M_2)} \sum_{\alpha_{MT}=1,2} \left(r_{\alpha_{MT}}\right)^4 \\
&\times \sum_{l=0}^{\infty}(2l+1) \sum_{m=-l}^{l} m \frac{(l-|m|)!}{(l+|m|)!} \Big\{ \varsigma_{l,\alpha_{MT}} \bar{a}^{P_2M_2N_2,\mathrm{k}L}_{lm,\alpha_{MT}} a^{P_1M_1N_1,\mathrm{k}L}_{lm,\alpha_{MT}} + \\
&\ddot{\varsigma}_{l,\alpha_{MT}} \bar{b}^{P_2M_2N_2,\mathrm{k}L}_{lm,\alpha_{MT}} b^{P_1M_1N_1,\mathrm{k}L}_{lm,\alpha_{MT}} + \dot{\varsigma}_{l,\alpha_{MT}} \Big[\bar{a}^{P_2M_2N_2,\mathrm{k}L}_{lm,\alpha_{MT}} b^{P_1M_1N_1,\mathrm{k}L}_{lm,\alpha_{MT}} + \\
&\bar{b}^{P_2M_2N_2,\mathrm{k}L}_{lm,\alpha_{MT}} a^{P_1M_1N_1,\mathrm{k}L}_{lm,\alpha_{MT}} \Big] \Big\} \times \exp\Big\{ i \Big[\Big(\mathrm{k}_{p_1} - \mathrm{k}_{p_2} - n(M_1 - M_2)\frac{\omega}{h} \Big) \times \\
&Z_{\alpha_{MT}} + n(M_1 - M_2)\Phi_{\alpha_{MT}} \Big] \Big\} \times \\
&\Big[C^{J,L}_{M_2N_2} J_{m-(L+nM_2)}\left(\kappa_{|L+nM_2|,N_2} R_{\alpha_{MT}}\right) \\
&+ C^{Y,L}_{M_2N_2} Y_{m-(L+nM_2)}\left(\kappa_{|L+nM_2|,N_2} R_{\alpha_{MT}}\right) \Big] \times \\
&\Big[C^{J,L}_{M_1N_1} J_{m-(L+nM_1)}\left(\kappa_{|L+nM_1|,N_1} R_{\alpha_{MT}}\right) + \\
&C^{Y,L}_{M_1N_1} Y_{m-(L+nM_1)}\left(\kappa_{|L+nM_1|,N_1} R_{\alpha_{MT}}\right) \Big],
\end{aligned} \quad (5.20)$$

and

$$\begin{aligned}
&\left\langle \Psi_{P_2M_2N_2}(\mathbf{r}\,|\,\mathrm{k}L)\beta \middle| H_{S-O} \middle| \Psi_{P_1M_1N_1}(\mathbf{r}\,|\,\mathrm{k}L)\alpha \right\rangle = \frac{n}{2c^2h}(-1)^{n(M_1+M_2)} \sum_{\alpha_{MT}=1,2} \left(r_{\alpha_{MT}}\right)^4 \\
&\times \sum_{l=0}^{\infty}(2l+1) \sum_{m=-l}^{l} \left[l(l+1) - m(m+1) \right] (-1)^{\frac{m+|m|}{2} + \frac{m+1+|m+1|}{2}} \\
&\times \left[\frac{(l-|m|)!(l-|m|+1)!}{(l+|m|)!(l+|m+1|)!} \right]^{1/2} \times \Big\{ \varsigma_{l,\alpha_{MT}} \bar{a}^{P_2M_2N_2,\mathrm{k}L}_{lm+1,\alpha_{MT}} a^{P_1M_1N_1,\mathrm{k}L}_{lm,\alpha_{MT}} +
\end{aligned}$$

$$\ddot{\varsigma}_{l,\alpha_{MT}} \bar{b}_{lm+1,\alpha_{MT}}^{P_2M_2N_2,kL} b_{lm,\alpha_{MT}}^{P_1M_1N_1,kL} + \dot{\varsigma}_{l,\alpha_{MT}}$$

$$\times\left[\bar{a}_{lm+1,\alpha_{MT}}^{P_2M_2N_2,kL} b_{lm,\alpha_{MT}}^{P_1M_1N_1,kL} + \bar{b}_{lm+1,\alpha_{MT}}^{P_2M_2N_2,kL} a_{lm,\alpha_{MT}}^{P_1M_1N_1,kL}\right]\Bigg\}\times$$

$$\exp\left\{i\left[\left(\mathrm{k}_{p_1} - \mathrm{k}_{p_2} - n(M_1 - M_2)\frac{\omega}{h}\right)Z_{\alpha_{MT}} +\right.\right.$$

$$\left.\left. n(M_1 - M_2)\Phi_{\alpha_{MT}}\right]\right\}\times\left[C_{M_2N_2}^{J,L} J_{m+1-(L+nM_2)}\left(\kappa_{|L+nM_2|,N_2} R_{\alpha_{MT}}\right)+\right.$$

$$\left. C_{M_2N_2}^{Y,L} Y_{m+1-(L+nM_2)}\left(\kappa_{|L+nM_2|,N_2} R_{\alpha_{MT}}\right)\right]\times$$

$$\left[C_{M_1N_1}^{J,L} J_{m-(L+nM_1)}\left(\kappa_{|L+nM_1|,N_1} R_{\alpha_{MT}}\right) + C_{M_1N_1}^{Y,L} Y_{m-(L+nM_1)}\left(\kappa_{|L+nM_1|,N_1} R_{\alpha_{MT}}\right)\right]. \quad (5.21)$$

Here we should not explain the meaning of symbols used in Eqs. (5.20) and (5.21); they are the same as in the previous section and in Chapter 3.

5.3.2.2. Carbon Nanotubes

As the first example, Fig. 5.3 shows the results of spin-dependent LACW calculations of the nanotube (11, 3) for $0 \le \mathrm{k} \le \pi/h$.

This is the chiral system without rotational symmetry having more than six hundred atoms per translational unit cell. However, due to the screw symmetry of the carbon nanotubes, the LACW method makes it easy to perform the relativistic calculations of the band structure of this giant tubule; moreover, the results can be presented in a very simple form, because there are only eight electrons per true unit cell and, therefore, only eight spin-dependent energy bands in the valence band and only two or several low-energy unoccupied spin-dependent approximately π-type curves in the conduction band. The spin-orbit coupling effect appears as the splitting of non-relativistic dispersion curves. As expected, the gaps formed due to the splitting are small, but they can be seen on an enlarged energy scale as shown in the insets of Fig. 5.3 for the π-bands, where the dispersion curves in the regions of the seven local maxima and minima (van Hove singularities) of the valence and conduction π-bands (A, B, ..., G) are presented. The arrows indicated on each van Hove singularity correspond to the direction of spin polarization relative to the nanotube's axis. The polarization is almost perfect (>0.99) and polarization directions are opposite for split pairs of bands. According to the relativistic LACW data, the (11, 3) nanotube is the semiconducting compound with the direct optical gap E_g = 0.662 eV at the Brillouin zone point B corresponding to the wave vector k = 2.395 a.u.$^{-1}$ = 0.24(π/h). At the points A, D, E, and F, the spin-orbit splitting of the conduction band is greater than the splitting of the valence band states; on the contrary, at the points B, C, and G, the splitting of the conduction band is weaker than that of the valence band. The spin-orbit gaps of electron and hole states vary in the ranges of 0.05-0.91 and 0.01-0.71 meV, respectively. At all these points of the Brillouin zone, the minimum gaps correspond to direct transitions between the spin states of the same polarization. The polarization is positive at points A, C, and F, but it is negative at points B, D, E, and G.

Figure 5.4 shows a character of spin-state polarization of all the valence and conduction band states both for k > 0 and k < 0.

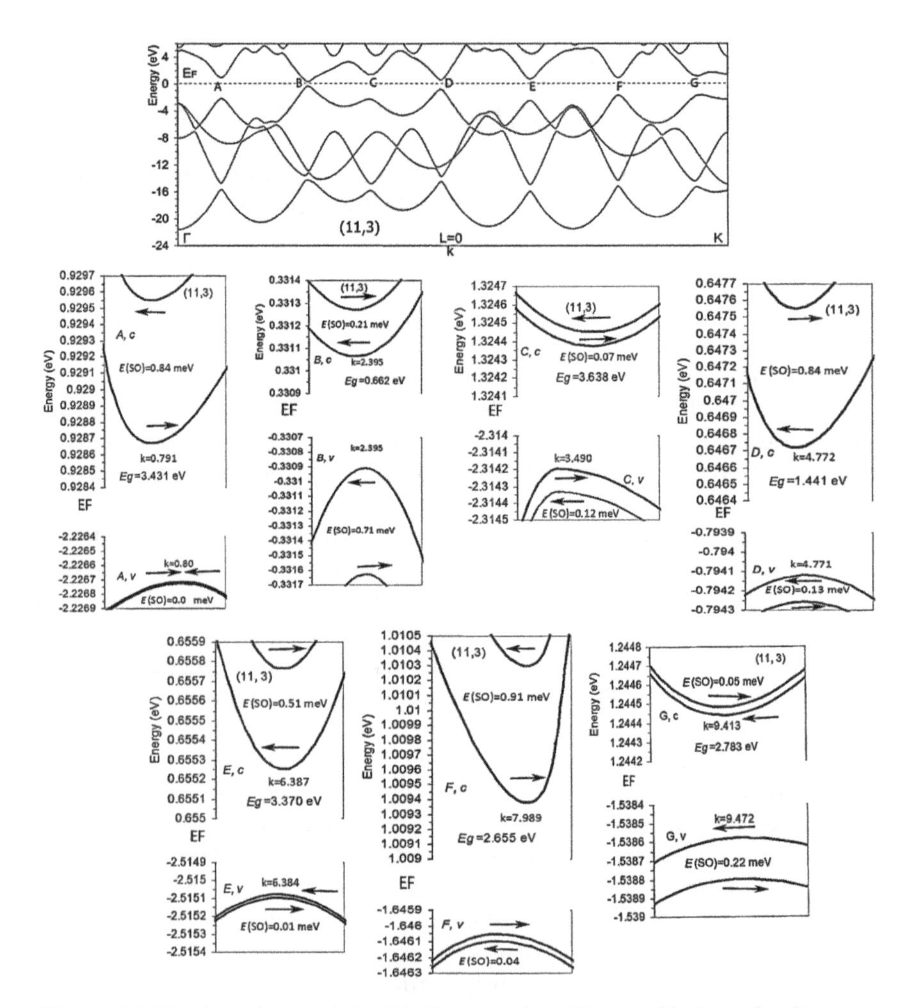

Figure 5.3. Band structure of the (11, 3) nanotubes. Here and below, the dispersion curves in the regions of valence (v) and conduction (c) bands local maxima and minima (A, B, ...) are presented in an enlarged energy scale; the arrows show the spin polarization direction relative to the tubule axis. $h = 0.315$ a.u., $k = \pi/h = 9.98$ a.u.$^{-1}$ for the Brillouin zone edge point K.

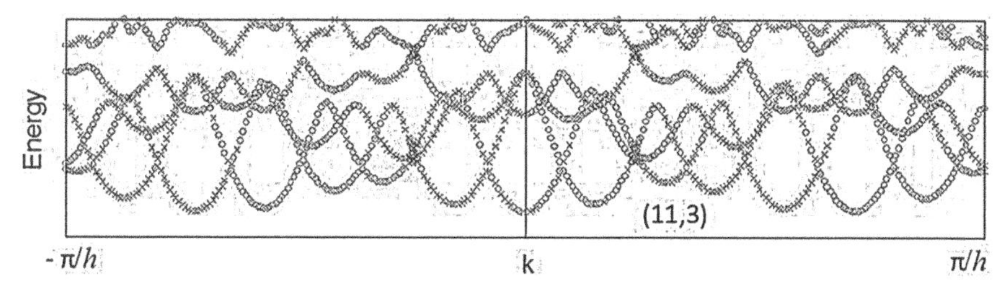

Figure 5.4. Polarization direction for the split pairs of energy states. The times sign corresponds to splitting $\overset{\rightarrow}{\leftarrow}$ with positive polarization of the higher level and negative polarization of the lower level; the circles sign points on the splitting $\overset{\leftarrow}{\rightarrow}$ with opposite polarizations.

One can see that the dispersion curves are anti symmetric with respect to substitution k for −k; they have the same energies, but opposite polarization of spins

$$E^{\rightarrow}_{\leftarrow}(\mathrm{k}) = E^{\leftarrow}_{\rightarrow}(-\mathrm{k})\,. \tag{5.22}$$

The (9, 6) nanotube is another example of chiral system. For this tubule having the rotational third order axis, Fig. 5.5 shows the values of band energies and spin-orbit gaps of the seven main low-energy van Hove singularities for $L = 0, 1, 2$ and $0 \leq \mathrm{k} \leq \pi/h$.

According to the simple π electron tight-binding theory, this tubule should be semimetallic with zero gap between the occupied and unoccupied π states at the point Γ (k = 0) for $L = 1$ or 2. Due to the carbon nanotubes curvature and relativistic effects, the conduction band minimum and valence band maximum shift from point k = 0 to the point k = 0.010, and a minigap equal to 47 meV between the π bonding and antibonding states opens. At this k point, the spin-orbit gaps are equal to 0.55 and 0.36 meV for the occupied and unoccupied levels, respectively.

Figure 5.6 shows that for $L = 0$ the dispersion curves at negative k values are described by Eq. (5.22); they are the same as at the positive k values, but spin polarization is opposite.

The situation is different for $L = 1$ and 2; namely, the states located at the points (k, L) and (−k, 3 − L) have the same energy and opposite polarization

$$E^{\rightarrow}_{\leftarrow}(\mathrm{k}, L) = E^{\leftarrow}_{\rightarrow}(-\mathrm{k}, n - L). \tag{5.23}$$

The (10, 5) tubule with C_5 axis is the chiral semiconductor with minimum gap $E_g = 0.660$ eV corresponding to electron transition at point F for $L = 2$ (Fig. 5.7).

The spin-orbit splitting of hole and electron states at this point are equal to the 0.28 and 0.56 meV. In the total Brillouin zone $-\pi/h < \mathrm{k} \leq \pi/h$, the dependencies of spin-bands on wave vector k are similar to those of

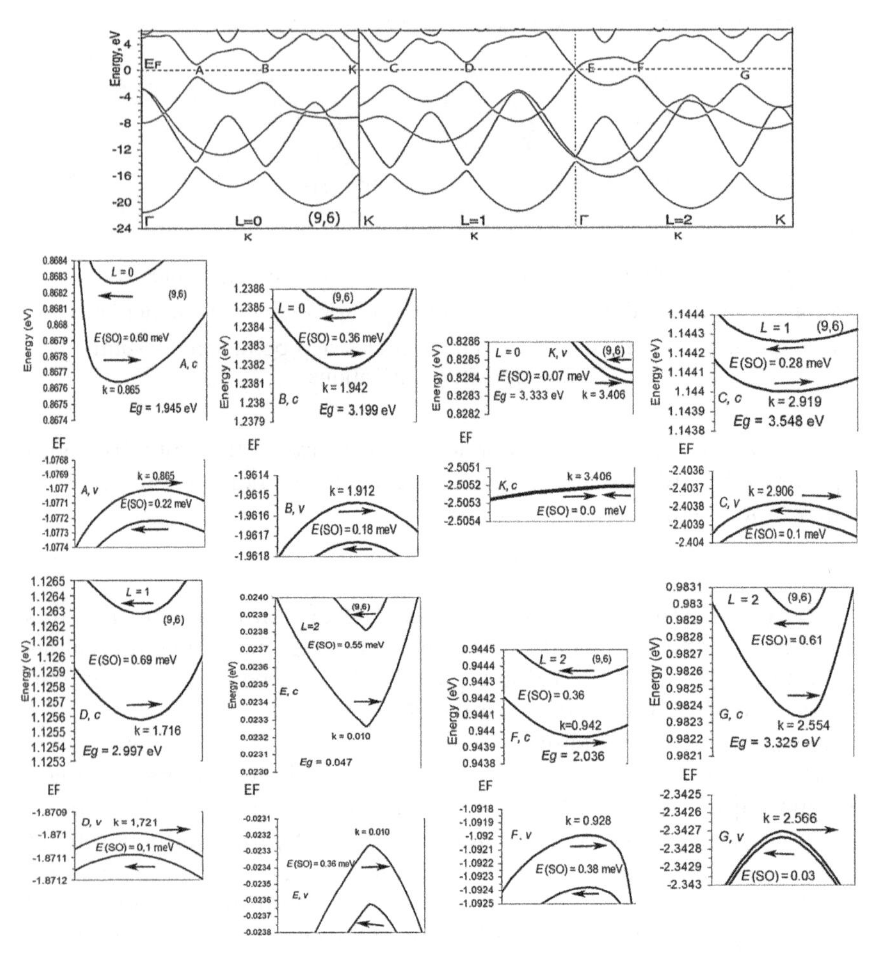

Figure 5.5. Band structure of the (9, 6) carbon nanotube. $h = 1.519$ a.u., $k = \pi/h = 3.408$ a.u.$^{-1}$ for point K.

the (9, 6) nanotube; they are described by Eqs. (5.22) and (5.23) for $L = 0$ and $L \neq 0$, respectively.

Figure 5.8 shows the energy bands of the nonchiral armchair (7,7) nanotubes.

For the armchair nanotube in the absence of the spin-orbit coupling, the π and π^* bands cross at the Fermi energy for $L = 0$ at point A near $k = 2\pi/(3h)$, but the gap $E_g = E_{S\text{-}O}$ equal to 0.44 meV opens between these bands due to the spin-orbit interaction. Each energy sub band for $L = 0$ has the twofold spin degeneracy and $E(k) = E(-k)$, because the nonchiral armchair tubule has an inversion symmetry. Due to the inversion, rotational, and time-reversal symmetries, the dispersion curves in the regions of the

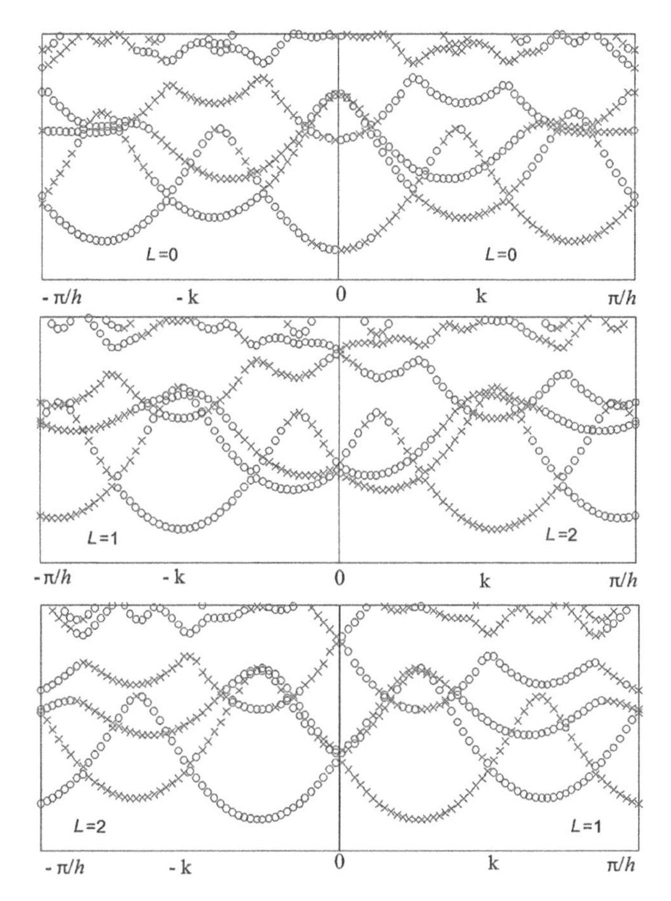

Figure 5.6. Polarization directions relative to the (9, 6) nanotube axis for the split pairs of energy states.

points B (L = 1) and C (L = 6), D (L = 2) and E (L = 5), F (L = 3) and G (L = 4) coincide, and the polarization of spins at these pairs is opposite. For $L \neq 0$, the rule (5.23) is performed.

Figure 5.9 shows the LACW data on the spin dependent band structure of achiral semiconducting zigzag (13, 0) nanotube.

As for the armchair tubule, the bands for $L = 0$ are degenerate and symmetric, and Eq. (5.23) is valid in the case of $L \neq 0$.

In conclusion, we note that the energies of the spin-orbit splitting of the inner p_σ and s bands also range from 0.01 to 1 meV.

5.3.2.3. Gold Nanotubes

In recent years, theoretical studies of gold nanotubes have been widely developed. There are multi-walled (Kondo and Takayanagi 2000, Bridges

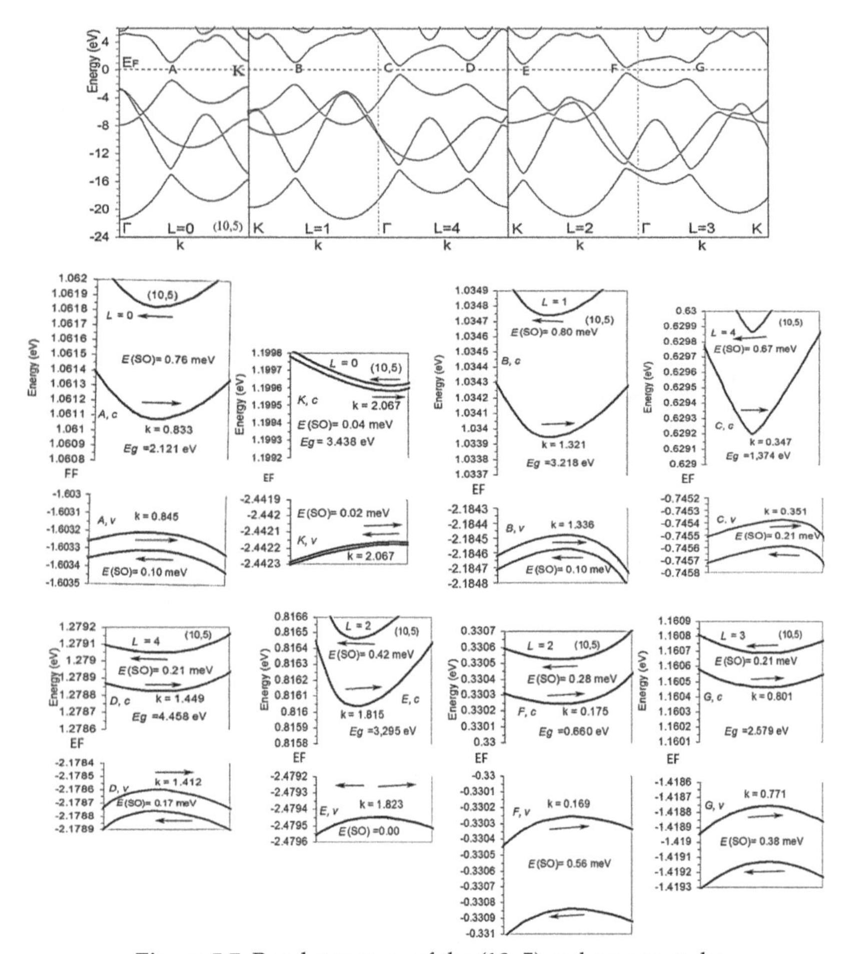

Figure 5.7. Band structure of the (10, 5) carbon nanotube. $h = 0.921$ a.u., $k = \pi/h = 2.067$ a.u.$^{-1}$ for point K.

et al. 2012, 2013) and single-walled gold nanotubes (Oshima et al. 2003). By the methods of transmission electron microscopy, single-walled tubes with a minimum diameter of 4 Å have been found; the thinnest observed single-walled gold nanotube (5, 3) is formed by a chain of five Au atoms folded around the axis of the tube.

The ideal single-walled gold nanotube is a plane curved into a cylinder laid out by regular hexagons, in the center of which there are gold atoms (Fig. 5.10).

The spatial arrangement of atoms in a single-walled gold nanotube is determined by the Au-Au bond length $d_{\text{Au-Au}} = 2.9$ Å and by two integer indices n_1 and n_2. While considering the rotational and screw symmetry, a minimum cell with only one Au atom can be defined.

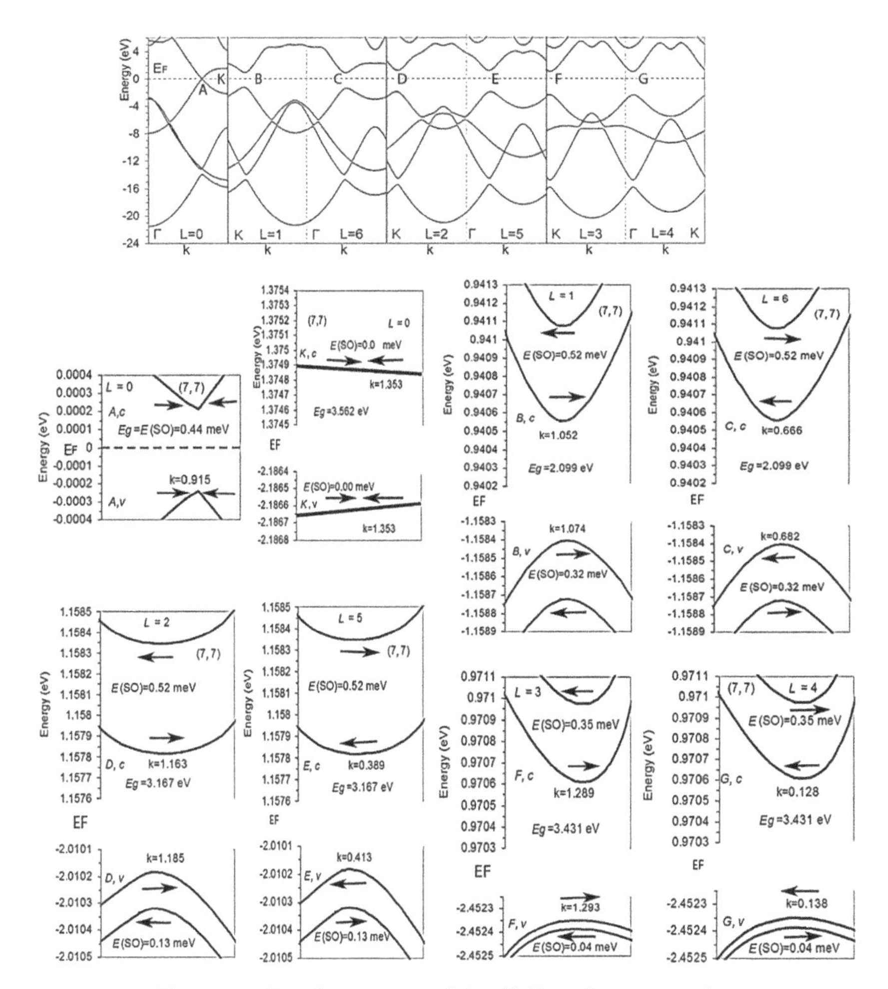

Figure 5.8. Band structure of the (7, 7) carbon nanotube. $h = 2.321$ a.u., $k = \pi/h = 1.353$ a.u.$^{-1}$ for point K.

Technological interest in gold nanotubes is due primarily to the possibility of using them for creating and combining elements of molecular electronics (Huang et al. 2003); therefore, much attention is paid to modeling their electronic, optical, electrical, and spin properties. It is reported that the noble metal helical nanotubes have distinct electronic and mechanical properties, quantum ballistic transport, and conduction channels originating from their helical structure (Oshima et al. 2006, del Valle et al. 2006, Sen et al. 2013, Ono and Hirose 2005, Shimada et al. 2011, Manrique et al. 2010).

The electronic band structure of gold nanotubes has been studied theoretically in several papers. Initially, the pseudopotential method and

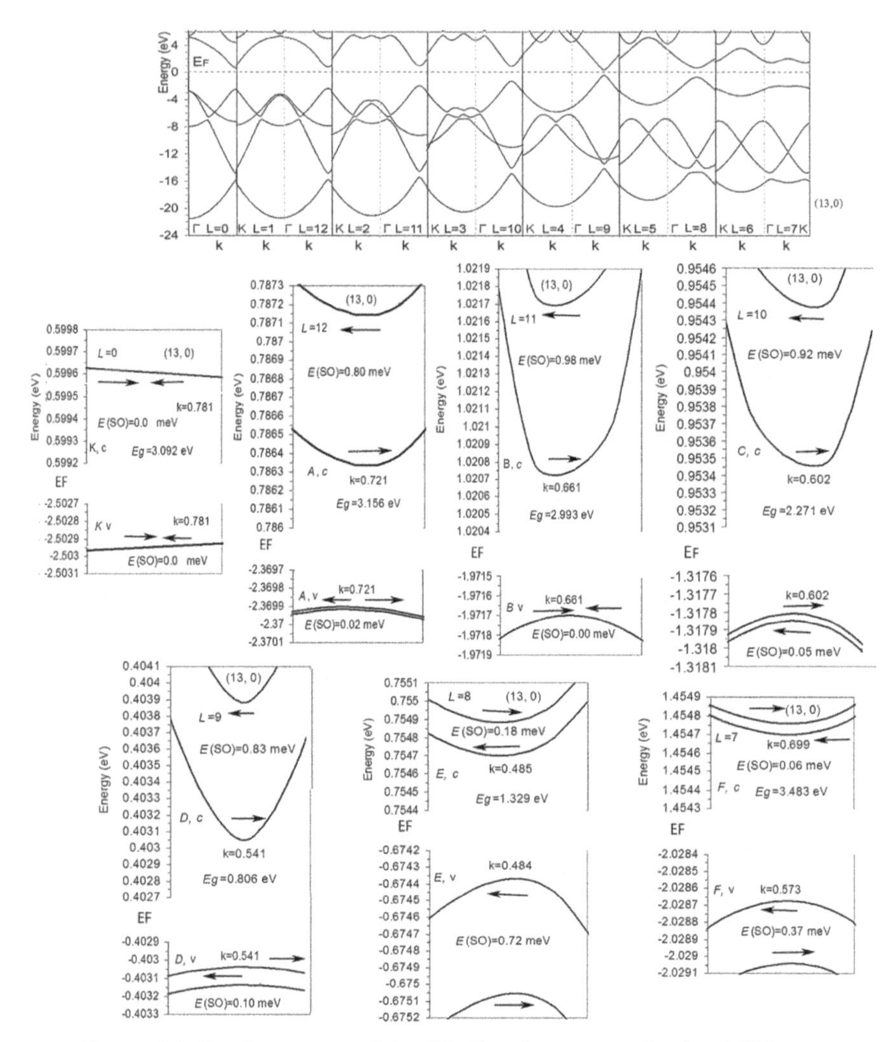

Figure 5.9. Band structure of the (13, 0) carbon nanotube. $h = 4.020$ a.u., $k = \pi/h = 0.781$ a.u.$^{-1}$ for point K.

the plane-waves basis were used to calculate the band structures and DOS of nanotubes (5, 3), (4, 0), and (6, 3) (Yang and Dong 2005). Later, the optical properties of gold nanotubes were studied and the nature and polarization dependences of the plasma resonance bands in gold nanotubes were interpreted using the quantum chemistry calculations (Zhang and Zhang 2014). The results of measuring the ballistic conductivity of gold nanotubes (Valle et al. 2006) were simulated using the Green's function method and atomic orbitals basic set (Oshima et al. 2002, 2006). Similarly, using *ab initio* calculations and the plane-wave basis set, the electrical characteristics of

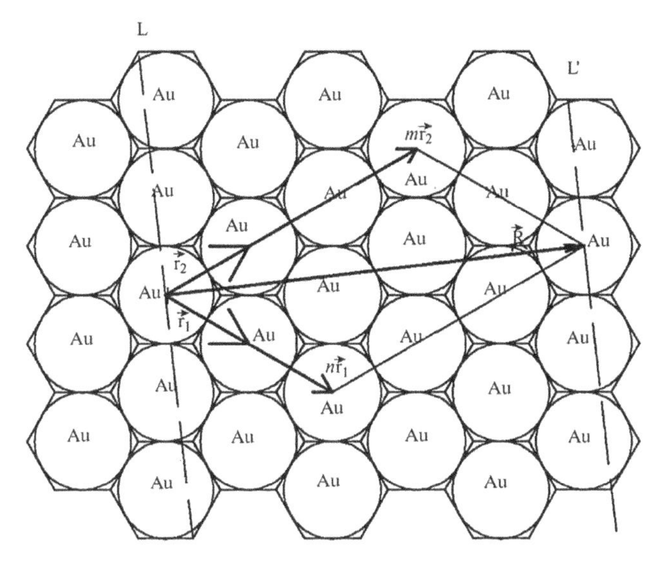

Figure 5.10. Rolling up of the sheet for producing Au (n, m) nanotube, $\mathbf{R} = n\mathbf{r}_1 + m\mathbf{r}_2$.

gold nanotubes as electrodes in contact with an organic molecule were studied in order to model the corresponding heterojunctions (Sen et al. 2013). The electronic conductivity of a gold nanotube suspended between two bulk gold electrodes was also investigated (Ono and Hirose 2005). With the help of analytical modeling and numerical calculations, the currents in the gold nanotubes are studied and a formation of magnetic fields inside the gold chiral tubes is predicted (Manrique et al. 2010).

In these theoretical studies, all calculations were carried out neglecting the spin-orbit interaction, but relativistic effects are very important for the quantitative description of the properties of heavy elements such as gold. For example, in a monoatomic gold layer with a hexagonal structure, the spin-orbit interaction induces a formation of the forbidden band with a width of 0.1 eV, and the gold layer from the metallic goes into semiconducting state unusual for the gold materials (Liu et al. 2017). For the first time, the electronic bands of gold nanotubes considering the effect of the spin-orbit contribution were calculated by the pseudopotential and LAPW methods in Yang et al (2008) and Mokrousov et al. (2005, 2006). In these studies, only the translational symmetry of nanotubes was taken into account, so the calculations were limited to the thinnest nanotubes (5, 3), (4, 0), and (6, 0) containing the minimum numbers of atoms in the unit cells. In the symmetrized LACW method, the calculations of any nanotube are possible, regardless of their diameter and chirality; the stable band structure was achieved using about 25 basic functions (Krasnov et al. 2019).

As an example, let us consider the results of calculating the spin-dependent band structure of the gold nanotube (5, 3) of radius 3.23 Å (Fig. 5.11).

For greater clarity, in the upper part of the figures, the results for the neighborhood of the Fermi level are presented on an enlarged scale. The dashed and solid lines correspond to the polarization of the α and β spins along the axis of the nanotube in the z and $-z$ directions. The results are presented in terms of 10 filled zones and one half-filled zone. (If only the translational symmetry was considered, there would be 1078 curves in the band structure.) The spin-orbit interaction appears as a strong splitting of non-relativistic dispersion curves. For the curves crossing the Fermi level, this splitting is equal to 0.5 eV and decrease with the transition to the inner states of the valence band. For the band located below −5 eV, the splitting is an order of magnitude smaller. The spin polarization directions are opposite for the split pairs of bands, and the spin polarization is almost ideal as in the case of the carbon analogs. Each intersection point of the

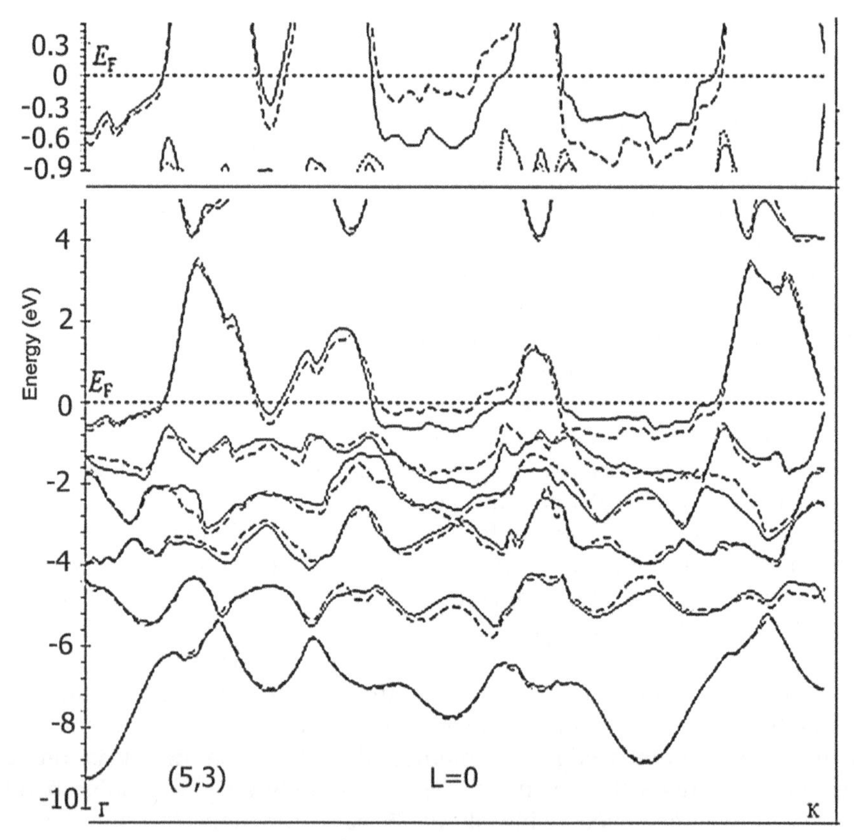

Figure 5.11. Band structure of gold nanotube (5, 3).

dispersion curve with the Fermi level corresponds to the ballistic transport channel. In the nanotube (5, 3), the calculations reveal seven such channels for the spins α and β.

Figures 5.12 and 5.13 show two more examples of the results of calculations of the spin-dependent band structure of chiral gold nanotubes that do not have rotational symmetry axes.

These are the tubes (8, 7) and (11, 3) with different values of the chirality angles, but almost the same radii 5.9 and 6.0 Å and translational cells formed by 338 and 326 Au atoms. An almost twofold increase in the radius and a decrease in the curvature of the cylindrical surface of these nanotubes in comparison with the tube (5, 3) lead to a noticeable decrease in the spin-orbit splitting; the maximum values of the spin-orbit gaps are equal to about 0.2 eV. The number of conduction channels in these tubes, on the contrary, is twice as large as in the (5, 3) tubule.

The transition to (18, 11) nanotube with an almost unchanged chirality angle is accompanied by a further about twofold increase in the tube radius up to 11.7 Å and a decrease in the spin-orbit splitting of the levels to a maximum of 0.1 eV due to a decrease in the curvature of the cylindrical surface (Fig. 5.14).

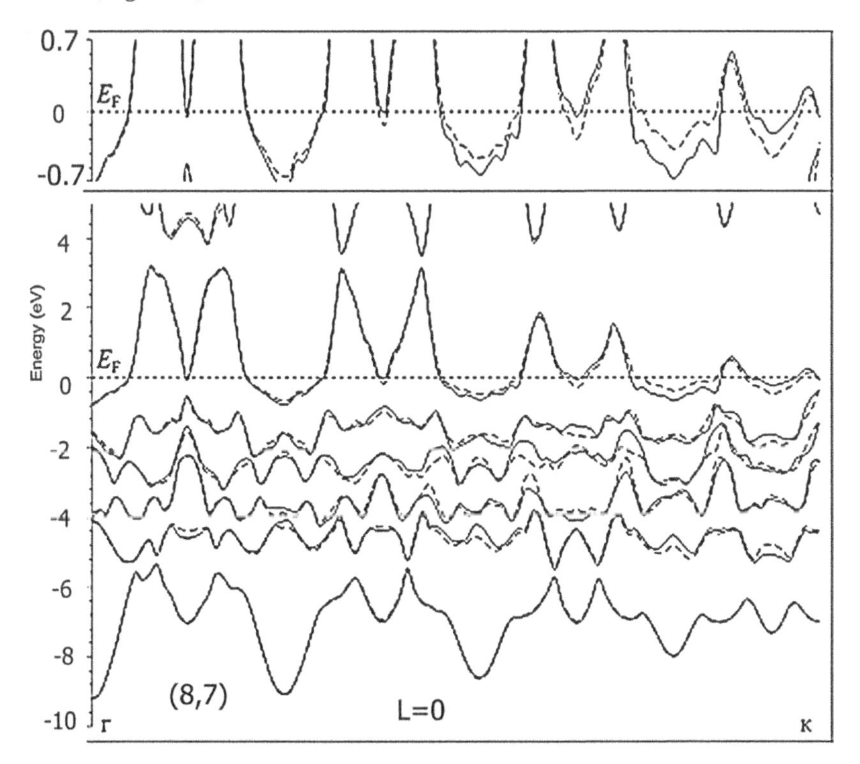

Figure 5.12. Band structure of gold nanotube (8, 7).

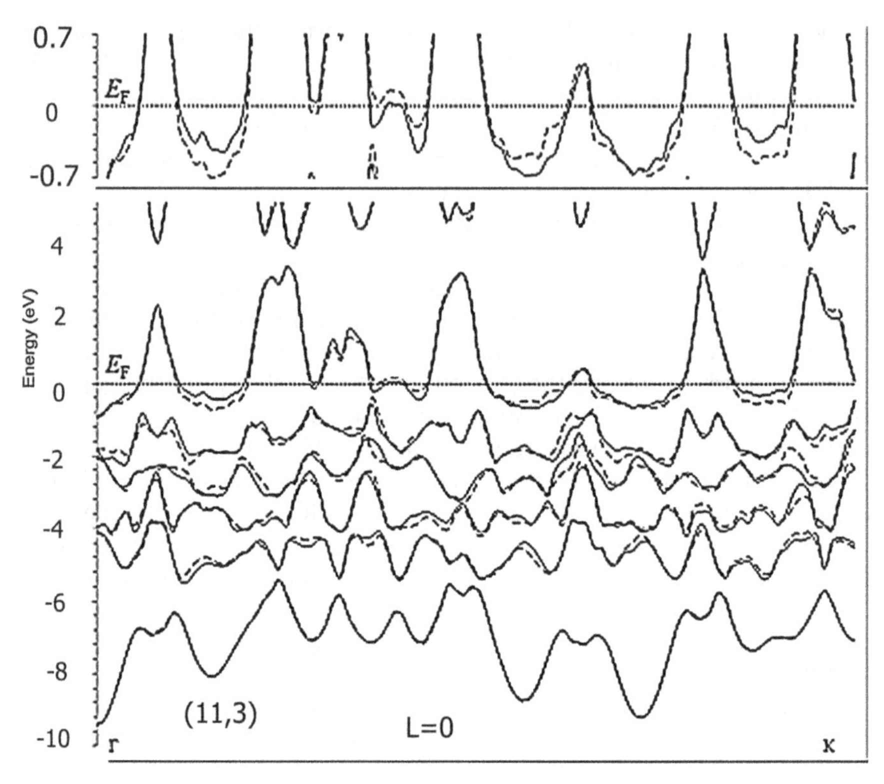

Figure 5.13. Band structure of gold nanotube (11, 3).

In this case, the double growth of the tube radius leads approximately to a doubling of the number of conduction channels in the tube (18, 11) in comparison with (8, 7) and (11, 3) system. The dispersion curves for the tubes (5, 3), (8, 7), (11, 3), and (18, 11) for negative values of $-(\pi/h) \leq k < 0$ are found using the Kramers theorem. They are anti symmetric with respect to the replacement of k by –k, i.e. have the same energy, but opposite spins (5.22).

The nanotube (10, 5) is another example of a chiral system, but there is also a rotational axis of the fifth order in addition to the screw axis in this tube. Figure 5.15 shows the dispersion curves for integer values of L in the interval from 0 to 4 and $0 \leq k \leq \pi/h$ for (10, 5) tube.

The radius of this tube (6.1 Å) almost does not differ from the radii of the tubes (8, 7) and (11, 3); therefore, the maximum energy of the spin-orbit gaps equal to 0.2 eV and the number of electron transport channels are practically the same as in the tubes (8, 7) and (11, 3).

Figures 5.16 and 5.17 show the results of calculations of the spin-dependent band structures of nonchiral tubes (8, 8) and (13, 0) with close radii (6.0 and 6.4 Å) and maximally different chirality angles (0 and 30°).

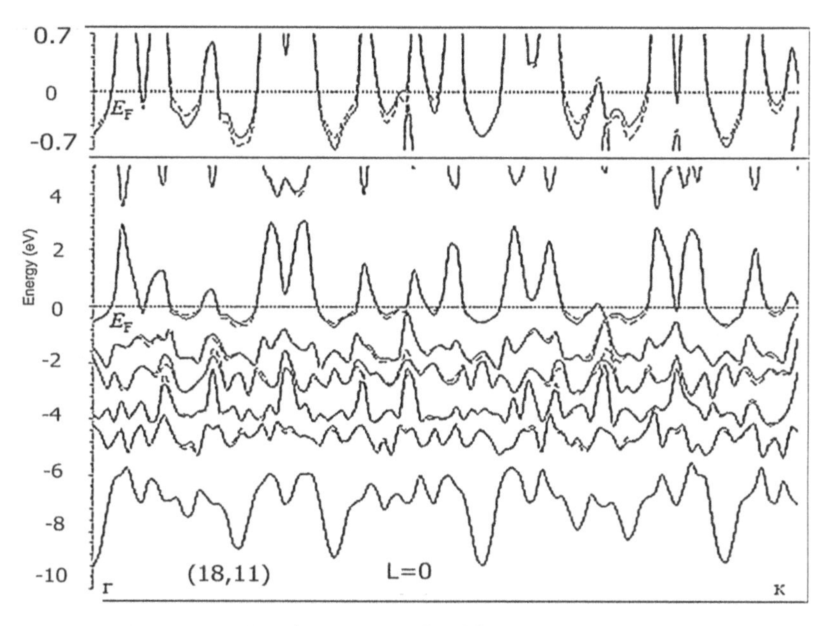

Figure 5.14. Band structure of gold nanotube (18, 11).

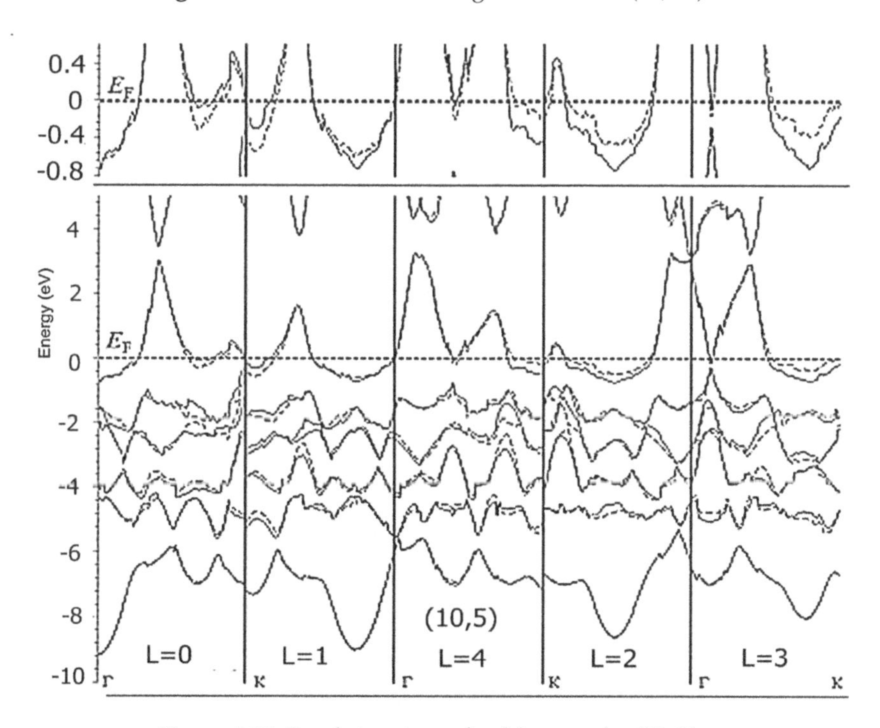

Figure 5.15. Band structure of gold nanotube (10, 5).

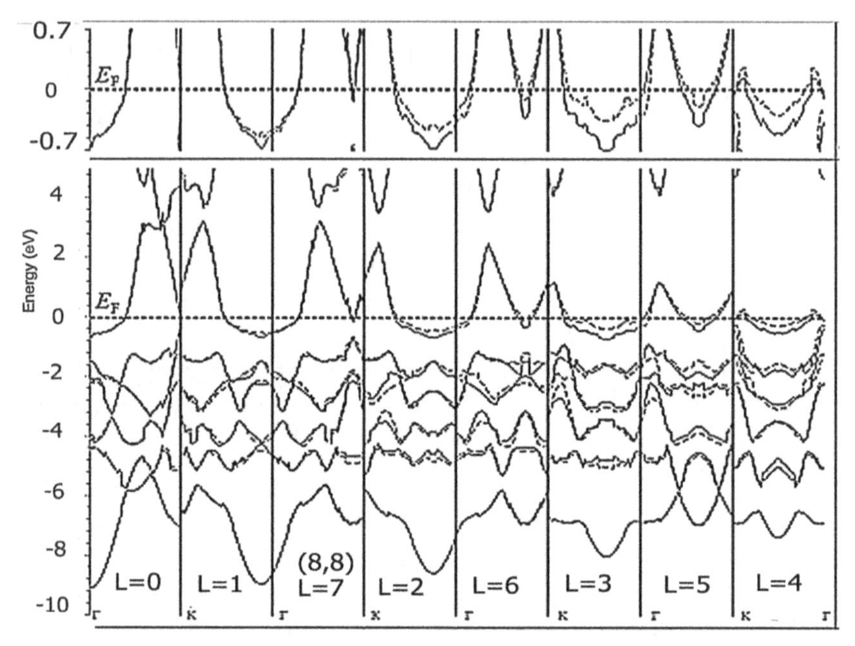

Figure 5.16. Band structure of gold nanotube (8, 8).

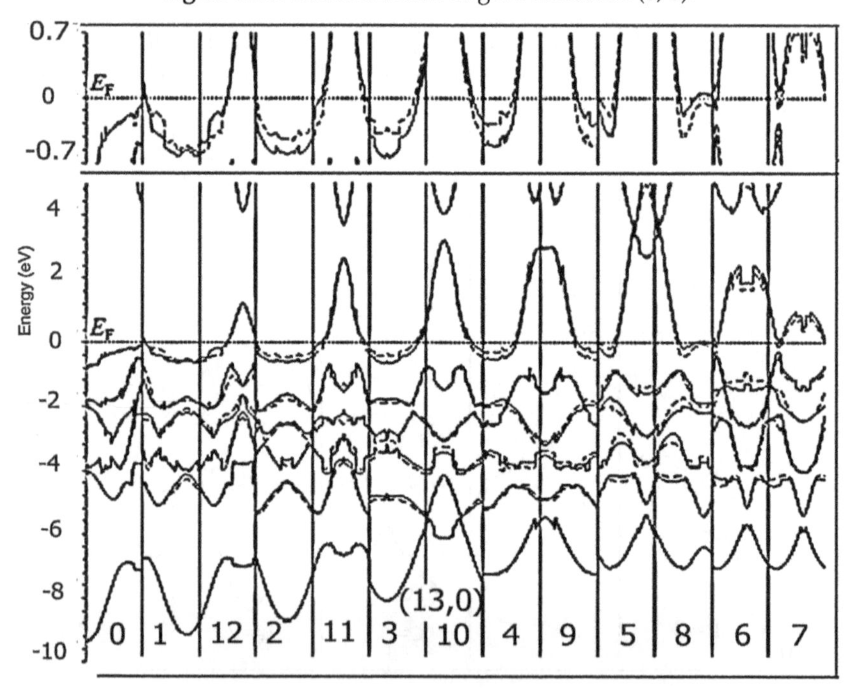

Figure 5.17. Band structure of gold nanotube (13, 0).

Comparison of these figures shows that it is the radius or curvature of the surface of the gold nanotube, but not its chirality that determines the energies of the spin-orbit gaps in the gold systems. For nanotubes (10, 5), (8, 8), and (13, 0) with rotational axes of orders $n = 5$, 8 and 13 for $L = 0$ and negative values of k, the dispersion curves have the same energy as for positive values k, but the spin direction is reversed, Eq. (5.22). When $L \neq 0$, the same energies, but opposite spin have possessed the states at the points (k, L) and (−k, $n - L$), Eq. (5.23). For a nanotube (8, 8) with inversion symmetry, the spin-orbit coupling does not split the electronic states at $L = 0$.

Thus, the spin-orbit interaction in gold nanotubes appears as the strong splitting of non-relativistic dispersion curves, which decreases upon transition to the inner states of the valence band and to nanotubes of larger radius. All the tubes have a metallic type of band structure, and the number of conduction channels increases with the nanotube's radius.

5.3.2.4. Platinum Nanotubes

Single-walled platinum nanotubes can be obtained starting from platinum nanowire and applying an electron-beam thinning method (Oshima et al. 2002, 2003). Such nanotubes have the same structure as the gold analogs; it is a triangular network with Pt atom rows helically coiling around the wire axis and Pt-Pt bond length equal to 2.825 Å. There has been no detailed study of a band structure of the platinum nanotubes, which are expected to have the electron properties quite different from those of the Pt bulk or surfaces (Andersen 1970, Bordoloit and Auluck 1983, Wern et al. 1985, Herrera-Suárez et al. 2012), because the Pt atoms in the nanotube have two times smaller coordination numbers. The calculations of Pt nanotubes are limited to works (Xiao and Wang 2006, Matanović et al. 2013, Hui et al. 2003, Konar and Gupta 2008). In the paper (Matanović et al. 2013), in the terms of a projected augmented wave method and DFT, the dispersion curves of achiral (6, 6) and (13, 13) Pt nanotubes for the Fermi energy region were calculated. In the work (Xiao and Wang 2006), the structures, magnetic moments, and DOS of finite length Pt nanotubes were calculated using a similar approach and plane-wave basic set. Finally, some data on the DOS of Pt (6, 4) and (5, 3) nanotube were obtained using the supercell cluster calculations and plane wave basis (Konar and Gupta 2008). However, in these calculations, the spin-orbit coupling terms were not taken into account in Hamiltonian, but these terms must give rise to a splitting of energy bands and formation of spin-orbit gaps. Platinum is a heavy metal; for its compounds, the spin-orbit interaction should be surely taken into consideration. It is the aim of this section to present the LACW spin-dependent electron structures of the single-walled Pt nanotubes.

Figure 5.18 shows the results (D'yachkov and Krasnov 2019) of calculations of the band structure of chiral tubule Pt (5, 3).

The results are presented in a very simple form with only 10 spin-dependant energy bands in the valence band and two low-energy curves in the conduction band. The Fermi level clearly separates the valence and conduction band curves. There is no crossing of the occupied and vacant dispersion curves, but only slight overlap of a top of valence and of a bottom of conduction states typical for semi metallic systems. In the band structure, the spin-orbit coupling appears as the large splitting of non-relativistic dispersion curves equal up to 0.5 eV for the bands in Fermi energy region.

Figure 5.19 shows the band structure of the (5, 5) nanotube presented in the repeated zone scheme.

Again, the curves for valence and conduction states are separated and the tubule is semimetallic with significant spin-orbit splitting. An optical gap up to 50 meV opens between the valence and conduction bands due to about 3% stretching this tubule.

The semimetal type of the band structure is also preserved for (8, 0) zigzag Pt tubule (Fig. 5.20).

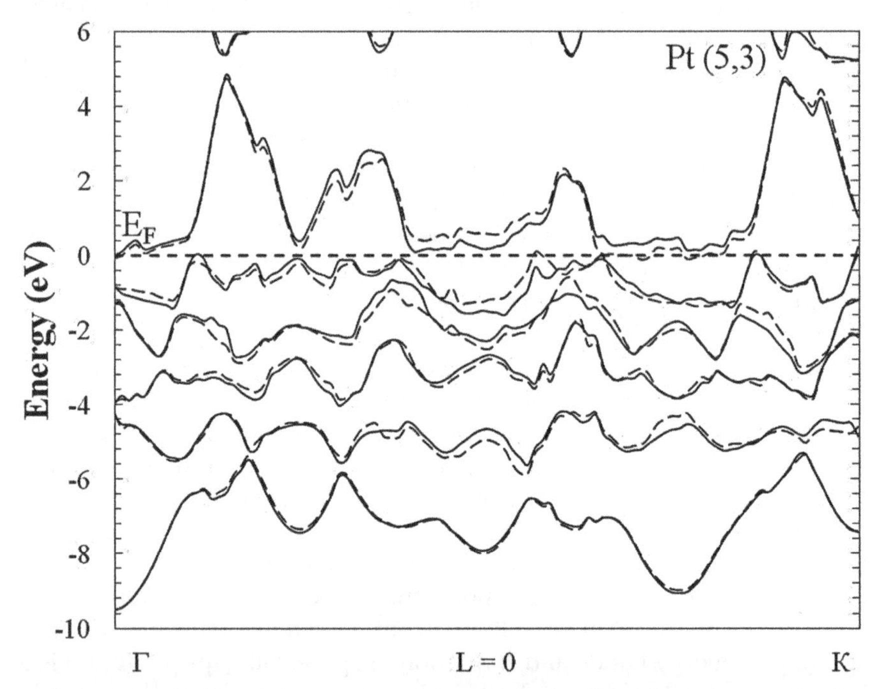

Figure 5.18. Electron structure of the chiral (5, 3) Pt nanotube. The dashed and solid lines correspond to the spin α and β. The dispersion curves in the Fermi energy region are presented in enlarged energy scale.

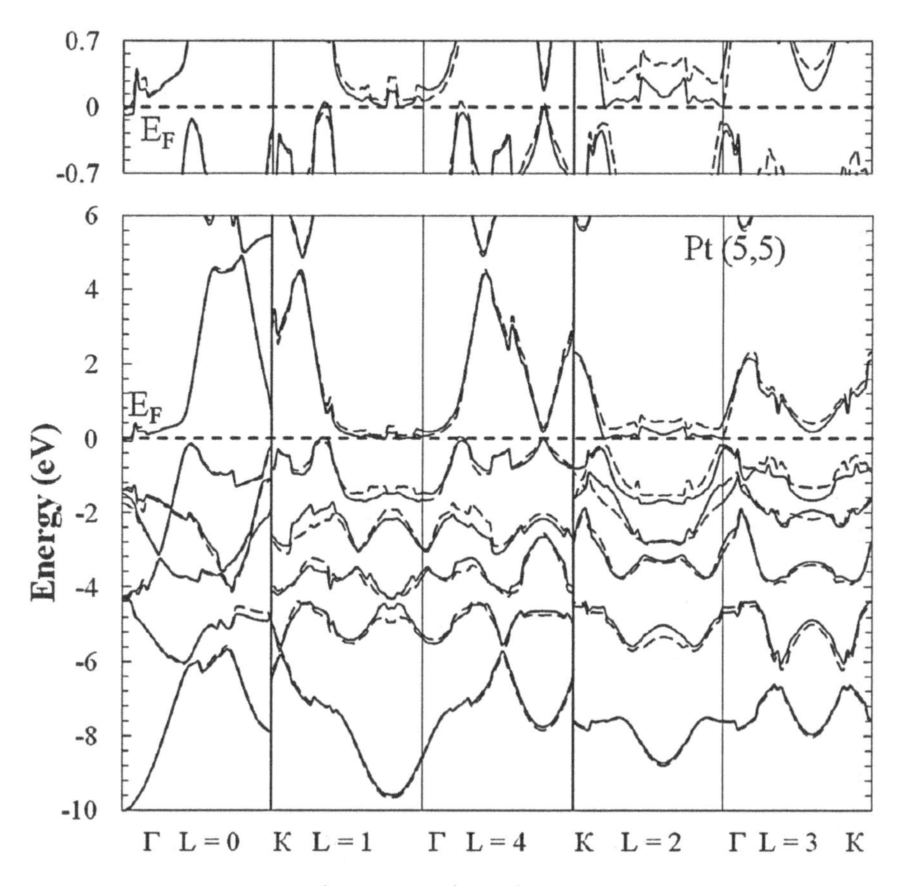

Figure 5.19. Band structure of armchair (5, 5) Pt nanotube.

5.3.3. Atomic Chains

In neglect of relativistic effects, the band structure of carbynes and transition metal chains was considered in Chapter 1. Here we consider these effects on the electronic properties of the A^NB^{8-N} atomic chains. It is known from the first-principles projected plane wave calculations, that these chains are stable, the binding energies of chains being close to the cohesive energies of the bulk materials with the same composition.

5.3.3.1. Secular Equation

The LACW technique presented above for the achiral nanotubes is easily rewritten for the case of the linear monoatomic chains having a more simple geometry (D'yachkov and Zaluev 2014). Again we start from the two-component Hamiltonian (5.1) with cylindrical MT potential, apply the basic functions corresponding to the free electron movement in the

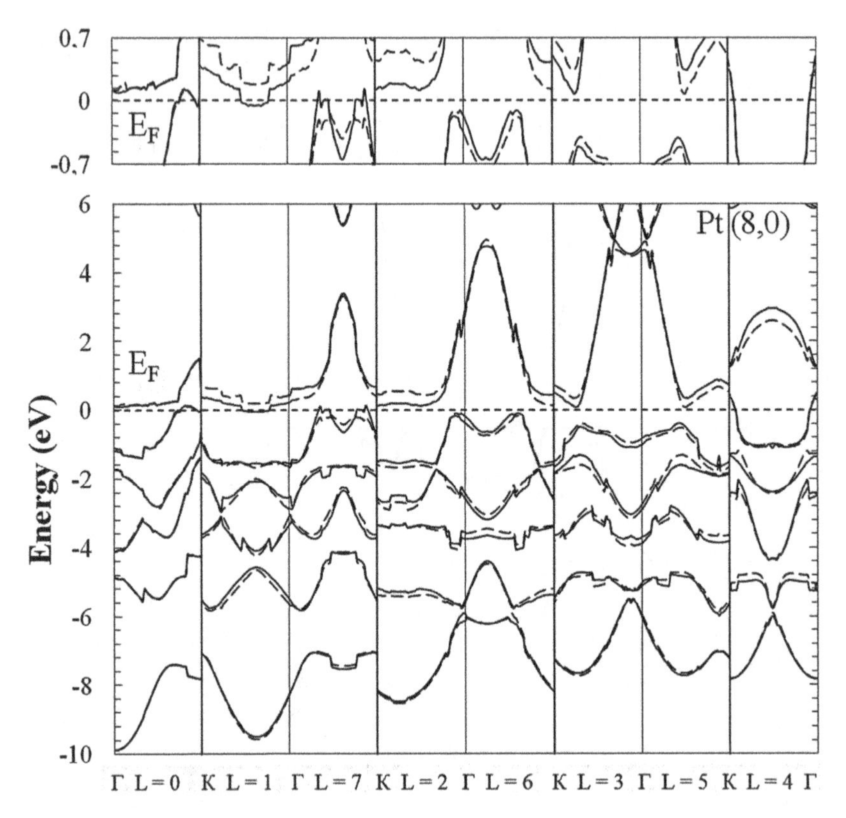

Figure 5.20. Electron structure of the zigzag (8, 0) Pt nanotube.

cylindrical potential well of the interspherical region sewn together with the solutions of the Schrödinger equation for MT regions expanded in spherical harmonics. Since the general cylindrical and local spherical coordinate systems coincide in the case of linear chains, it follows that the nonzero values of the $A^{PMN,k}_{lm,\alpha_{MT}}$ and $B^{PMN,k}_{lm,\alpha_{MT}}$ coefficients correspond to $m = M$

$$A^{PMN,k}_{lM,\alpha_{MT}} = r^2_{\alpha_{MT}} D^{PMN,k}_{lM,\alpha_{MT}} a^{PMN,k}_{lM,\alpha_{MT}}, \tag{5.24}$$

$$B^{PMN,k}_{lM,\alpha_{MT}} = r^2_{\alpha_{MT}} D^{PMN,k}_{lM,\alpha_{MT}} b^{PMN,k}_{lM,\alpha_{MT}}. \tag{5.25}$$

As a result, the sums over m disappear in the equations for the matrix elements of the spin-orbit interaction (5.11), (5.15) and these equations take the simpler form

$$\left\langle \Psi^k_{P_2M_2N_2}(\mathbf{r},\alpha) \middle| H_{S-O} \middle| \Psi^k_{P_1M_1N_1}(\mathbf{r},\alpha) \right\rangle = \delta_{M_2M_1} \frac{\pi}{c^2\Omega} \left\{ \left| J'_{M_1}\left(\kappa_{|M_1|,N_2} a\right) J'_{M_1} \right. \right.$$

$$\left(\kappa_{|M_1|,N_1}a\right)\Big|\Big\}^{-1}\times\sum_{\alpha_{MT}} r^4_{\alpha_{MT}}\sum_{l=|M_1|}^{\infty}(2l+1)M_1\frac{(l-|M_1|)!}{(l+|M_1|)!}$$
$$\times\exp\left\{i\left[\left(K_{p_2}-K_{p_1}\right)Z_{\alpha_{MT}}\right]\right\}\left\{\varsigma_{l,\alpha_{MT}}\left(\overline{a}^{P_2M_1N_2,k}_{lM_1,\alpha_{MT}}\right)\right.$$
$$\times\left(a^{P_1M_1N_1,k}_{lM_1,\alpha_{MT}}\right)+\ddot{\varsigma}_{l,\alpha_{MT}}\left(\overline{b}^{P_2M_1N_2,k}_{lM_1,\alpha_{MT}}\right)\left(b^{P_1M_1N_1,k}_{lM_1,\alpha_{MT}}\right)+\dot{\varsigma}_{l,\alpha_{MT}}$$
$$\left.\left[\left(\overline{a}^{P_2M_1N_2,k}_{lM_1,\alpha_{MT}}\right)\left(b^{P_1M_1N_1,k}_{lM_1,\alpha_{MT}}\right)+\left(\overline{b}^{P_2M_1N_2,k}_{lM_1,\alpha_{MT}}\right)\left(a^{P_1M_1N_1,k}_{lM_1,\alpha_{MT}}\right)\right]\right\} \tag{5.26}$$

and

$$\left\langle\Psi^k_{P_2M_2N_2}(\mathbf{r},\beta)\middle|H_{S-O}\middle|\Psi^k_{P_1M_1N_1}(\mathbf{r},\alpha)\right\rangle=\delta_{M_2-M_1,M_1}\,sign(M_1)\frac{\pi}{c^2\Omega}$$
$$\times\left\{\left|J'_{M_2}\left(\kappa_{|M_2|,N_2}a\right)J'_{M_1}\left(\kappa_{|M_1|,N_1}a\right)\right|\right\}^{-1}\times\sum_{\alpha_{MT}} r^4_{\alpha_{MT}}\sum_{l=M_1}^{\infty}\left[l(l+1)-M_1(M_1+1)\right]$$
$$\times\left[\frac{(l-M_1)!(l-|M_1+1|)!}{(l+|M_1|)!(l+|M_1+1|)!}\right]^{1/2}\left\{\varsigma_{l,\alpha_{MT}}\left(\overline{a}^{P_2M_2N_2,k}_{lm+1,\alpha_{MT}}\right)a^{P_1M_1N_1,k}_{lm,\alpha_{MT}}+\right.$$
$$\ddot{\varsigma}_{l,\alpha_{MT}}\left(\overline{b}^{P_2M_2N_2,k}_{lm+1,\alpha_{MT}}\right)\left(b^{P_1M_1N_1,k}_{lm,\alpha_{MT}}\right)+\dot{\varsigma}_{l,\alpha_{MT}}\left[\left(\overline{a}^{P_2M_2N_2,k}_{lm+1,\alpha_{MT}}\right)\left(b^{P_1M_1N_1,k}_{lm,\alpha_{MT}}\right)+\left(\overline{b}^{P_2M_2N_2,k}_{lm+1,\alpha_{MT}}\right)\right.$$
$$\left.\left.a^{P_1M_1N_1,k}_{lm,\alpha_{MT}}\right]\right\}\times\exp\left\{i\left[\left(K_{p_2}-K_{p_1}\right)Z_{\alpha_{MT}}\right]\right\} \tag{5.27}$$

where $sign(M_1) = 1$ for $M_1 \geq 0$ and $sign(M_1) = -1$ for $M_1 < 0$.

Based on previous LACW calculations of nanotubes, the barrier radii $a(A^{IV})$ for the fourth group chains were set equal to the arithmetic mean value of the atomic covalent and van der Waals radii. For the partially ionic chains, the corresponding radii $a(A^NB^{8-N})$ were taken as being the same as for the covalent analogs: $a(A^NB^{8-N}) = a(A^{IV})$. A good convergence of the relativistic band structures is obtained using 45–50 basic functions for the monoatomic chains and using 70–80 functions in the case of chains with two atoms per unit cell.

5.3.3.2. Group IV Chains

Figures 5.21 and 5.22 show the spin-dependent energy bands of cumulenic and polyynic carbynes.

Because of the rotational symmetry of the chains, there are σ- and π-type dispersion curves in the band structures of these systems. The two-fold orbitally degenerate π bands correspond to the semiclassical clockwise and anticlockwise rotational motion of electrons around the chain axis. In the absence of spin-orbit interaction with the two possible directions of spin, the π bands would be the fourfold degenerate ones. The spin and orbital motions of electrons are coupled, thereby splitting the fourfold degeneracy clearly seen in the insets of the figure. However, each energy

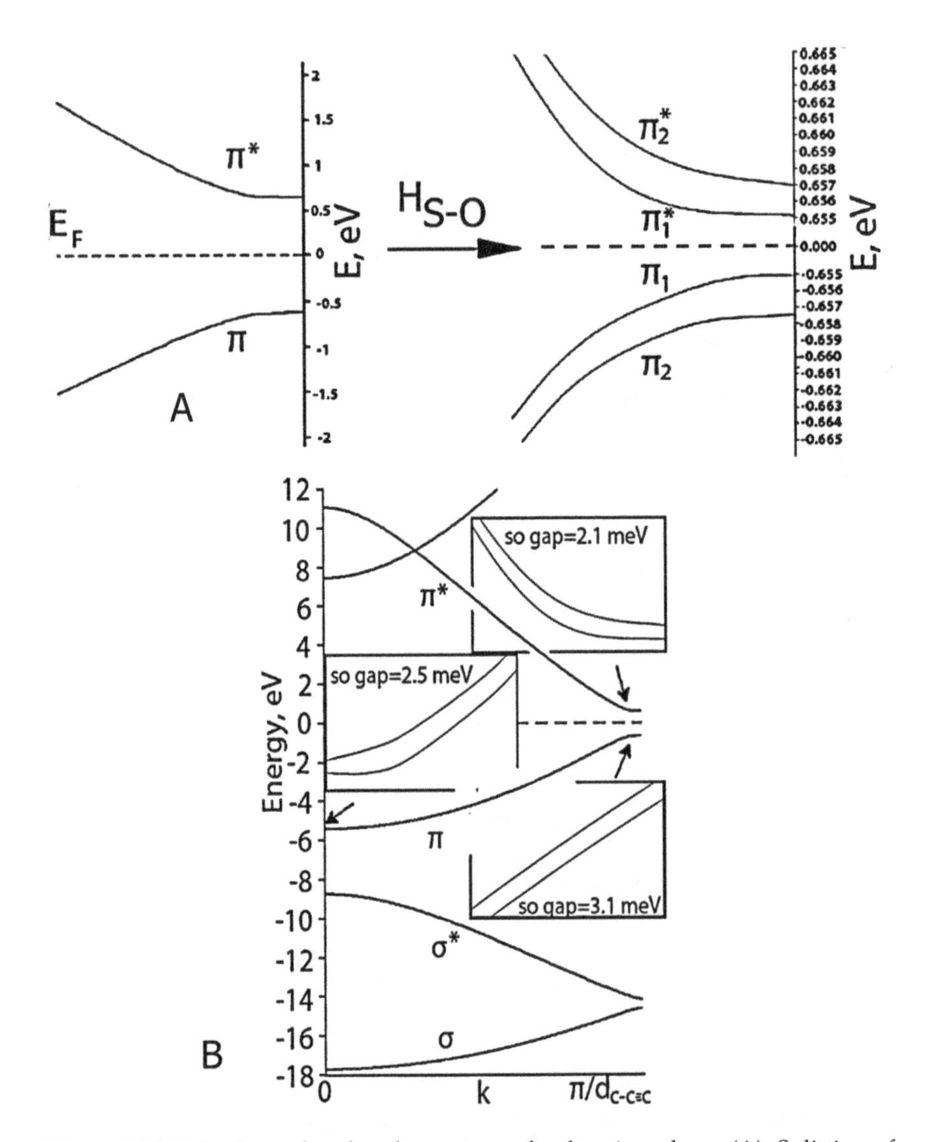

Figure 5.21. Spin-dependent band structure of polyynic carbyne (A). Splitting of levels in the gap region is shown on an enlarged scale (B).

π sub band still has the twofold degeneracy; Kramers theorem on time-reversal symmetry alongside the inversion symmetry of chains preserve the spin degeneracy, the spin polarization direction between degenerate two bands being opposite to each other. Obviously, there is no spin-orbit splitting of the orbitally non-degenerate σ bands.

The polyynic carbyne has the semiconducting-type band structure with a gap between bonding (π) and antibonding (π*) states equal to 1.14

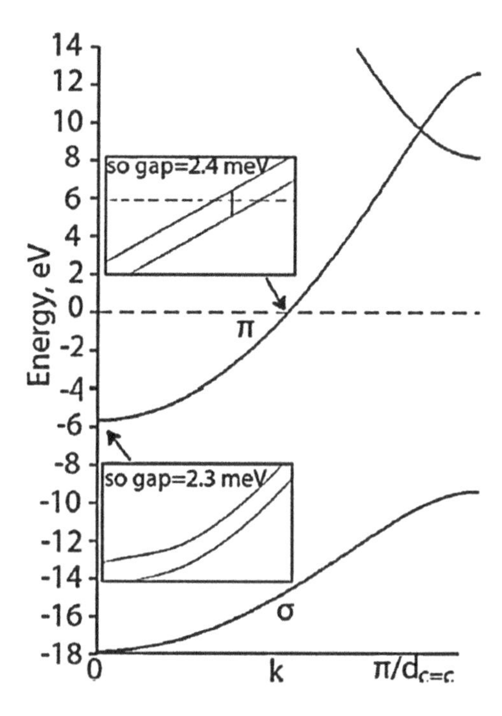

Figure 5.22. Spin-dependent band structure of cumulenic carbyne.

eV (Fig. 5.21). The spin-orbit interaction results in the splitting of these bands. Particularly, at the Brillouin zone boundary, the splitting induced by the spin-orbit interaction term is different for the highest valence band (3.1 meV) and the lowest conduction band (2.1 meV). For the absence of spin-orbit interaction, the metallic cumulenic carbyne would have one π band crossing the Fermi level at the center of the Brillouin zone. Due to the spin-orbit coupling, the band structure becomes more complex; the Fermi level crosses not one, but the two bands. The present calculation shows a direct optical gap equal to 2.4 meV at the Fermi level near the point $k = \pi/2d_{C-C}$, but the metallic nature for the cumulenic system is not broken by the spin-orbit coupling. The spin-orbit gaps in carbynes are about 2 or 3 times smaller than the spin-orbit splitting (6 meV) in the carbon atom.

The obtained spin-orbit gaps equal to 2–3 meV are in reasonable agreement with the semiclassical model and inverse diameter dependence of the spin-orbit splitting in carbon nanotubes. For example, in the case of the small-diameter (4, 4) nanotube with radius equal to 2.7 Å, the spin-orbit gap is equal to 0.54 meV. For carbyne, a radius of electron orbit is approximately 3.5 times smaller; it is about the carbon atomic radius r_C = 0.77 Å. Therefore, the spin-orbit gaps in carbynes are expected to be about 2 meV, in agreement with the numerical LACW calculations.

Figure 5.23 shows the band structures of all the group IV covalent chains calculated using the relativistic LACW method in comparison with non-relativistic *ab initio* LCAO data (D'yachkov et al. 2015); the relativistic LACW band structure calculations were performed for the geometries optimized using the LCAO method.

The generic band structures of these atomic wires are composed of the σ(s) bonding, σ*(p_z) antibonding, and one π band which is bonding at $k < \pi/(2d)$ and antibonding between $k = \pi/(2d)$ and $k = \pi/d$. Both the LACW and LCAO techniques predict that the chains of Si, Ge, Sn, and Pb are metallic. However, there is a great difference between the relativistic and non-relativistic band structures. The spin–orbit splitting energy $\Delta_{S\text{–}O}$ varies between about 1.5 meV and 0.45 eV for the C and Sn chains, respectively. In the case of the chain of the heaviest element Pb, $\Delta_{S\text{–}O}$ is already equal to 4 eV.

As is well known, in addition to the spin-orbit interaction, the Darwin H_{Dar} and mass-velocity H_{m-v} corrections are also referred to as the relativistic effects. As different from the spin-orbit coupling, which may split the degenerated levels and cause the mixing of levels, the H_{Dar} and H_{m-v} corrections are invariant under the operations of the single group and therefore do not split the levels, but these terms may mix levels of the same single-group symmetry and shift the levels. The effects of the H_{Dar}

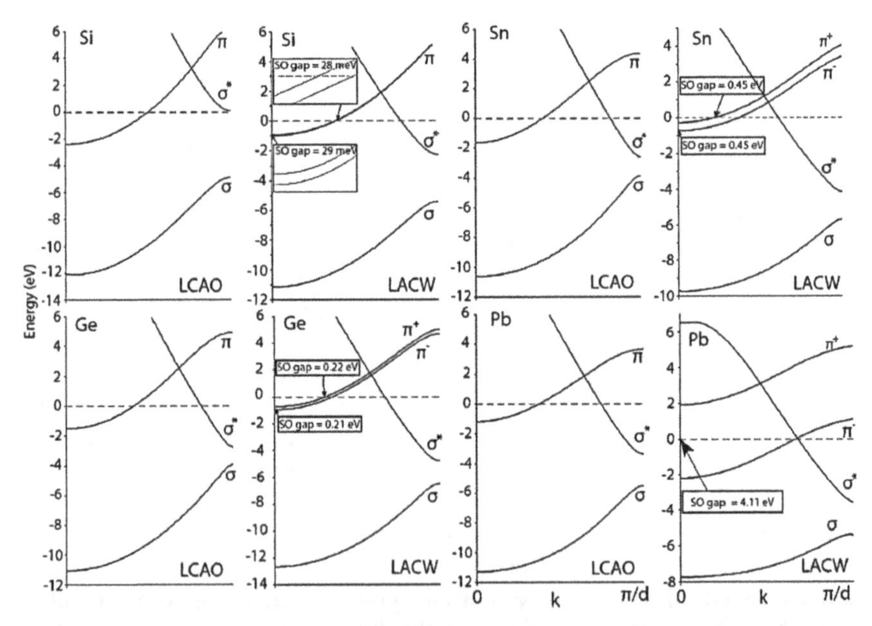

Figure 5.23. Band structures of the linear group IV chains calculated using the non-relativistic LCAO and relativistic LACW methods. The zero of energy is set at the Fermi level.

and H_{m-v} terms on the band structure of atomic chains was also studied using the LACW technique (D'yachkov et al. 2015). The method for calculating these corrections for atomic chains is given in the Appendix 1, Eqs. A6 and A10. The results show that the mass-velocity corrections shift all the valence band levels in the lower energy region. In the C and Si chains, the shifts are equal to 2–5 and 10–30 meV for different k points and are possibly negligible, but the mass-velocity shifts increase up to 0.6, 2.2, and 3.7 eV in the cases of the Ge, Sn, and Pb atomic wires. The Darwin corrections are several times smaller in comparison to the mass-velocity contributions, and can somewhat shift the energy levels to the high or low-energy region.

In the Fermi energy region, there are important differences between the band structures of the carbon and other group IV chains. In the case of the C chain, the π bands cross the Fermi level at $k = \pi/2d$ giving rise to Peierls' distortion, whereas the Si, Ge, Sn, and Pb chains behave differently; in addition to the π bands, there are the σ^* band dips below the Fermi level near the zone edge preventing Peierls' dimerization.

5.3.3.3. $A^{III}B^{V}$ and $A^{II}B^{VI}$ Chains

Figures 5.24 and 5.25 show that the transition from the covalent chains to the partially ionic ones is accompanied by a sharp change in the band structure.

For example, the carbon chain with all bond lengths equal to 1.27 Å has a metal type electronic structure, but the boron nitride chain with almost the same bond lengths of 1.30 Å is an insulator with an optical gap corresponding to the transition between the occupied π and vacant π^* states at the edge of the Brillouin zone and equal to 6 and 8 eV in the LCAO and LACW calculations, respectively. Qualitatively, the differences between the electronic bands of the covalent and partially ionic chains are explained by the presence of the antisymmetric components of the electron potential in the boron nitride chain. This is also the reason for a similar splitting of the s bands in the BN chain. The transition from the atomic BN chain to the AlP, GaAs, and InSb chains is accompanied by a decrease in the ionic character of the chemical bond leading to a gradual decrease in the $\pi - \pi^*$ gap and shift of the $\sigma_1{}^*$ bands in the region of lower energies. These effects lead to the fact that the AlP chain is a semiconductor with an indirect band gap equal to 2.2 eV (LACW) or 3.5 eV (LCAO).

According to the LCAO calculations, the chain GaAs is a semiconductor with a band gap equal to 0.5 eV; the gap is due to a transition between the π state on the edge of the Brillouin zone and the σ state in the center. It is metal according to the relativistic LACW data; the Fermi level crosses the π^+ zone, a formation of which is associated with spin-orbit splitting. Both methods predict that the InSb chain has a metal type band structure due

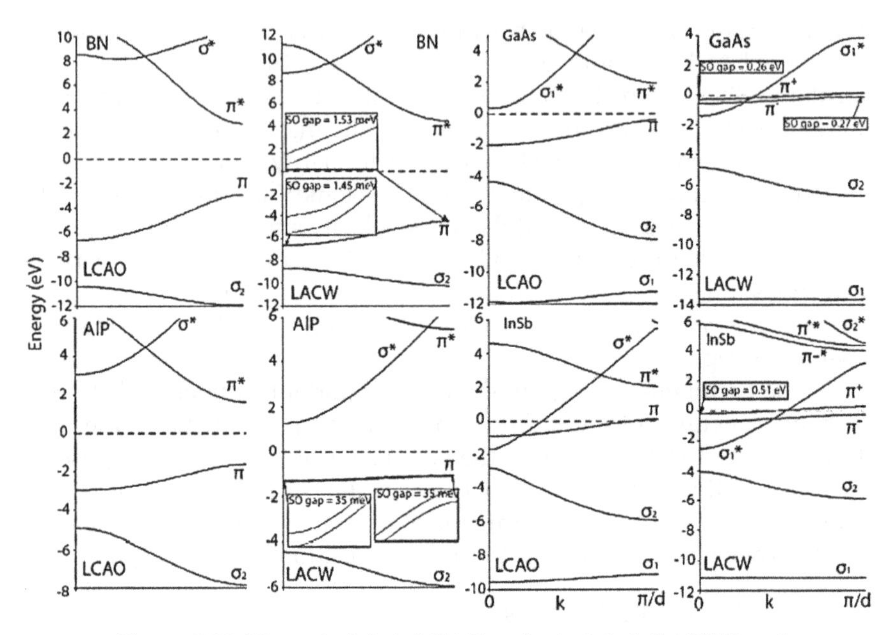

Figure 5.24. Non-relativistic LCAO and relativistic LACW band structures of the $A^{III}B^{V}$ chains.

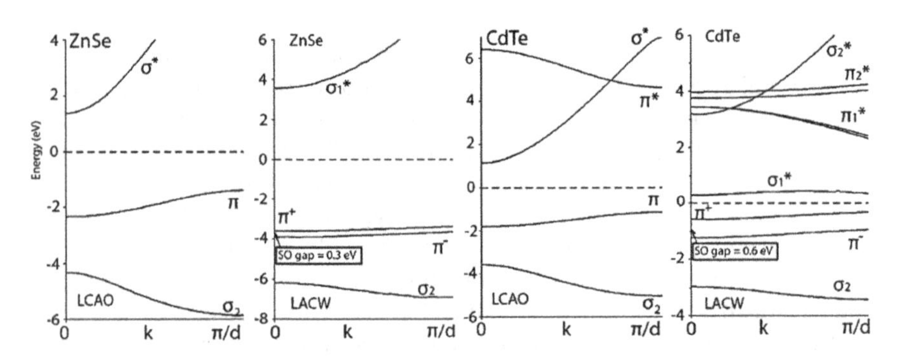

Figure 5.25. Non-relativistic LCAO and relativistic LACW band structures of $A^{II}B^{VI}$ chains.

to the intersection of the π and σ* bands; the spin-orbit interaction splits the π states, forms the π^+ and π^- bands, and noticeably complicates the band structure and DOS in the Fermi level region. The chemical bonding in the compounds $A^{II}B^{VI}$ is much more ionic than in $A^{III}B^{V}$; therefore, the antisymmetric component of the electron potential in the $A^{III}B^{V}$ chains is stronger, and, in the case of compounds from the same horizontal row in the periodic table, the transition from the $A^{III}B^{V}$ chains to the $A^{II}B^{VI}$ ones is accompanied by a sharp increase in the band gaps. For example, Figs.

5.24 and 5.25 show that, in the GaAs chain, the gap is absent or equal to about 0.5 eV, based on different data, but it is large in the ZnSe chain. All methods indicate the metal type of the band structure of the InSe chain, but the CdTe is an insulator with band gap equal to 2 and 0.8 eV in the non-relativistic LCAO and relativistic LACW calculations with due account of relativistic corrections, respectively.

It is interesting to underscore that the presence of the ionic component of the electronic potential dramatically changes the band structure in the transition from the covalent to partially ionic chains, although the ionicity almost does not affect the energy of spin-orbit splitting.

Appendix: Darwin and Mass-velocity Corrections

We present here the formulas for calculating the Darwin and mass-velocity terms in atomic wires using the LACW technique (Dyachkov et al. 2015). In terms of Rydberg atomic units, the general two-component Hamiltonian can be written as (Conklin et al. 1965)

$$H = -\Delta + V + \frac{1}{c^2}\sigma \cdot \left[(\nabla V) \times \mathbf{p} \right] + \frac{1}{2c^2}\nabla^2 V - \frac{1}{c^2} p^4. \tag{A1}$$

The first two terms correspond to the non-relativistic Hamiltonian operator and the third term is the operator of the spin–orbit interaction H_{S-O}, which may split the degenerated levels and cause the mixing of levels, thus altering the band picture in comparison with non-relativistic results. Methods for considering the spin-orbit coupling are described in Chapter 5. The fourth and fifth terms are the Darwin H_{Dar} and mass-velocity H_{m-v} corrections which are invariant under the operations of the single group and therefore do not split levels, but may mix levels of the same single-group symmetry and shift non-relativistic levels. In terms of the spinor basic functions, the matrix elements of the Hamiltonian (A1) can be written as

$$\begin{aligned} \left\langle \Psi^0_{n_2,k}(\mathbf{r})\chi_2 \middle| H \middle| \Psi^0_{n_1,k}(\mathbf{r})\chi_1 \right\rangle = E^0_{n_1,k}\delta_{n_1,n_2} &+ \left\langle \Psi^0_{n_2,k}(\mathbf{r})\chi_2 \middle| H_{S-O} \middle| \Psi^0_{n_1,k}(\mathbf{r})\chi_1 \right\rangle \\ &+ \delta_{\chi_2,\chi_1} \left\langle \Psi^0_{n_2,k}(\mathbf{r}) \middle| H_{Dar} \middle| \Psi^0_{n_1,k}(\mathbf{r}) \right\rangle \\ &+ \delta_{\chi_2,\chi_1} \left\langle \Psi^0_{n_2,k}(\mathbf{r}) \middle| H_{m-v} \middle| \Psi^0_{n_1,k}(\mathbf{r}) \right\rangle \end{aligned} \tag{A2}$$

The H_{m-v} and H_{Dar} parts of Eq. (A2) are completely determined by the matrix elements between the eigenfunctions of the non-relativistic Hamiltonian.

Following the description given in ref. (Conklin et al. 1965), note that the Hermitian character of the operator p requires that

$$\left\langle \Psi^0_{n_2,k}(\mathbf{r}) \middle| H_{m-v} \middle| \Psi^0_{n_1,k}(\mathbf{r}) \right\rangle = -\frac{1}{c^2}\left\langle \Psi^0_{n_2,k}(\mathbf{r}) \middle| p^4 \middle| \Psi^0_{n_1,k}(\mathbf{r}) \right\rangle$$
$$= -\frac{1}{c^2}\left\langle p^2\Psi^0_{n_2,k}(\mathbf{r}) \middle| p^2\Psi^0_{n_1,k}(\mathbf{r}) \right\rangle$$
$$= -\frac{1}{c^2}\left\langle \left[H_0 - V(\mathbf{r})\right]\Psi^0_{n_2,k}(\mathbf{r}) \middle| \left[H_0 - V(\mathbf{r})\right]\Psi^0_{n_1,k}(\mathbf{r}) \right\rangle$$
$$= -\frac{1}{c^2}\left\langle \left[E^0_{n_2,k} - V(\mathbf{r})\right]\Psi^0_{n_2,k}(\mathbf{r}) \middle| \left[E^0_{n_1,k} - V(\mathbf{r})\right]\Psi^0_{n_1,k}(\mathbf{r}) \right\rangle. \quad \text{(A3)}$$

Because of the approximate potential $V(\mathbf{r}) = 0$ in the region outside the MT spheres, taking into account the equation $\Psi^0_{n,k}(\mathbf{r}) = \sum_{PMN} a^{kn}_{PMN}\Psi^k_{PMN}(r)$, we have

$$\left\langle \Psi^0_{n_2,k}(\mathbf{r}) \middle| H_{m-v} \middle| \Psi^0_{n_1,k}(\mathbf{r}) \right\rangle =$$
$$-\frac{1}{c^2}\left\{ E^0_{n_2,k}E^0_{n_1,k}\delta_{n_2,n_1} + \sum_{\alpha_{MT}} \int_{\Omega_{\alpha_{MT}}} F^{\alpha_{MT}}_{n_2,n_1,k}(r)\bar{\Psi}^0_{n_2,k}(r)\Psi^0_{n_1,k}(r)\,dr \right\}$$
$$= -\frac{1}{c^2}\left\{ E^0_{n_2,k}E^0_{n_1,k}\delta_{n_2,n_1} + \sum_{P_2M_2N_2}\sum_{P_1M_1N_1} \bar{a}^{n_2,k}_{P_2M_2N_2}a^{n_1,k}_{P_1M_1N_1} \times \right.$$
$$\left. \sum_{\alpha_{MT}} \int_{\Omega_{\alpha_{MT}}} F^{\alpha_{MT}}_{n_2,n_1,k}(r)\bar{\Psi}^{P_2M_2N_2,k}_{\alpha_{MT}}(r)\Psi^{P_1M_1N_1,k}_{\alpha_{MT}}(r)\,dr \right\}, \quad \text{(A4)}$$

where

$$F^{\alpha_{MT}}_{n_2,n_1,k}(r) \equiv \left[E^0_{n_1,k} - V_{\alpha_{MT}}(r)\right]\left[E^0_{n_2,k} - V_{\alpha_{MT}}(r)\right] - E^0_{n_1,k}E^0_{n_2,k}. \quad \text{(A5)}$$

Finally, considering the analytical form of the basic function $\Psi^{PMN,k}_{\alpha_{MT}}$ in the region of MT spheres one obtains the final formula for calculating the mass-velocity matrix elements. Particularly, for the linear atomic chains we have

$$\left\langle \Psi^0_{n_2,k}(\mathbf{r}) \middle| H_{m-v} \middle| \Psi^0_{n_1,k}(\mathbf{r}) \right\rangle = -\frac{1}{c^2}\left\{ E^0_{n_2,k}E^0_{n_1,k}\delta_{n_2,n_1} + \right.$$
$$\frac{\pi}{\Omega}\sum_{P_2M_2N_2}\sum_{P_1M_1N_1} \delta_{M_2,M_1}\bar{a}^{n_2,k}_{P_2M_2N_2}a^{n_1,k}_{P_1M_1N_1}\left[J'_{M_1}\left(\kappa_{|M_1|N_2}a\right)J'_{M_1}\left(\kappa_{|M_1|N_1}a\right)\right]^{-1}$$
$$\times \sum_{\alpha_{MT}} r^4_{\alpha_{MT}} \exp\left[i\left(k_{P_1} - k_{P2}\right)Z_{\alpha_{MT}}\right] \sum_{l=|M_1|}^{\infty} (2l+1)\frac{(l-|M_1|)!}{(l+|M_1|)!} \times$$
$$\left[\mu_{n_2,n_1,k,l,\alpha_{MT}}\bar{a}^{P_2M_1N_2,k}_{lM_1,\alpha_{MT}}a^{P_1M_1N_1,k}_{lM_1,\alpha_{MT}} + \ddot{\mu}_{n_2,n_1,k,l,\alpha_{MT}}\bar{b}^{P_2M_1N_2,k}_{lM_1,\alpha_{MT}}b^{P_1M_1N_1,k}_{lM_1,\alpha_{MT}} + \right.$$

$$\dot{\mu}_{n_2,n_1,k,l,\alpha_{MT}} \left(\bar{a}^{P_2M_1N_2,k}_{lM_1,\alpha_{MT}} b^{P_1M_1N_1,k}_{lM_1,\alpha_{MT}} + \bar{b}^{P_2M_1N_2,k}_{lM_1,\alpha_{MT}} a^{P_1M_1N_1,k}_{lM_1,\alpha_{MT}} \right) \Big] \Bigg\}. \tag{A6}$$

Here

$$\mu_{n_2,n_1,k,l,\alpha_{MT}} = \int_0^{r_\alpha^{MT}} F^{\alpha_{MT}}_{n_2,n_1,k}(r) u^2_{l,\alpha_{MT}}(r) r^2 dr,$$

$$\dot{\mu}_{n_2,n_1,k,l,\alpha_{MT}} = \int_0^{r_\alpha^{MT}} F^{\alpha_{MT}}_{n_2,n_1,k}(r) u_{l,\alpha_{MT}}(r) \dot{u}_{l,\alpha_{MT}}(r) r^2 dr,$$

$$\ddot{\mu}_{n_2,n_1,k,l,\alpha_{MT}} = \int_0^{r_\alpha^{MT}} F^{\alpha_{MT}}_{n_2,n_1,k}(r) \dot{u}^2_{l,\alpha_{MT}}(r) r^2 dr.$$

The Darwin term may be rewritten in a more convenient form by using the following relations (Conklin et al. 1965)

$$\int \nabla \cdot \left[\bar{\Phi}\Phi'(\nabla V) \right] dv = \int_{\text{Encled surface}} \bar{\Phi}\Phi' \nabla V \cdot d\sigma = 0, \tag{A7}$$

$$\nabla \cdot \left[\bar{\Phi}\Phi'(\nabla V) \right] = \bar{\Phi}\Phi' \left(\nabla^2 V \right) + (\nabla V) \cdot (\nabla \bar{\Phi}) \Phi' + (\nabla V) \cdot (\nabla \Phi') \bar{\Phi}, \tag{A8}$$

where the integrals are taken over the volume defined by the periodic boundary conditions. Combining these equations leads to the result, with $\Phi = \Psi^0_{n_2,k}(r)$ and $\Phi' = \Psi^0_{n_1,k}(r)$:

$$\frac{1}{2c^2} \left\langle \Psi^0_{n_2,k}(\mathbf{r}) \middle| \nabla^2 V(r) \middle| \Psi^0_{n_1,k}(\mathbf{r}) \right\rangle = -\frac{1}{2c^2} \sum_{P_2M_2N_2} \sum_{P_1M_1N_1} \bar{a}^{n_2,k}_{P_2M_2N_2} a^{n_1,k}_{P_1M_1N_1} \times$$
$$\sum_{\alpha_{MT}} \int_{\Omega_{\alpha_{MT}}} \left\{ \bar{\Psi}^{P_2M_2N_2,k}_{\alpha_{MT}}(r) \left[\nabla V_{\alpha_{MT}}(r) \right] \left[\nabla \Psi^{P_1M_1N_1,k}_{\alpha_{MT}}(r) \right] + \right.$$
$$\left. \left[\nabla \bar{\Psi}^{P_2M_2N_2,k}_{\alpha_{MT}}(r) \right] \left[\nabla V_{\alpha_{MT}}(r) \right] \Psi^{P_1M_1N_1,k}_{\alpha_{MT}}(r) \right\} dr. \tag{A9}$$

Taking the analytical form of function $\Psi^{PMN,k}_{\alpha_{MT}}$ into account, we have for the linear atomic chains

$$\left\langle \Psi^0_{n_2,k}(\mathbf{r}) \middle| H_{Dar} \middle| \Psi^0_{n_1,k}(\mathbf{r}) \right\rangle = -\frac{\pi}{2c^2\Omega} \sum_{P_2M_2N_2} \sum_{P_1M_1N_1} \delta_{M_1,M_2} \bar{a}^{n_2,k}_{P_2M_2N_2} a^{n_1,k}_{P_1M_1N_1}$$
$$\times \left[J'_{M_1}\left(\kappa_{|M_1|N_2} a \right) J'_{M_1}\left(\kappa_{|M_1|N_1} a \right) \right]^{-1} \sum_{\alpha_{MT}} r^4_{\alpha_{MT}} \exp\left[i\left(k_{P_1} - k_{P_2} \right) Z_{\alpha_{MT}} \right] \times$$
$$\sum_{l=|M_1|}^{\infty} (2l+1) \frac{(l-|M_1|)!}{(l+|M_1|)!} \left\{ 2\varsigma_{l,\alpha_{MT}}(u,u') \bar{a}^{P_2M_1N_2,k}_{lM_1,\alpha_{MT}} a^{P_1M_1N_1,k}_{lM_1,\alpha_{MT}} \right.$$
$$+ 2\varsigma_{l,\alpha_{MT}}(\dot{u},\dot{u}') \bar{b}^{P_2M_1N_2,k}_{lM_1,\alpha_{MT}} b^{P_1M_1N_1,k}_{lM_1,\alpha_{MT}} + \left[\varsigma_{l,\alpha_{MT}}(u,\dot{u}') + \varsigma_{l,\alpha_{MT}}(\dot{u},u') \right]$$
$$\times \left[\bar{a}^{P_2M_1N_2,k}_{lM_1,\alpha_{MT}} b^{P_1M_1N_1,k}_{lM_1,\alpha_{MT}} + \bar{b}^{P_2M_1N_2,k}_{lM_1,\alpha_{MT}} a^{P_1M_1N_1,k}_{lM_1,\alpha_{MT}} \right] \tag{A10}$$

Here

$$\varsigma_{l,\alpha_{MT}}\left(u,u'\right)=\int_0^{r_\alpha^{MT}} u_{l,\alpha_{MT}}\left(r\right)\frac{dV_{\alpha_{MT}}\left(r\right)}{dr}u'_{l,\alpha_{MT}}\left(r\right)r^2dr,$$

$$\varsigma_{l,\alpha_{MT}}\left(\dot{u},\dot{u}'\right)=\int_0^{r_\alpha^{MT}} \dot{u}_{l,\alpha_{MT}}\left(r\right)\frac{dV_{\alpha_{MT}}\left(r\right)}{dr}\dot{u}'_{l,\alpha_{MT}}\left(r\right)r^2dr,$$

$$\varsigma_{l,\alpha_{MT}}\left(u,\dot{u}'\right)=\int_0^{r_\alpha^{MT}} u_{l,\alpha_{MT}}\left(r\right)\frac{dV_{\alpha_{MT}}\left(r\right)}{dr}\dot{u}'_{l,\alpha_{MT}}\left(r\right)r^2dr,$$

$$\varsigma_{l,\alpha_{MT}}\left(\dot{u},u'\right)=\int_0^{r_\alpha^{MT}} \dot{u}_{l,\alpha_{MT}}\left(r\right)\frac{dV_{\alpha_{MT}}\left(r\right)}{dr}u'_{l,\alpha_{MT}}\left(r\right)r^2dr.$$

References

Ajayan, P.M. 2004. Nanotechnology: How does a nanofibre grow? Nature 427: 402-403.

Ajayan, P.M., C. Colliex, J.M. Lambert, P. Bernier, L. Barbedette, M. Tence et al. 1994. Growth of manganese filled carbon nanofibers in the vapor phase. Phys. Rev. Lett. 72: 1722.

Ajiki, H. and T. Ando. 1993. Electronic states of carbon nanotubes. J. Phys. Soc. Jpn. 62: 1255-1266.

Ajiki, H. and T. Ando. 1996. Energy bands of carbon nanotubes in magnetic fields. J. Phys. Soc. Jpn. 65: 505-514.

Amer, M.R., A. Bushmaker and S.B. Cronin. 2012. The influence of substrate in determining the band gap of metallic carbon nanotubes. Nano Lett. 12: 4843.

Amer, M.R., S.-W. Chang, R. Dhall, J. Qiu and S.B. Cronin. 2013. Zener tunneling and photocurrent generation in quasi-metallic carbon nanotube *pn*-devices. Nano Lett. 13: 5129.

Anantram, M.P. and T.R. Govindan. 1998. Conductance of carbon nanotubes with disorder: A numerical study. Phys. Rev. B 58: 4882.

Andersen, O.K. 1970. Electronic structure of the fcc transition metals Ir, Rh, Pt and Pd. Phys. Rev. B 2: 883-907.

Andersen, O.K. 1975. Linear methods in band theory. Phys. Rev. B 12: 3060-3083.

Ando, T. 2000. Spin-orbit interaction in carbon nanotubes. J. Phys. Soc. Jpn. 69: 1757.

Ando, T. 2005. Theory of electronic states and transport in carbon nanotubes. J. Phys. Soc. Jpn. 74: 777-817.

Ando, T., T. Nakanishi and M. Igami. 1999. Effective-mass theory of carbon nanotubes with vacancy. J. Phys. Soc. Jpn. 68: 3994.

Ataca, C., S. Cahangirov, E. Durgun, Y.-R. Jang and S. Ciraci. 2008. Structural, electronic, and magnetic properties of 3d transition metal monatomic chains: First-principles calculations. Phys. Rev. B 77: 214413.

Avramescu, M.-L., P.E. Rasmussen and M. Chénier. 2016. Determination of metal impurities in carbon nanotubes sampled using surface wipes. J. Anal. Methods Chem. Article ID 3834292, 10 p.

Bachilo, S.M., M.S. Strano, C. Kittrell, R.H. Hauge, R.E. Smalley and R.B. Weisman 2002. Structure-assigned optical spectra of single-walled carbon nanotubes. Science 298: 2361-2366.

Bacsa, R.R., E. Flahaut, Ch. Laurent, A. Peigney, S. Aloni, P. Puech et al. 2003. Narrow diameter double-wall carbon nanotubes: Synthesis, electron microscopy and inelastic light scattering. New J. Phys. 5: 131.

Bai, Y., R. Zhang, X. Ye, Z. Zhu, H. Xie, B. Shen et al. 2018. Carbon nanotube bundles with tensile strength over 80 GPa. Nature Nanotechnology 13: 589-595.

Baňacký, P., J. Noga and V. Szöcs 2013. Electronic structure of single-wall silicon nanotubes and silicon nanoribbons: Helical symmetry treatment and effect of dimensionality. Adv. Condens. Matter. Phys. 374371.

Bandow, S., K. Hirahara, T. Hiraoka, G. Chen and P.C. Eklund. 2004. Turning peapods into double-walled carbon nanotubes. MRS Bull. 29: 260-264.

Bandow, S., M. Takizawa, K. Hirahara, M. Yudasaka and S. Iijima. 2001. Raman scattering study of double-wall carbon nanotubes derived from the chains of fullerenes in single-wall carbon nanotubes. Chem. Phys. Lett. 337: 48.

Barraza-Lopez, S., P.M. Albrecht and J.W. Lyding. 2009. Carbon nanotubes on partially depassivated n-doped Si (100) – (2×1): H substrates. Phys. Rev. B 80: 045415.

Beeby, J.L. 1967. The electronic structure of disordered system. Proc. R. Soc. London, Ser. A 302: 113.

Berthe, M., S. Yoshida, Y. Ebine, K. Kanazawa, A. Okada, A. Taninaka et al. 2007. Reversible defect engineering of single-walled carbon nanotubes using scanning tunneling microscopy. Nano Lett. 7: 3623.

Bi, Y. 2008. Controlled synthesis of pentagonal gold nanotubes at room temperature. Nanotechnology 19: 275306.

Blase, X., L.X. Benedict, E.L. Shirley and S.G. Louie. 1994. Hybridization effects and metallicity in small radius carbon nanotubes. Phys. Rev. Lett. 72: 1878-1881.

Bobenko, N.G., V.E. Egorushkin, N.V. Melnikova and A.N. Ponomarev. 2014. Are carbon nanotubes with impurities and structure disorder metals or semiconductors? Physica E 60: 11-16.

Bochkov, I.A. and P.N. D'yachkov. 2013. Electronic structure of carbon nanotubes with copper core. Russ. J. Inorg. Chem. 58: 800-802.

Bockrath, M., W. Liang, D. Bozovic, J.H. Hafner, C.M. Lieber, M. Tinkham et al. 2001. Resonant electron scattering by defects in single-walled carbon. Science 291: 283.

Bolshakov, A.D., A.M. Mozharov, G.A. Sapunov, I.V. Shtrom, N.V. Sibirev, V.V. Fedorov et al. 2018. Dopant-stimulated growth of GaN nanotube-like nanostructures on Si(111) by molecular beam epitaxy. Beilstein J. Nanotechnol. 9: 146-154.

Bonard, J.M., J.P. Salvetat, T. Stockli, W.A. de Heer, L. Forro and A. Chatelain. 1998. Field emission from single-wall carbon nanotube films. Appl. Phys. Lett. 73: 918-920.

Bordoloit, A.K. and S. Auluck. 1983. Electronic structure of platinum. J. Phys. F: Met. Phys. 13: 2101-2105.

Borondocs, F., K. Kamarás, M. Nikolou, D.B. Tanner, Z.H. Chen and A.G. Rinzle. 2006. Charge dynamics in transparent single-walled carbon nanotube films from optical transmission measurements. Phys. Rev. B 74: 045431.

Braspenning, P.J., R. Zeller, A. Lodder and P.H. Dederichs. 1984. Self-consistent cluster calculations with correct embedding for 3*d*, 4*d*, and some *sp* impurities in copper. Phys. Rev. B 29: 703.

Bridges, C.R., P.M. DiCarmine and D.S. Seferos. 2012. Gold nanotubes as sensitive, solution-suspendable refractive index reporters. Chem. Mater. 24(6): 963-965.

Bridges, C.R., P.M. DiCarmine, A. Fokina, D. Huesmann and D.S. Seferos. 2013. Synthesis of gold nanotubes with variable wall thicknesses. J. Mater. Chem. A1: 1127-1133.

Bulaev, D., B. Trauzettel and D. Loss. 2008. Spin-orbit interaction and anomalous spin relaxation in carbon nanotube quantum dots. Phys. Rev. B 77: 235301.

Bulusheva, L.G., A.V. Okotrub, D.A. Romanov and D. Tomanek. 1998. Electronic structure of (*n*,0) zigzag carbon nanotubes: Cluster and crystal approach. J. Phys. Chem. A 102: 975-981.

Bulusu, S. and X.C. Zeng. 2006. Structures and relative stability of neutral gold clusters: Au_n (n = 15-19). J. Chem. Phys. 125: 154303.

Burda, C., X.B. Chen, R. Narayanan and M.A. El-Sayed. 2005. Chemistry and properties of nanocrystals of different shapes. Chem. Rev. 105: 1025-1102.

Bushmaker, A.W., V.V. Deshpande, S. Hsieh, M.W. Bockrath and Stephen B. Cronin. 2009. Large modulations in the intensity of Raman-scattered light from pristine carbon nanotubes. Phys. Rev. Lett. 103: 067401.

Bussi, G., J. Menèndez, J. Ren, M. Canonico and E. Molinari. 2005. Quantum interferences in the Raman cross section for the radial breathing mode in metallic carbon nanotubes. Phys. Rev. B 71: 041404(R).

Cabria, I., J.W. Mintmire and C.T. White. 2003. Metallic and semiconducting narrow carbon nanotubes. Phys. Rev. B 67: 121406(R).

Cao, Q., J. Tersoff, D.B. Farmer, Y. Zhu and S.-J. Han. 2017. Carbon nanotube transistors scaled to a 40-nanometer footprint. Science 356: 1369-1372.

Carlsson, J.M. 2006. Curvature and chirality dependence of the properties of point defects in nanotubes. Phys. Status Solidi B 243: 3452-3457.

Carroll, D. L., Ph. Redlich, X. Blase, J.-C. Charlier, S. Curran, P.M. Ajayan et al. 1998. Effects of nanodomain formation on the electronic structure of doped carbon nanotubes. Phys. Rev. Lett. 81: 2332.

Chalifoux, W. and R. Tykwinski. 2010. Synthesis of polyynes to model the sp-carbon allotrope carbine. Nature Chemistry 2: 967-971.

Chang, S.-W., J. Hazra, M. Amer, R. Kapadia and S.B. Cronin. 2015. A comparison of photocurrent mechanisms in quasi-metallic and semiconducting carbon nanotube pn-junctions. ACS Nano 9: 11551-11556.

Charlier, J.-C., Ph. Lambin and T.W. Ebbesen. 1996a. Electronic properties of carbon nanotubes with polygonized cross sections. Phys. Rev. B 54: R8377.

Charlier, J.-C., T.W. Ebbesen and P. Lambin. 1996b. Structural and electronic properties of pentagon-heptagon pair defects in carbon nanotubes. Phys. Rev. B 53: 11108.

Chen, G., S. Bandow, E.R. Margine, C. Nisoli, A.N. Kolmogorov, Vincent H. Crespi et al. 2003. Chemically doped double-walled carbon nanotubes: Cylindrical molecular capacitors. Phys. Rev. Lett. 90: 257403.

Chen, I.-W.P., M.-D. Fu, W.-H. Tseng, J.-Y. Yu, S.-H. Wu, C.-J. Ku et al. 2006. Conductance and stochastic switching of ligand-supported linear chains of metal atoms. Angew. Chem., Int. Ed. 45: 5814-5818.

Chen, Y., J. Zou, S.J. Campbell and G. Le Caer. 2004. Boron nitride nanotubes: Pronounced resistance to oxidation. Appl. Phys. Lett. 84: 2430-2432.

Chibotaru, L.F., S.A. Bovin and A. Ceulemans. 2002. Bend-induced insulating gap in carbon nanotubes. Phys. Rev. B 66: 161401(R).

Chico, L., M.P. Lopez-Sancho and M.C. Muñoz. 1998. Carbon-nanotube-based quantum dot. Phys. Rev. Lett. 81: 1278.

Chiko, L., M.P. López-Sancho and M.C. Muñoz. 2004. Spin splitting induced by spin-orbit interaction in chiral nanotubes. Phys. Rev. Lett. 93: 176402.

Chico, L., V.H. Crespi, L.X. Benedict, S.G. Louie and M.L. Cohen. 1996. Pure carbon nanoscale devices: Nanotube heterojunctions. Phys. Rev. Lett. 76: 971.

Chiko, L., M.P. López-Sancho and M.C. Muñoz. 2009. Curvature-induced anisotropic spin-orbit splitting in carbon nanotubes. Phys. Rev. B 79: 235423.

Choi, H.J. and J. Ihm. 1999. Exact solutions to the tight-binding model for the conductance of carbon nanotubes. Solid State Commun. 111: 385-390.

Choi, H.J., J. Ihm, S.G. Louie and M.L. Cohen. 2000. Defects, quasibound states, and quantum conductance in metallic carbon nanotubes. Phys. Rev. Lett. 84: 2917.

Choi, S. 2014. Efficient antennas for terahertz and optical frequencies. A dissertation. University of Michigan.

Chopra, N.G., R.J. Luyken, K. Cherrey, V.H. Crespi, M.L. Cohen, S.G. Louie et al. 1995. Boron nitride nanotubes. Science 269: 966-967.

Christ, K.V. and H.R. Sadeghpour. 2007. Energy dispersion in graphene and carbon nanotubes and molecular encapsulation in nanotubes. Phys. Rev. B 75: 195418.

Chung, D.-S., S.H. Park, H.W. Lee, J.H. Choi, S.N. Cha and J.W. Kim. 2002. Carbon nanotube electron emitters with a gated structure using backside exposure processes. Appl. Phys. Lett. 80: 4045-4047.

Clogston, A.M. 1962. Impurity states in metals. Phys. Rev. 125: 439.

Collins, P.G., A. Zettl, H. Bando, A. Thess and R.E. Smalley. 1997. Nanotube nanodevice. Science 278: 100-102.

Conklin, J.B., L.E. Johnson and G.W. Pratt. 1965. Energy bands in PbTe. Phys. Rev. 137: 4A, 1283.

Craighead, H.G. 2000. Nanoelectromechanical systems. Science 290: 1532-1535.

Crespi, V.H., M.L. Cohen and A. Rubio. 1997. *In situ* band gap engineering of carbon nanotubes. Phys. Rev. Lett. 79: 2093.

Cretu, O., A.R. Botello Mendez, I.M. Janowska, C. Pham-Huu, J.-C. Charlier and F. Banhart. 2013. Electrical transport measured in atomic carbon chains. Nano Lett. 13: 3487-3493.

Cui, X., M. Freitag, R. Martel, L. Brus and P. Avouris. 2003. Controlling energy-level alignments at carbon nanotube/Au contacts. Nano Lett. 3: 783.

D'yachkov, E.P. and P.N. D'yachkov. 2016. The effect of 3d-metal dopants on the electronic structure of carbon nanotubes. Russ. J. Inorg. Chem. 61: 726-730.

D'yachkov, E.P., L.O. Khoroshavin, I.A. Bochkov, E.M. Kol'tsova and P.N.

D'yachkov, P. and D. Kutlubaev. 2012. Spin-orbit gaps in armchair nanotubes calculated using the linear augmented cylindrical wave method. IOP Conf. Ser.: Mater. Sci. Eng. 38: 012003.

D'yachkov, P. and D. Makaev. 2015. *Ab initio* spin-dependent band structures of carbon nanotubes. Intern. J. Quantum Chem. 116: 316-324.

D'yachkov, P.N and I.A. Bochkov. 2018. *Ab initio* band structure of quasi-metallic carbon nanotubes for terahertz applications. Computer Model. New Technol. 22: 1-19.

D'yachkov, P.N. 1997. LAPW studies of electronic interactions in fullerene tubes doped with transition metals. Zeitschr. fur Phys. Chem. 200: 165-169.

D'yachkov, P.N. 2004. Augmented waves for nanomaterials. pp. 191-212. *In*: H.S. Nalwa [ed.]. Encyclopedia of Nanoscience and Nanotechnology. Vol. 1. American Scientific Publishers, Valencia, CA.

D'yachkov, P.N. 2015. Electronic structure of a gold nanotube. Russ. J. Inorg. Chem. 60: 947-949.

D'yachkov, P.N. 2016. Linear augmented cylindrical wave method for nanotubes electronic structure. Intern. J. Quantum Chem. 116: 174-188.

D'yachkov, P.N. and B.S. Kuznetsov. 2004. Linear augmented spherical wave method for spherical clusters. Doklady Phys. Chem. 395: 57-61.

D'yachkov, P.N. and D.V. Kirin. 1999. Extension of the linear augmented-cylindrical-wave method to the electronic structure of nanotubes with an interior hole. Doklady Phys. Chem. (Engl. Trans.) 369: 326-333.

D'yachkov, P.N. and D.V. Kirin. 2001. Linearized augmented-cylindrical-wave method and its applications to band structure of nanotubes. pp. 273-280. *In*: S. Belucci [ed.]. Proceedings of the School and Workshop on Nanotubes and Nanostructures, Vol. 74. Italian Physical Society, Bologna, Italy.

D'yachkov, P.N. and D.V. Makaev. 2005a. Electronic structure of embedded carbon nanotubes. Phys. Rev. B 71: 081101(R).

D'yachkov, P.N. and D.V. Makaev. 2005b. Linear augmented cylindrical wave method for embedded carbon nanotubes. Doklady Phys. Chem. 402: 109-114.

D'yachkov, P.N. and D.V. Makaev. 2006. Linear augmented cylindrical wave method for calculating the electronic structure of double-wall carbon nanotubes. Phys. Rev. B 74: 155442.

D'yachkov, P.N. and D.V. Makaev. 2007. Account of helical and rotational symmetries in the linear augmented cylindrical wave method for calculating the electronic structure of nanotubes: Towards the *ab initio* determination of the band structure of a (100, 99) tubule. Phys. Rev. B 76: 195411.

D'yachkov, P.N. and D.V. Makaev. 2009. Electronic structure of BN nanotubes with intrinsic defects NB and BN and isoelectronic substitutional impurities PN, AsN, SbN, InB, GaB, and AlB. J. Phys. Chem. Solids 70: 180-185.

D'yachkov, P.N. and H. Hermann. 2004. Electronic structure and interband transitions of semiconducting carbon nanotubes. J. Appl. Phys. 95: 399-401.

D'yachkov, P.N. and O.M. Kepp. 2000. Linear augmented cylindrical wave method for nanotubes: Band structure of $[Cu@Cu]_{\infty}$. pp. 77-82. *In*: D. Tomanek and

R.J. Enbody [eds.]. Science and Application of Nanotubes. Kluwer, Academic/ Plenum, New York, USA.

D'yachkov, P.N. and V.A. Zaluev. 2014. Spin–orbit gaps in carbynes. J. Phys. Chem. C 118: 2799-2803.

D'yachkov, P.N. and D.O. Krasnov. 2019. Electronic and transport properties of deformed platinum nanotubes calculated using relativistic linear augmented cylindrical wave method.Chem. Phys. Lett. 720: 15–18.

D'yachkov, P.N., D.Z. Kutlubaev and D.V. Makaev. 2010. Linear augmented cylindrical wave Green's function method for electronic structure of nanotubes with substitutional impurities. Phys. Rev. B 82: 035426.

D'yachkov, P.N., H. Hermann and D.V. Kirin. 2002. Electronic structure and interband transitions of metallic carbon nanotubes. Appl. Phys. Lett. 81: 5228-5230.

D'yachkov, P.N., O.M. Kepp and A.V. Nikolaev. 1998. Augmented cylindrical wave method in the theory of electronic structure of quantum nanowires. Macromol. Symp. 136: 17-25.

D'yachkov, P.N., O.M. Kepp and A.V. Nikolaev. 1999. Linearized augmented-cylindrical-wave method in the electronic structure theory of nanowires. Doklady Chem. (Engl. Trans.) 365: 67-72.

D'yachkov, P.N., V.A. Zaluev, S.N. Piskunov and Y.F. Zhukovskii. 2015. Comparative analysis of the electronic structures of mono- and bi-atomic chains of IV, III–V and II–VI group elements calculated using the DFT LCAO and LACW methods. RSC Adv. 5: 91751-91759.

D'yachkov, P.N. 2014. The effect of 3d-metal intercalation on the electronic structure of metallic and semiconducting nanotubes. Russ. J. Inorg. Chem. 59: 682-687.

Dai, J.Y., J.M. Lauerhaas, A.A. Seltlur and R.P.H. Chang. 1996. Synthesis of carbon-encapsulated nanowires using polycyclic aromatic hydrocarbon precursors. Chem. Phys. Lett. 258: 547.

Daniel, M.C. and D. Astruc. 2004. Gold nanoparticles: Assembly, supramolecular chemistry, quantum-size-related properties, and applications toward biology, catalysis, and nanotechnology. Chem. Rev. 104: 293-346.

Davydov, A.S. 1965. Quantum Mechanics, 2nd ed. Pergamon Press, Oxford, New York.

De Heer, W.A., A. Chatelain and D. Ugarte. 1995. A carbon nanotube field-emission electron source. Science 270: 1179 1180.

De Volder, M.F.L., S.H. Tawfick, R.H. Baughman and A.J. Hart. 2013. Carbon nanotubes: Present and future commercial applications. Science 339: 535-539.

Dederichs, P.H., S. Lounis and R. Zeller. 2006. The Korringa-Kohn-Rostoker (KKR) Green Function Method II. Impurities and clusters in the bulk and on surface. pp. 279. *In*: J. Grotendorst, S. Blügel and D. Marx [eds.]. Computational Nanoscience: Do It Yourself! NIC Series, Vol. 31, edited by John von Neumann Institute for Computing, Jülich.

Dekker, C. 1999. Carbon nanotubes as molecular quantum wires. Phys. Today 52: 22-28.

Dekker, C. 2018. How we made the carbon nanotube transistor. Nature Electronics 1: 518.

Del Valle, M., C. Tejedor and G. Cuniberti. 2006. Scaling of the conductance in gold nanotubes. Phys. Rev. B 74: 045408.

Deshpande, V.V., B. Chandra, R. Caldwell, D.S. Novikov, J. Hone and M. Bockrath. 2009. Mott insulating state in ultraclean carbon nanotubes. Science 323: 106.

Dubay, O., G. Kresse and H. Kuzmany. 2002. Phonon softening in metallic nanotubes by a Peierls-like mechanism. Phys. Rev. Lett. 88: 235506.

Ebbesen, T.W. 1996. Carbon nanotubes. Phys. Today 49(6): 26.

Ebbesen, T.W. and P.M. Ajayan. 1992. Large-scale synthesis of carbon nanotubes. Nature 358: 220-222.

Endo, M., H. Muramatsu, T. Hayashi, Y.A. Kim, M. Terrones and M.S. Dresselhaus. 2005. Nanotechnology: 'Buckypaper' from coaxial nanotubes. Nature London 433: 476.

Enyashin, A.N., G. Seifert and A.L. Ivanovskii. 2004. Electronic, structural, and thermal properties of a nanocable consisting of carbon and BN nanotube. JETP Lett. 80: 608-611.

Espinosa-Soria, A. and A. Martínez. 2016. Transverse spin and spin-orbit coupling in silicon waveguides. IEEE Photonics Technol. Lett. 28: 1561-1564.

Ezawa, M. 2012. Dirac theory and topological phases of silicon nanotube. Eur. Phys. Lett. 98: 67001.

Ezawa, M. 2012. Quantum hall effects in silicene. J. Phys. Soc. Jpn. 81: 064705.

Fan, Y., B.R. Goldsmith and P.G. Collins. 2005. Identifying and counting point defects in carbon nanotubes. Nature Mater. 4: 906.

Farberovich, O.V., A. Yaresko, K. Kikoin and V. Fleurov. 2008. Electronic structure of transition-metal impurities in semiconductors: Cu in GaP. Phys. Rev. B 78: 085206.

Fenoglio, I., G. Greco, M. Tomatis, J. Muller, E. Raymundo-Pinero, F. Béguin et al. 2008. Structural defects play a major role in the acute lung toxicity of multiwall carbon nanotubes: Physicochemical aspects. Chem. Res. Toxicol. 21: 1690.

Fink, J.H. and Ph. Lambin. 2001. Electron spectroscopy studies of carbon nanotubes. pp. 247-272. *In*: M.S. Dresselhaus, G. Dresselhaus and Ph. Avouris [eds.]. Carbon Nanotubes, Topics. Appl. Phys. 80. Springer-Verlag, Berlin, Heidelberg.

Flensberg, K. and C. Marcus. 2010. Bends in nanotubes allow electric spin control and coupling. Phys. Rev. B 81: 195418.

Frank, S., P. Poncharal, Z.L. Wang and W.A. de Heer. 1998. Carbon nanotube quantum resistors. Science 280: 1744-1746.

Freitag, M. 2008. Doped defects tracked down. Nature Mater. 7: 840-844.

Fu, K., R. Zannoni, S.H. Chan, S.H. Adams, J. Nicholson, E. Polizzi et al. 2008. Terahertz detection in single wall carbon nanotubes. Appl. Phys. Lett. 92: 033105.

Galitski, V., B. Karnakov, V. Kogan and V. Galitski, Jr. 2013. Exploring Quantum Mechanics: A Collection of 7001 Solved Problems for Students, Lecturers, and Researchers. Oxford University Press, Oxford.

Galitskii, V.M., B.M. Karnakov and V.I. Kogan. 1984. Zadachi po Kvantovoi Mekhanike (Problems in Quantum Mechanics). Nauka, Moscow.

Gelin, M.F. and I.V. Bondarev. 2016. One-dimensional transport in hybrid metal-semiconductor nanotube systems. Phys. Rev. B 93: 115422.

Ghedjatti, A., Y. Magnin, F. Fossard, G. Wang, H. Amara, E. Flahaut et al. 2017. Structural properties of double-walled carbon nanotubes driven by mechanical interlayer coupling. ACS Nano 11: 4840-4847.

Golovacheva, A.Yu. and P.N. D'yachkov. 2005. Effect of intrinsic defects on the electronic structure of BN nanotubes. JETP Letters 82: 737-741.

Gordillo, M.C. 2006. Role of vacancies in the adsorption of quantum noble gases inside a bundle of carbon nanotubes. Phys. Rev. Lett. 96: 216102.

Green, A.A. and M.C. Hersam. 2011. Properties and application of double-walled carbon nanotubes sorted by outer-wall electronic type. ACS Nano. 5: 1459-1467.

Grünzel, T., Y.J. Lee, K. Kuepper and J. Bachmann. 2013. Preparation of electrochemically active silicon nanotubes in highly ordered arrays. Beilstein J. Nanotechnol. 4: 655-664.

Grzelczak, M., M.A. Correa-Duarte and L.M. Liz-Marzán. 2006. Carbon nanotubes encapsulated in wormlike hollow silica shells. Small 2: 1174-1177.

Guan, L., K. Suenaga, S. Okubo, T. Okazaki and S. Iijima. 2008. Metallic wires of lanthanum atoms inside carbon nanotubes. J. Am. Chem. Soc. 130: 2162-2163.

Guimarães, F.S.M., D.F. Kirwan, A.T. Costa, R.B. Muniz, D.L. Mills and M.S. Ferreira. 2010. Carbon nanotube: A low-loss spin-current waveguide. Phys. Rev. B 81: 153408.

Gülseren, O., T. Yildirim and S. Ciraci. 2002. Systematic *ab initio* study of curvature effects in carbon nanotubes. Phys. Rev. B65: 153405.

Gunlycke, D., J.H. Jefferson, S.W.D. Bailey, C.J. Lambert, D.G. Pettifor and G.A.D. Briggs. 2006. Zener quantum dot spin filter in a carbon nanotube. J. Phys.: Condens. Matter. 18: S843–S849.

Guo, G.Y. and J.C. Lin. 2005. Systematic *ab initio* study of the optical properties of BN nanotubes. Phys. Rev. B71: 165402.

Gupta, B.K., G. Kedawat, P. Kumar, S. Singh, S.R. Suryawanshi, N. Agrawal et al. 2018. High-performance field emission device utilizing vertically aligned carbon nanotubes-based pillar architectures. AIP. Advances 8: 015117.

Hagen, A. and T. Hertel. 2003. Quantitative analysis of optical spectra from individual single-wall carbon nanotubes. Nano Lett. 3: 383-388.

Hamada, N.S., Sawada and A. Oshiyama. 1992. New one-dimensional conductors: Graphitic microtubule. Phys. Rev. Lett. 68: 1579-1581.

Hartmann, R.R. and M.E. Portnoi. 2015. Terahertz transitions in quasi-metallic carbon nanotubes. IOP Conf. Ser.: Mater. Sci. Engin. 79: 012014.

Hartmann, R.R. and M.E. Portnoi. 2016. Exciton states in narrow-gap carbon nanotubes. AIP Conf. Proc. 1705: 020046.

Hartmann, R.R., J. Kono and M.E. Portnoi. 2014. Terahertz science and technology of carbon nanomaterials. Nanotechnology 25: 322001.

Haruta, M. 2005. Catalysis: Gold rush. Nature 437: 1098-1099.

Hashimoto, A., K. Suenaga, A. Gloter, K. Urita and S. Iijima. 2004. Direct evidence for atomic defects in graphene layers. Nature London 430: 870.

Hashimoto, A., K. Suenaga, K. Urita, T. Shimada, T. Sugai, S. Bandow et al. 2005. Atomic correlation between adjacent graphene layers in double-wall carbon nanotubes. Phys. Rev. Lett. 94: 045504.

He, H., J. Klinovski, M. Foster and A. Lerf. 1998. A new structural model for graphite oxide. Chem. Phys. Lett. 287: 53-56.

He, X., H. Htoon, S.K. Doorn, W.H.P. Pernice, F. Pyatkov, R. Krupke et al. 2018. Carbon nanotubes as emerging quantum-light sources. Nature Materials 17: 663-670.

Herrera-Suárez, H.J., A. Rubio–Ponce and D. Olguín. 2012. Electronic band structure of platinum low-index surfaces: Ab initio and tight-binding study. II Revista Mexicana de Física 58: 46-54.

Hertel, T., A. Hagen, V. Talalaev, K. Arnold, F. Hennrich, M. Kappes et al. 2005. Spectroscopy of single- and double-wall carbon nanotubes in different environments. Nano Lett. 5: 511-514.

Hever, A., J. Bernstein and O. Hod. 2012. Structural stability and electronic properties of sp^3 type silicon nanotubes. J. Chem. Phys. 137: 214702.

Hohenberg, P. and W. Kohn. 1964. Inhomogeneous electron gas. Phys. Rev. 136: B864-B871.

Hu, Y.H. 2011. Bending effect of sp-hybridized carbon (carbyne) chains on their structures and properties. J. Phys. Chem. C 115: 1843-1850.

Huang, D., F. Liao, S. Molesa, D. Redinger and V. Subramanian. 2003. Plastic-compatible low resistance printable gold nanoparticle conductors for flexible electronics. J. Electrochem. Soc. 150: 412-417.

Huertas-Hernando, D., F. Guinea and A. Brataas. 2006. Spin-orbit coupling in curved graphene, fullerenes, nanotubes, and nanotube caps. Phys. Rev. B 74: 155426.

Hui, L., F. Pederiva, W. Guanghou and W. Baolin. 2003. Structural calculation and properties of one-dimensional Pt materials. Chem. Phys. Lett. 381: 94-101.

Hutchings, G.J. and A.S.K. Hashmi. 2006. Gold catalysis. Angew. Chem. Int. Ed. Engl. 45: 7896-7936.

Hutchison, J.L., N.A. Kiselev, E.P. Krinichnaya, A.V. Krestinin, R.O. Loutfy, A.P. Morawsky et al. 2001. Double-walled carbon nanotubes fabricated by a hydrogen arc discharge method. Carbon 39: 761.

Igami, M., T. Nakanishi and T. Ando. 1999. Conductance of carbon nanotubes with a vacancy. J. Phys. Soc. Jpn. 68: 716.

Iijima, S. 1991. Helical microtubules of graphitic carbon. Nature 354: 56-58.

Ilani, S. and P.L. McEuen. 2010. Electron transport in carbon nanotubes. Annu. Rev. Condens. Matter Phys. 1: 1-25.

Itkis, M.E., S. Niyogi, M.E. Meng, M.A. Hamon, H. Hu and R.C. Haddon. 2002. Spectroscopic study of the fermi level electronic structure of single-walled carbon nanotubes. Nano Lett. 2: 155.

Izumida, W., K. Sato and R. Saito. 2009. Spin-orbit interaction in single wall carbon nanotubes: Symmetry adapted tight-binding calculation and effective model analysis. J. Phys. Soc. Jpn. 78: 074707.

Jaspersen, T.S., K. Grove-Rusmussen, J. Paaske, K. Muraki, T. Fujisawa, J. Nigård et al. 2011. Gate-dependent spin-orbit coupling in multielectron carbon nanotubes. Nature Physics 7: 348.

Jensen, A., J.R. Hauptmann, J. Nygerd, J. Sadowski and P.E. Lindelof. 2004. Hybrid devices from single wall carbon nanotubes epitaxially grown into a semiconductor heterostructure. Nano Lett. 4: 349-352.

Jensen, K., J. Weldon, H. Garcia and A. Zettl. 2007. Nanotube radio. Nano Lett. 7: 3508.

Jensen, K., K. Kim and A. Zettl. 2008. An atomic-resolution nanomechanical mass sensor. Nature Nanotechnol. 3: 533.

Jepsen, O., J. Madsen and O.K. Andersen. 1978. Band structure of thin films by the linear augmented-plane-wave method. Phys. Rev. B 18: 605.

Jhang, S.H., M. Marganska, Y. Skuorsky, D. Preusche, B. Witkamp, M. Grifony et al. 2010. Spin-orbit interaction in chiral carbon nanotubes probed in pulsed magnetic fields. Phys. Rev. B 82: 041404.

Jin, C.H., H.P. Lan, L.M. Peng, K. Suenaga and S. Iijima. 2009. Deriving carbon atomic chains from graphene. Phys. Rev. Lett. 102: 205501.

Jorio, A., R. Saito, J.H. Hafner, C.M. Lieber, M. Hunter, T. McClure et al. 2001. Structural (*n*, *m*) determination of isolated single-wall carbon nanotubes by resonant Raman scattering. Phys. Rev. Lett. 86: 1118-1121.

Joselevich, E. 2006. Twisting nanotubes: From torsion to chirality. ChemPhysChem. 7: 1405-1407.

Kampfrath, T., K. von Volkmann, C.M. Aguirre, P. Desjardins, R. Martel, M. Krenz et al. 2008. Mechanism of the far-infrared absorption of carbon-nanotube films. Phys. Rev. B 101: 267403.

Kane, C.L. and E.J. Mele. 1997. Size, shape, and low energy electronic structure of carbon nanotubes. Phys. Rev. Lett. 78: 1932.

Kazaoui, S., N. Minami and H. Yamawaki. 2000. Pressure dependence of the optical absorption spectra of single-walled carbon nanotube films. Phys. Rev. B 62: 1643.

Kepp, O.M. and P.N. D'yachkov. 1999. Band structure of metal chains revealed by an added cylinder wave method. Doklady Akad. Nauk 365: 354-359.

Kim, J.H., T.V. Pham, J.H. Hwang, C.S. Kim and M.J. Kim. 2018. Boron nitride nanotubes: Synthesis and applications. Nano Convergence 5: 17.

Kim, Y.A., H. Muramatsu, T. Hayashi, M. Endo, M. Terrones and M.S. Dresselhaus. 2004. Thermal stability and structural changes of double-walled carbon nanotubes by heat treatment. Chem. Phys. Lett. 398: 87-92.

Kirin, D.V. and P.N. D'yachkov. 2001. Electronic structure of carbyne and $C_{1-x}(BN)_x$ ($x = 0, 1, 1/3$) and GaAs nanotubes as determined by the full-potential linear augmented-cylindrical-wave method. Doklady Phys. Chem. 380: 227-233.

Knupfer, M. 2001. Electronic properties of carbon nanostructures. Surf. Sci. Rep. 42: 1-74.

Koelling, D.D. and B.N. Harmon. 1977. A technique for relativistic spin-polarized calculations. J. Phys. C: Solid State Phys. 10: 3107.

Koelling, D.D. and G.O. Arbman. 1975. Use of energy derivative of the radial solution in an augmented plane wave method: Application to copper. J. Phys. F: Met. Phys. 5: 2041-2054.

Kohn, W. and L.J. Sham. 1965. Self-consistent equations including exchange and correlation effects. Phys. Rev. 140: A1133-A1138.

Kohn, W. and N. Rostoker. 1954. Solution of the Schrödinger equation in periodic lattices with an application to metallic lithium. Phys. Rev. 94: 1111.

Komorowski, P.G. and M.G. Cottam. 2017. Electronic modes in carbon nanotubes with single and double impurity sites. Intern. J. Mod. Phys. B 31: 1750220.

Konar, S. and B.C. Gupta. 2008. Density functional study of single-wall and double-wall platinum nanotubes. Phys. Rev. B 78: 235414.

Kondo, Y. and K. Takayanagi. 2000. Synthesis and characterization of helical multi-shell gold nanowires. Science 289: 606-608.

Kong, J., N.R. Franklin, C. Zhou, M.G. Chapline, S. Peng, K. Cho et al. 2000. Nanotube molecular wires as chemical sensors. Science 287: 622.

Korn, G.A. and T.M. Korn. 1961. Mathematical Handbook for Scientists and Engineers. McGraw-Hill, New York, USA.

Korringa, J. 1947. On the calculation of the energy of a bloch wave in a metal. Physica (Amsterdam) 13: 392.

Kostov, M.K., E.E. Santiso, A.M. George, K.E. Gubbins and M.B. Nardelli. 2005. Dissociation of water on defective carbon substrates. Phys. Rev. Lett. 95: 136105.

Kostyrko, T., M. Bartkowiak and G.D. Mahan. 1999a. Reflection by defects in a tight-binding model of nanotubes. Phys. Rev. B 59: 3241.

Kostyrko, T., M. Bartkowiak and G.D. Mahan. 1999b. Localization in carbon nanotubes within a tight-binding model. Phys. Rev. B 60: 10735.

Kotakoski, J., A.V. Krasheninnikov, Y. Ma, A.S. Foster, K. Nordlund and R.M. Nieminen. 2005. B and N ion implantation into carbon nanotubes: Insight from atomistic simulations. Phys. Rev. B 71: 205408.

Krakauer, H., M. Posternak and A.J. Freeman. 1979. Linearized augmented plane-wave method for the electronic band structure of thin films. Phys. Rev. B 19: 1706-1719.

Krasheninnikov, A. 2001. Predicted scanning tunneling microscopy images of carbon nanotubes with atomic vacancies. Solid State Commun. 118: 361.

Krasheninnikov, A.V. 2007. Irradiation-induced phenomena in carbon nanotubes. pp. 250. *In*: V.A. Basiuk and E.V. Basiuk [eds.]. Chemistry of Carbon Nanotubes. Blackwell, Oxford.

Krasheninnikov, A.V. and F. Banhart. 2007. Engineering of nanostructured carbon materials with electron or ion beams. Nature Mater. 6: 723.

Krasheninnikov, A.V., K. Nordlund, M. Sirviö, E. Salonen and J. Keinonen. 2001. Formation of ion-irradiation-induced atomic-scale defects on walls of carbon nanotubes. Phys. Rev. B 63: 245405.

Krasnov, D.O., L.O. Khoroshavin and P.N. D'yachkov. 2019. Spin-orbital interaction in single-wall gold nanotubes. Russ. J. Inorg. Chem. 64: 1.

Kudryavtsev, Yu.P., M.B. Evsyukov and M.B. Guseva. 1993. Carbyne—the third allotropic form of carbon. Russ. Chem. Bull. 3: 399-413.

Kuemmeth, F., H. Churchill, P. Herring and C. Marcus. 2010. Carbon nanotubes for coherent spintronics. Mater. Today 13: 18-26.

Kuemmeth, F., S. Ilani, D.C. Ralph and P.L. McEuen. 2008. Coupling of spin and orbital motion of electrons in carbon nanotubes. Nature (London) 452: 448.

Kürti, J., V. Zólyomi, M. Kertesz, G. Sun, R.H. Baughman and H. Kuzmany. 2004. Individualities and average behavior in the physical properties of small diameter single-walled carbon nanotubes. Carbon 42: 971-978.

Kuznetsova, A., J.T. Jates Jr., J. Liu and R.E. Smalley. 2000. Physical adsorption of xenon in open single walled carbon nanotubes: Observation of a quasi-one-dimensional confined Xe phase. J. Chem. Phys. 112: 9590-9598.

Kwon, Y.-K. and D. Tomanek. 1998. Electronic and structural properties of multiwall carbon nanotubes. Phys. Rev. B 58: R16001.

Laird, E.A., F. Kuemmeth, G.A. Steele, K. Grove-Rasmussen, J. Nygård, K. Flensberg et al. 2015. Quantum transport in carbon nanotubes. Rev. of Mod. Phys. 87: 703-764.

Lambin, Ph., A. Fonseca, J.P. Vigneron, J.B. Nagy and A.A. Lucas. 1995. Structural and electronic properties of bent carbon nanotubes. Chem. Phys. Lett. 245: 85-89.

Lambin, Ph., L. Philippe, J.C. Charlier and J.P. Michenaud. 1994. Electronic band structure of multilayered carbon tubules. Comput. Mater. Sci. 2: 350-356.

Larina, E.V., V.I. Chmyrev, V.M. Skorikov, P.N. D'yachkov and D.V. Makaev. 2008. Band structure of silicon carbide nanotubes. Inorg. Mater. 44: 823-830.

Levshov, D.I., R. Parret, H.-N. Tran, T. Michel, T.T. Cao, V.C. Nguyen et al. 2017. Photoluminescence from an individual double-walled carbon nanotube. Phys. Rev. B 96: 195410.

Li, F., S.G. Chou, W. Ren, J.A. Gardecki, A.K. Swan, B.B. Goldberg et al. 2003. Identification of the constituents of double-walled carbon nanotubes using Raman spectra taken with different laser-excitation energies. J. Mater. Res. 18: 1251-1258.

Li, W.Z., J.G. Wen, Y. Tu and Z.F. Ren. 2001. Effect of gas pressure on the growth and structure of carbon nanotubes by chemical vapor deposition. Appl. Phys. A: Mater. Sci. Process. 73: 259-264.

Li, W.Z., S.S. Xie, L.X. Qian, B.H. Chung, B.S. Zou, W.Y. Zhou et al. 1996. Large-scale synthesis of aligned carbon nanotubes. Science 274: 1701-1703.

Li, Z.M., Z.K. Tang, H.J. Liu, N. Wang, C.T. Chan, R. Saito et al. 2001. Polarized absorption spectra of single-walled 4 Å carbon nanotubes aligned in channels of an $AlPO_4^{-5}$ single crystal. Phys. Rev. Lett. 87: 127401.

Lim, J., R. Loþez and R. Aguado. 2011. Josephson current in carbon nanotubes with spin-orbit interaction. Phys. Rev. Lett. 107: 196801.

Lin, H., J. Lagoute, V. Repain, C. Chacon, Y. Girard, J.-S. Lauret et al. 2010. Many-body effects in electronic bandgaps of carbon nanotubes measured by scanning tunnelling spectroscopy. Nature Mater. 9: 235.

Liu, A.Y., R.M. Wentzcovich and M.L. Cohen. 1989. Atomic arrangement and electronic structure of BC_2N. Phys. Rev. B 39: 1760-1765.

Liu, H. 2011. Band structures of carbon nanotube with spin-orbit coupling interaction. Physica B 406: 104.

Liu, H.J. and C.T. Chan. 2002. Properties of 4 Å carbon nanotubes from first-principles calculations. Phys. Rev. B 66: 115416.

Liu, N., S. Jin, L. Guo, G. Wang, H. Shao and L. Chen. 2017. Two-dimensional semiconducting gold. Phys. Rev. B 95: 155311.

Liu, X., T. Pichler, M. Knupfer, M. Golden, J. Fink, H. Kataura et al. 2001. Electronic structure and optical properties of single wall carbon nanotubes and C60 peapods. AIP Conf. Proc. 591: 266-303.

Loiseau, A., N. Demoncy, O. Stéphan, C. Colliex and H. Pascard. 2000. Filling carbon nanotubes using an ARC discharge. pp. 1-16. *In*: D. Tomanek and R.J.

Enbody [eds.]. Science and Application of Nanotubes. Kluwer, New York, USA.

Loucks, T.L. 1965. Relativistic electronic structure in crystals. I. Theory. Phys. Rev. 139: 4A, 1333.

Ma, L., D. Guan, F. Wang and C. Yuan. 2018. Environmental emissions from chemical etching synthesis of silicon nanotube for lithium ion battery applications. J. Manuf. Mater. Process. 2: 11.

MacDonald, A.H., W.E. Pickettll and D.D. Koelling. 1980. A linearised relativistic augmented-plane-wave method utilizing approximate pure spin basis functions. J. Phys. C: Solid St. Phys. 13: 2675.

Machón, M., S. Reich, C. Thomsen, D. Sánchez-Portal and P. Ordejón. 2002. Ab initio calculations of the optical properties of 4Å-diameter single-walled nanotubes. Phys. Rev. B 66: 155410.

Maciel, I.O., N. Anderson, M.A. Pimenta, A. Hartschuh, H. Qian, M. Terrones et al. 2008. Electron and phonon renormalization near charged defects in carbon nanotubes. Nature Mater. 7: 878.

Makaev, D.V. and P.N. D'yachkov. 2006. Band structure and optical transitions in semiconducting double-wall carbon nanotube. JETP Lett. 84: 335.

Mann, D., Y.K. Kato, A. Kinkhabwala, E. Pop and J. Cao. 2007. Electrically driven thermal light emission from individual single-walled carbon nanotubes. Nature Nanotechnol. 2: 33.

Manrique, D.Z., J. Cserti and C.J. Lambert. 2010. Chiral currents in gold nanotubes. Phys. Rev. B 81: 073103.

Mantsch, H.H. and D. Naumann. 2010. Terahertz spectroscopy: The renaissance of far infrared spectroscopy. J. Mol. Struct. 964: 1.

Marinopoulos, A.G., L. Reining, A. Rubio and N. Vast. 2003. Optical and loss spectra of carbon nanotubes: Depolarization effects and intertube interactions. Phys. Rev. Lett. 91: 046402.

Matanović, I., P.R.C. Kent, F.H. Garzon and N.J. Henson. 2013. Density functional study of the structure, stability and oxygen reduction activity of ultrathin platinum nanowires. J. Electrochem. Soc. 160: F548-F553.

Mavropoulos, P. and N. Papanikolaou. 2006. The Korringa-Kohn-Rostoker (KKR) green function method. I. Electronic structure of periodic systems. pp. 131-155. *In*: J. Grotendorst, S. Blügel and D. Marx [eds.]. Computational Nanoscience: Do It Your-self! NIC Series. Vol. 31. John von Neumann Institute for Computing, Jülich.

Minot, E.D., Y. Yaish, V. Sazonova and P.L. McEuen. 2004. Determination of electron orbital magnetic moments in carbon nanotubes. Nature (London) 428: 536.

Mintmire, J.W. and C.T. White. 1998. Universal density of states for carbon nanotubes. Phys. Rev. Lett. 81: 2506-2509.

Mintmire, J.W., B.I. Dunlop and C.T. White. 1992. Are fullerene tubules metallic? Phys. Rev. Lett. 68: 631-634.

Mintmire, J.W., D.H. Robertson and C.T. White. 1993. Properties of fullerene nanotubules. J. Phys. Chem. Solids 54: 1835-1840.

Miyake, T. and S. Saito. 2003. Quasiparticle band structure of carbon nanotubes. Phys. Rev. B 68: 155424.

Miyake, T. and S. Saito. 2005. Band-gap formation in (*n*, 0) single-walled carbon nanotubes (*n* = 9, 12, 15, 18): A first-principles study. Phys. Rev. B 72: 073404.

Miyamoto, Y., A. Rubio, M.L. Cohen and S.G. Louie. 1994. Chiral tubules of hexagonal BC_2N. Phys. Rev. 50: 4976-4979.

Miyamoto, Y., S.G. Louie, M.L. 1996. Cohen Chiral Conductivities of Nanotubes. Phys. Rev. Lett. 76: 2121.

Miyamoto, Y., S. Saito and D. Tomanek. 2001. Electronic interwall interactions and charge redistribution in multiwall nanotubes. Phys. Rev. B 65: 041402R.

Modi, A., N. Koratkar, E. Lass, B. Wei and P.M. Ajayan. 2003. Miniaturized gas ionization sensors using carbon nanotubes. Nature London 424: 171.

Mokrousov, Y., G. Bihlmayer and S. Blügel. 2005. Full-potential linearized augmented plane-wave method for one-dimensional systems: Gold nanowire and iron monowires in a gold tube. Phys. Rev. B 72: 045402.

Mokrousov, Y., G. Bihlmayer, S. Heinze and S. Blügel. 2006. Giant magnetocrystalline anisotropies of 4d transition-metal monowires. Phys. Rev. Lett. 96: 147201.

Mueller, T., M. Kinoshita, M. Steiner, V. Perebeinos, A.A. Bol and D.B. Farmer. 2010. Efficient narrow-band light emission from a single carbon nanotube p-n diode. Nature Nanotechnol. 5: 27.

Nemilentsau, A.M., G.Y. Slepyan, S.A. Maksimenko, O.V. Kibis and M.E. Portnoi. 2010. Terahertz Radiation from Carbon Nanotubes. pp. 1-15. *In*: K.D. Sattler [ed.]. The Handbook of Nanophysics. V. 4. Nanotubes and Nanowires. London and New York

Neophytou, N., S. Ahmed and G. Klimeck. 2007. Influence of vacancies on metallic nanotube transport properties. Appl. Phys. Lett. 90: 182119.

Nevidomskyy, A.H., G. Csányi and M.C. Payne. 2003. Chemically active substitutional nitrogen impurity in carbon nanotubes. Phys. Rev. Lett. 91: 105502.

Nikulkina, A.V. and P.N. D'yachkov. 2004. Electronic structure of nitrogen, boron, and oxygen doped nanotubes. Russ. J. Inorg. Chem. 49: 430-436.

Odom, T.W., J.-L. Huang, C.L. Cheung and C.M. Lieber. 2000. Magnetic clusters on single-walled carbon nanotubes: The Kondo effect in a one-dimensional host. Science 290: 1549.

Odom, T.W., J.-L. Huang, Ph. Kim and Ch.M. Lieber. 1998. Atomic structure and electronic properties of single-walled carbon nanotubes. Nature 391: 62-64.

Ohm, C., C. Stampfer, J. Splettstoesser and M. Wegewijs. 2012. Readout of carbon nanotube vibrations based on spin-phonon coupling. Appl. Phys. Lett. 100: 143103.

Ohnishi, H., Y. Kondo and K. Takayanagi. 1998. Quantized conductance through individual rows of suspended gold atoms. Nature 395: 780-783.

Okada, S., A. Oshiyama and S. Saito. 2000. Nearly free electron states in carbon nanotube bundles. Phys. Rev. B 62: 7634-7638.

Ono, T. and K. Hirose. 2005. First-principles study of electron-conduction properties of helical gold nanowires. Phys. Rev. Lett. 94: 206806.

Osadchii, A.V., E.D. Obraztsova, S.V. Terekhov and V. Yu. Yurov. 2003. Modeling of electronic density of states for single-wall carbon and boron nitride nanotubes. JETP Lett. 77: 405-410.

Oshima, Y., A. Onga and K. Takayanagi. 2003. Helical gold nanotube synthesized at 150 K. Phys. Rev. Lett. 91: 205503.

Oshima, Y., H. Koizumi, K. Mouri, H. Hirayama, K. Takayanagi and Y. Kondo. 2002. Evidence of a single-wall platinum nanotube. Phys. Rev. B 65: 121401(R).

Oshima, Y., K. Mouri, H. Hirayama and K. Takayanagi. 2006. Quantized electrical conductance of gold helical multishell nanowires. J. Phys. Soc. Jpn. 75: 053705.

Östling, D., D. Tomanek and A. Rosèn. 1997. Electronic structure of single-wall, multiwall and filled carbon nanotubes. Phys. Rev. B 55: 13980.

Osváth, Z., G. Vértesy, L. Tapasztó, F. Wéber, Z.E. Horváth, J. Gyulai et al. 2006. Scanning tunneling microscopy investigation of atomic-scale carbon nanotube defects produced by Ar^+ ion irradiation. Mater. Sci. Eng. C 26: 1194.

Osváth, Z., L. Tapasztó, G. Vértesy, A.A. Koós, Z.E. Horváth, J. Gyulai et al. 2007. STM imaging of carbon nanotube point defects. Phys. Stat. Solidi A 204: 1825.

Ouyang, M., J.-L. Huang, C.L. Cheung and C.M. Lieber. 2001. Energy gaps in "metallic" single-walled carbon nanotubes. Science 292: 702-704.

Paĺyi, A., P. Struck, M. Rudner, K. Flensberg and G. Burkard. 2012. Spin-orbit induced strong coupling of a single spin to a nanomechanical resonator. Phys. Rev. Lett. 108: 206811.

Park, M.H., M.G. Kim and J. Joo. 2009. Silicon nanotube battery anodes. Nano Lett. 9: 3844-3847.

Pekker, A. and K. Kamara. 2008. Wide range optical studies on various single-walled carbon nanotubes: Origin of the low-energy gap. Phys. Rev. B 84: 075475.

Peng, L.-M. 2018. A new stage for flexible nanotube devices. Nature Electronics 1: 158-159.

Piquini, P., R.J. Baierle, T.M. Schmidt and A. Fazzio. 2005. Formation energy of native defects in BN nanotubes: An ab initio study. Nanotechnology 16: 827-831.

Portnoi, M.E., O.V. Kibis and M.R. da Costa. 2008. Terahertz applications of carbon nanotubes. Superlattices Microstruct. 43: 399.

Postma, H.W.Ch., T. Teepen, Z. Yao, M. Grifoni and C. Dekker. 2001. Carbon nanotube single-electron transistors at room temperature. Science 293: 76.

Qiu, M., Zh. Zhang, Zh. Fan, X. Deng and J. Pan. 2011. Transport properties of a squeezed carbon monatomic ring: A route to a negative differential resistance device. J. Phys. Chem. C 115: 11734-11737.

Reich, S., C. Thomsen and P. Ordejón. 2002. Electronic band structure of isolated and bundled carbon nanotubes. Phys. Rev. B 65: 155411.

Ren, W.C., F. Li, J. Chen, S. Bai and H.M. Cheng. 2002. Morphology, diameter distribution and Raman scattering measurements of double-walled carbon nanotubes synthesized by catalytic decomposition of methane. Chem. Phys. Lett. 359: 196-202.

Rinkiö, M., A. Johansson, G.S. Paraoanu and P. Torma. 2009. High-speed memory from carbon nanotube field-effect transistors with high-κ gate dielectric. Nano Lett. 9: 643.

Robinson, J.A., E.S. Snow, S.C. Badescu, T.L. Reinecke and F.K. Perkins. 2006. Role of defects in single-walled carbon nanotube chemical sensors. Nano Lett. 6: 1747.

Rodrigues, V., J. Bettini, P.C. Silva and D. Ugarte. 2003. Evidence for spontaneous spin-polarized transport in magnetic nanowires. Phys. Rev. Lett. 91: 096801.

Roy, A., T. Pandey, N. Ravishankar and A.K. Singh. 2013. Single-crystalline ultrathin gold nanowires: Promising nanoscale interconnects. AIP Adv. 3: 032131.

Rubio, A., D. Sánchez-Portal, E. Artacho, P. Ordejón and J.M. Soler. 1999. Electronic states in a finite carbon nanotube: A one-dimensional quantum box. Phys. Rev. Lett. 82: 35203523.

Rubio, A., J. Corkill and M.L. Cohen. 1994. Theory of graphitic boron nitride nanotubes. Phys. Rev. B 49: 5081-5084.

Saito, R., A. Jorio, J.H. Hafner, C.M. Lieber, M. Hunter, T. McClure et al. 2001. Chirality-dependent G-band Raman intensity of carbon nanotubes. Phys. Rev. B 64: 085312.

Saito, R., G. Dresselhaus and M.S. Dresselhaus. 1993. Electronic structure of double-layer graphene tubules. J. Appl. Phys. 73: 494.

Saito, R., G. Dresselhaus and M.S. Dresselhaus. 1998. Physical Properties of Carbon Nanotubes. Imperial College Press, London.

Saito, R., M. Fujita, G. Dresselhaus and M.S. Dresselhaus. 1992. Electronic structure of chiral graphene tubules. Appl. Phys. Lett. 60: 2204-2206.

Saito, R., M. Fujita, G. Dresselhaus and M.S. Dresselhaus. 1992. Electronic structure of graphene tubules based on C_{60}. Phys. Rev. B 46: 1804-1811.

Santavicca, D.F. and D.E. Prober. 2008. 33rd Intern. Conf. on Infrared, Millimeter and Terahertz Waves. IEEE, Pasadena, CA.

Sanvito, S., Y.-K. Kwon, D. Tomanek and C.J. Lambert. 2000. Fractional quantum conductance in carbon nanotubes. Phys. Rev. Lett. 84: 1974-1977.

Satishkumar, B.C., A. Govindaraj, K.R. Harikumar, J.-P. Zhang, A.K. Cheetham, and C.N.R. Rao 1999. Boron-carbon nanotubes from the pyrolysis of C_2H_2-B_2H_6 mixtures. Chem. Phys. Lett. 300: 473-477.

Schiff, L.I. 1949. Quantum Mechanics. McGraw-Hill, New York, USA.

Seltlur, A.A., J.M. Lauerhaas, J.Y. Dai and R.P.H. Chang. 1996. Cu-filled carbon nanotubes by simultaneous plasma-assisted copper incorporation. Appl. Phys. Lett. 69: 345.

Sen, A., C.-J. Lin and C.-C. Kaun. 2013. Single-molecule conductance through chiral gold nanotubes. J. Phys. Chem. C 117: 13676-13680.

Sen, R., B.C. Satishkumar, A. Govindaraj, K.R. Harikumar, G. Raina, J.-P. Zhang et al. 1998. B–C–N, C–N and B–N nanotubes produced by the pyrolysis of precursor molecules over Co catalysts. Chem. Phys. Lett. 287: 671-676.

Shan, B. and K. Cho. 2004. *Ab initio* study of Schottky barriers at metal-nanotube contacts. Phys. Rev. B 70: 233405.

Shan, B. and K. Cho. 2005. First-principles study of work functions of single wall carbon nanotubes. Phys. Rev. Lett. 94: 236602.

Shen, C., A.H. Brozena and Y.H. Wang. 2011. Double-walled carbon nanotubes: Challenges and opportunities. Nanoscale 2: 503-518.

Shimada, T., Y. Ishii and T. Kitamura. 2011. *Ab initio* study of ferromagnetic single-wall nickel nanotubes. Phys. Rev. B 84: 165452.

Shtogun, Y.V. and L.M. Woods. 2009a. Electronic structure modulations of radially deformed single wall carbon nanotubes under transverse external electric fields. J. Phys. Chem. C 113: 4792-4796.

Shtogun, Y.V. and L.M. Woods. 2009b. Electronic and magnetic properties of deformed and defective single wall carbon nanotubes. Carbon 47: 3252-3262.

Shtogun, Y.V. and L.M. Woods. 2010. Mechanical properties of defective single wall carbon nanotubes. J. Appl. Phys. 107: 061803.

Shulaker, M.M., G. Hills, N. Patil, H. Wei, H.-Y. Chen, H.-S. Wong et al. 2013. Carbon nanotube computer. Nature 501: 526-530.

Singh, D.J. 1994. Planewaves, Pseudopotentials and the LAPW Method. Kluwer, Boston, USA.

Singh, D.J. and L. Nordstrom. 2006. Planewaves, Pseudopotentials, and the LAPW Method. 2nd ed. Springer New York.

Skylaris, C.-K., P.D. Haynes, A. Mostofi and M.C. Payne. 2005. Introducing ONETEP: Linear-scaling density functional simulations on parallel computers. J. Chem. Phys. 122: 084119.

Slater, J.C. 1937. Wave functions in a periodic potential. Phys. Rev. 51: 846–851.

Slater, J.C. 1974. The Self-consistent Field for Molecules and Solids, Quantum Chemistry of Molecules and Crystals, Vol. 4. McGraw-Hill, New York.

Soares, G.P. and S. Guerini. 2011. Structural and electronic properties of impurities on boron nitride nanotube. J. Mod. Phys. 2: 857-863.

Song, T., J. Xia and J.-H. Lee. 2010. Arrays of sealed silicon nanotubes as anodes for lithium ion batteries. Nano Lett. 10: 1710-1716.

Stepanyuk, V.S., A. Szasz, A.A. Katsnelson, A.V. Kozlov and O.V. Farberovich. 1990. Application of LAPW and green-function methods to calculation of electronic structure of crystal defects. Z. Phys. B: Condens. Matter 81: 391-398.

Stephan, O., P.M. Ajayan, C. Colliex, P. Redlich and J.M. Lambert. 1994. Doping graphitic and carbon nanotube structures with boron and nitrogen. Science 266: 1683-1685.

Sugai, T., H. Yoshida, T. Shimada, T. Okazaki, H. Shinohara and S. Bandow. 2003. New synthesis of high-quality double-walled carbon nanotubes by high-temperature pulsed arc discharge. Nano Lett. 3: 769-773.

Sun, G., J. Kürti, M. Kertesz and R.H. Baughman. 2003. Variations of the geometries and band gaps of single-walled carbon nanotubes and the effect of charge injection. J. Phys. Chem. B 107: 6924.

Sun, L., F. Banhart, A.V. Krasheninnikov, J.A. Rodríguez-Manzo, M. Terrones and P.M. Ajayan. 2006. Carbon nanotubes as high-pressure cylinders and nanoextruders. Science 312: 1199.

Szczerba, W., M. Radtke, U. Reinholz, H. Riesemeier, R. Fenger, K. Radem et al. 2011. Au–Au bond length expansion in CTAB stabilized gold nanoparticles. Radiat. Phys. Chem. doi:10.1016/j.radphyschem.

Tang, Z.K., L. Zhang, N. Wang, X.X. Zhang, G.H. Wen, G.D. Li et al. 2001. Superconductivity in 4 angstrom single-walled carbon nanotubes. Science 292: 2462-2465.

Tans, S.J., A.R.M. Verschueren and C. Dekker. 1998. Room-temperature transistor based on a single carbon nanotube. Nature London 393: 49.

Tans, S.J., M.H. Devoret, R.J.A. Groeneveld and C. Dekker. 1997. Individual single-wall carbon nanotubes as quantum wires. Nature London 386: 474.

Tapasztó, L., P. Nemes-Incze, Z. Osváth, M.C. Bein, Al. Darabont and L.P. Biró. 2008. Complex superstructure patterns near defect sites of carbon nanotubes and graphite. Physica E 40: 2263.

Tatar, R.C. and S. Rabii. 1982. Electronic properties of graphite: A unified theoretical study. Phys. Rev. B 25: 4126-4141.

Tavazza, F., S. Barzilai, D.T. Smith and L.E. Levine. 2013. The increase in conductance of a gold single atom chain during elastic elongation. J. Appl. Phys. 113: 054316.

Terrones, M., A.M. Benito, C. Mantega-Diego and W.K. Hsu. 1996. Pyrolytically grown $B_xC_yN_z$ nanomaterials: Nanofibres and nanotubes. Chem. Phys. Lett. 257: 576-582.

Tien, L.-G., C.-H. Tsai, F.-Y. Li and M.-H. Lee. 2005. Band-gap modification of defective carbon nanotubes under a transverse electric field. Phys. Rev. B 72: 245417.

Tolvanen, A., G. Buchs, P. Ruffieux, P. Gröning, O. Gröning and A.V. Krasheninnikov. 2009. Modifying the electronic structure of semiconducting single-walled carbon nanotubes by Ar^+ ion irradiation. Phys. Rev. B 79: 125430.

Tony, V.C.S., C.H. Voon, C.C. Lee, B.Y. Lim and S.C.B. Gopinath. 2017. Effective synthesis of silicon carbide nanotubes by microwave heating of blended silicon dioxide and multi-walled carbon nanotube. Mater. Research 20: 1658-1668.

Tsukagoshi, K., B.W. Alphenaar and H. Ago. 1999. Coherent transport of electron spin in a ferromagnetically contacted carbon nanotube. Nature 401: 572-574.

Tu, Y., Z.P. Huang, D.Z. Wang, J.G. Wen and Z.F. Ren. 2002. Growth of aligned carbon nanotubes with controlled site density. Appl. Phys. Lett. 80: 4018-4020.

Ugarte, D., A. Chatelain and W.A. de Heer. 1996. Nanocapillarity and chemistry in carbon nanotubes. Science 274: 1897-1899.

Ugarte, D., T. Stockli, J.-M. Bonard, A. Châtelain and W.A. de Heer. 1998. Filling carbon nanotubes. Appl. Phys. 67: 101.

Vadapalli, R.K. and J.W. Mintmire. 2006. Endohedral carbon chains in chiral single-wall carbon nanotubes. Int. J. Quantum Chem. 106: 2324-2330.

Valle, M., C. Tejedor and G. Cuniberti. 2006. Scaling of the conductance in gold nanotubes. Phys. Rev. B 74: 045408.

Wang, K.Y., A.M. Blackburn, H.F. Wang, J. Wunderlich and D.A. Williams. 2013. Spin and orbital splitting in ferromagnetic contacted single-wall carbon nanotube devices. Appl. Phys. Lett. 102: 093508.

Wang, W.Y., J.G. Xu, Y.G. Zhang and G.X. Li. 2017. First-principles study of electronic structure of Si nanotubes. Comput. Chem. 5: 159-171.

Wang, X., L.Y. Lu, Y.K. Kato and E. Pop. 2007. Electrically driven light emission from hot single-walled carbon nanotubes at various temperatures and ambient pressures. Appl. Phys. Lett. 91: 261102.

Watson, G.N. 1966. Treatise on the Theory of Bessel Functions, 2nd ed. Cambridge University Press.

Wern, H., R. Courths, G. Leschik and S. Hüfner. 1985. On the band structure of silver and platinum from angle-resolved photoelectron spectroscopy (ARPS) measurements. Z. Phys. B: Condensed Matter 60: 293-310.

White, C.T., D.H. Robertson and J.W. Mintmire. 1993. Helical and rotational symmetries of nanoscale graphitic tubules. Phys. Rev. B 47: 5485-5489.

Wildöer, J.W.G., L.C. Venerma, A.G. Rinzler, R.E. Smolley and C. Dekker. 1998. Electronic structure of atomically resolved carbon nanotubes. Nature 391: 59-62.

Wirth, I., S. Eisebitt, G. Kann and W. Eberhardt. 2000. Statistical analysis of the electronic structure of single-wall carbon nanotubes. Phys. Rev. B 61: 5719-5723.

Wu, H., G. Chan and J.W. Choi. 2012. Stable cycling of double-walled silicon nanotube battery anodes through solid-electrolyte interphase control. Nature Nanotechnology 7: 310-315.

Wunsch, B. 2009. Few-electron physics in a nanotube quantum dot with spin-orbit coupling. Phys. Rev. B 79: 235408.

Xiang, H.J., J. Yang, J.G. Hou and Q. Zhu. 2003. First-principles study of small-radius single-walled BN nanotubes. Phys. Rev. B 68: 035427.

Xiao, L. and L. Wang. 2006. Density functional theory study of single-wall platinum nanotubes. Chem. Phys. Lett. 430: 319-322.

Xiao, L., Z. Chen, C. Feng, L. Liu, Z.-Q. Bai, Y. Wang et al. 2008. Flexible, stretchable, transparent carbon nanotube thin film loudspeakers. Nano Lett. 8: 4539.

Yang, X. and J. Dong. 2005. Geometrical and electronic structures of the (5, 3) single-walled gold nanotube from first-principles calculations. Phys. Rev. B 71: 233403.

Yang, X., J. Zhou, H. Weng and J. Dong. 2008. Spin-orbit interaction in Au structures of various dimensionalities. Appl. Phys. Lett. 92: 023115.

Yanson, A.I., G.R. Bollinger, H.E. van den Brom, N. Agrait and J.M. van Ruitenbeek. 1998. Formation and manipulation of a metallic wire of single gold atoms. Nature 395: 783-785.

Yao, Z., H.W.Ch. Postma, L. Balents and C. Dekker. 1999. Carbon nanotube intramolecular junctions. Nature 402: 273.

Yarzhemsky, V.G. and C. Battocchio. 2011. The structure of gold nanoparticles and Au based thiol self-organized monolayers. Russ. J. Inorg. Chem. 56: 2147-2159.

Zeller, R. and P.H. Dederichs. 1979. Electronic structure of impurities in Cu, calculated self-consistently by Korringa-Kohn-Rostoker Green's-function method. Phys. Rev. Lett. 42: 1713.

Zhang, K. and H. Zhang. 2014. Plasmon coupling in gold nanotube assemblies: Insight from a time-dependent density functional theory (TDDFT) calculation. J. Phys. Chem. C 118: 635-641.

Zhang, S., L. Kang, X. Wang, L. Tong, L. Yang, Z. Wang et al. 2017. Arrays of horizontal carbon nanotubes of controlled chirality grown using designed catalysts. Nature 543: 234-238.

Zhang, Y., H. Gu, K. Suenaga and S. Iijima. 1997. Heterogeneous growth of B-C-N nanotubes by laser ablation. Chem. Phys. Lett. 279: 264-269.

Zhao, X., Y. Ando, Y. Liu, M. Jinno and T. Suzuki. 2003. Carbon nanowire made of a long linear carbon chain inserted inside a multiwalled carbon nanotube. Phys. Rev. Lett. 90: 187401.

Zhi, C., Y. Bando, C. Tang and D. Golberg. 2010. Boron nitride nanotubes. Mater. Sci. Engineering 70(3-6): 92-111.

Zhong, Z., N.M. Gabor, J.E. Sharping, A.L. Gaeta and P.L. McEuen. 2008. Terahertz time-domain measurement of ballistic electron resonance in a single-walled carbon nanotube. Nature Nanotechnol. 3: 201.

Zhou, J., Q. Liang and J. Dong. 2009. Asymmetric spin-orbit coupling in single-walled carbon nanotubes. Phys. Rev. B 79: 195427.

Zólyomi, V. and J. Kürti. 2004. First-principles calculations for the electronic band structures of small diameter single-wall carbon nanotubes. Phys. Rev. B 70: 085403.

Index